1,000,000 Books

are available to read at

www.ForgottenBooks.com

Read online
Download PDF
Purchase in print

ISBN 978-1-5277-6579-5
PIBN 10889072

This book is a reproduction of an important historical work. Forgotten Books uses
state-of-the-art technology to digitally reconstruct the work, preserving the original format
whilst repairing imperfections present in the aged copy. In rare cases, an imperfection in
the original, such as a blemish or missing page, may be replicated in our edition. We do,
however, repair the vast majority of imperfections successfully; any imperfections that
remain are intentionally left to preserve the state of such historical works.

Forgotten Books is a registered trademark of FB &c Ltd.
Copyright © 2018 FB &c Ltd.
FB &c Ltd, Dalton House, 60 Windsor Avenue, London, SW19 2RR.
Company number 08720141. Registered in England and Wales.

For support please visit www.forgottenbooks.com

1 MONTH OF FREE READING

at

www.ForgottenBooks.com

By purchasing this book you are eligible for one month membership to ForgottenBooks.com, giving you unlimited access to our entire collection of over 1,000,000 titles via our web site and mobile apps.

To claim your free month visit: www.forgottenbooks.com/free889072

* Offer is valid for 45 days from date of purchase. Terms and conditions apply.

English
Français
Deutsche
Italiano
Español
Português

www.forgottenbooks.com

Mythology Photography **Fiction**
Fishing Christianity **Art** Cooking
Essays Buddhism Freemasonry
Medicine **Biology** Music **Ancient Egypt** Evolution Carpentry Physics
Dance Geology **Mathematics** Fitness
Shakespeare **Folklore** Yoga Marketing
Confidence Immortality Biographies
Poetry **Psychology** Witchcraft
Electronics Chemistry History **Law**
Accounting **Philosophy** Anthropology
Alchemy Drama Quantum Mechanics
Atheism Sexual Health **Ancient History**
Entrepreneurship Languages Sport
Paleontology Needlework Islam
Metaphysics Investment Archaeology
Parenting Statistics Criminology
Motivational

TEXTBOOK OF EVOLUTION AND GENETICS

THE MACMILLAN COMPANY
NEW YORK · BOSTON · CHICAGO · DALLAS
ATLANTA · SAN FRANCISCO

MACMILLAN & CO., Limited
LONDON · BOMBAY · CALCUTTA
MELBOURNE

THE MACMILLAN COMPANY
OF CANADA, Limited
TORONTO

TEXTBOOK OF
EVOLUTION AND GENETICS

BY

ARTHUR WARD LINDSEY

PROFESSOR OF ZOOLOGY IN DENISON UNIVERSITY

New York
THE MACMILLAN COMPANY
1929

ALL RIGHTS RESERVED, INCLUDING THE RIGHT OF REPRODUCTION
IN WHOLE OR IN PART IN ANY FORM

COPYRIGHT, 1929,
BY THE MACMILLAN COMPANY

Set up and electrotyped.
Published January, 1929.

SET UP AND ELECTROTYPED BY T. MOREY & SON
PRINTED IN THE UNITED STATES OF AMERICA BY
THE BERWICK & SMITH GO.

TO MY WIFE
WINIFRED WOOD LINDSEY

PREFACE

For nearly three quarters of a century the occurrence of evolution among organisms has been widely accepted by scientists. It is only natural that the work of Darwin, which made possible this acceptance, should have colored the beliefs of the period through which we have since been passing, and that the apparent opposition of his views and those of Lamarck should establish the basis for theoretical consideration of the processes of evolution. Initial views in any subject are likely to determine the trend of human thought upon that subject.

In the case of evolution it was obvious even when the *Origin of Species* was published that there were many difficulties to be surmounted before we could know even approximately how the wonderful adjustments of evolution were consummated. It was possible then as now to recognize the existence of natural selection, and to some a wider application of this principle appeared logical then than now seems justified. The Lamarckians offered their explanations as opposed to the Darwinian point of view and *vice versa*, and there is nothing so stimulating as conflict, although it is of doubtful productiveness.

When Mendel's discoveries were taken up early in the twentieth century and the new science of genetics arose it seemed that we might expect new concrete information regarding evolution, for in genetics we come as near as possible to the raw materials of evolution. The bearing of genetics on the larger problem of evolutionary processes is, indeed, of the utmost importance, but in the first quarter of the century it has made little if any impression upon the established treatment of evolutionary problems. The Lamarckian and Darwinian points of view still determine the course of a vast majority of writings on this subject.

It is difficult for most people to depart entirely from a point of view once learned as true. The printed page is probably the most potent influence in establishing an initial belief. Give a class a textbook which leans ever so slightly toward one opinion and no matter how vigorously an instructor may assert the opposite view, a majority of students will accept the word of the textbook.

These facts have impressed themselves so strongly upon me during my years of teaching that they have led to the production of this book. Our treatment of evolutionary processes and our methods of investigation have been at an impasse for many years. If we continue to teach the old point of view, can we expect to progress with reasonable rapidity beyond our old limits of knowledge? We have taught with unwarranted emphasis upon some of the factors in evolution, in spite of the fact that modern biology shows very clearly that many things must enter into the evolution of organisms. Such emphasis can hardly lead to great discoveries.

In this volume I have attempted first to present the materials of evolution in such a way that their true logical relationship is clear to the student, second to give a concise account of the fundamental principles of genetics, and finally to sum up the theoretical matter of the subject and to present a logical analysis of the factors bearing upon evolutionary theory. Since nothing is so interesting to man as himself, the bearing of all material upon the human species has been treated as fully as seems warranted.

The book has been written for students who desire a sound introduction to the subject and not merely such an elementary account as is presented very adequately in most textbooks of biology. The material contained in it has been used in my own classes for students who have previously completed a course in general zoölogy or biology and for a few students of marked ability without such prior training. While the facts presented must often be unfamiliar to such students, their significance should be evident with the brief treatment given to them here. The work is not designed for entertainment but for serious instruction in a difficult field of biology, although I have yet to find a student to whom the subject is not intensely interesting.

Acknowledgments for the use of illustrations are made where the figures appear, but I wish to extend my thanks again to all who have assisted in this way. My deepest gratitude is also due to my wife, Winifred Wood Lindsey, for her intelligent criticism of many scientific points, for assistance in the formulation of the manuscript, and for invaluable aid in the laborious work of proof reading.

PREFACE

If the book shall succeed in imparting to other adventurers in science the impartial attitude toward the problems of evolution which I believe to be essential to future progress it will have justified its preparation.

A. W. LINDSEY

GRANVILLE, OHIO,
December, 1928.

CONTENTS

CHAPTER	PAGE
I. INTRODUCTION	1
II. THE HISTORY OF EVOLUTION	5

RELATIONSHIPS OF ORGANISMS

III. THE RELATIONSHIP OF EXISTING ORGANISMS
    1. CLASSIFICATION 19

IV. THE RELATIONSHIP OF EXISTING ORGANISMS (Continued)
    2. EMBRYOLOGY OF VERTEBRATES . . . 39

V. THE RELATIONSHIP OF EXISTING ORGANISMS (Continued)
    3. COMPARATIVE ANATOMY OF VERTEBRATES . 68

VI. THE RELATIONSHIP OF EXISTING ORGANISMS (Continued)
    4. PHYSIOLOGY 90

VII. EVIDENCES OF EVOLUTION
    1. EXISTING ORGANISMS 105

VIII. EVIDENCES OF EVOLUTION (Continued)
    2. THE GEOLOGICAL RECORD 119

IX. EVOLUTION OF THE VERTEBRATES 144

X. ELEPHANTS, HORSES, AND CAMELS 164

XI. THE EVOLUTION OF MAN 187

THE PROCESS OF EVOLUTION

A. *The Foundation*

XII. ADAPTATION	210
XIII. THE BASIS OF ADAPTATION	247

B. *Genetics*

XIV. THE FOUNDATIONS OF GENETICS	264
XV. MENDELIAN HEREDITY	276

CHAPTER		PAGE
XVI.	THE CHROMOSOME THEORY OF HEREDITY	289
XVII.	GENES AND CHARACTERS	305
XVIII.	THE DETERMINATION OF SEX	325
XIX.	THE PRACTICAL VALUE OF GENETICS	341
XX.	HEREDITY IN MAN	359
XXI.	EUGENICS	374

C. *Theories of Evolution*

XXII.	NATURAL SELECTION	385
XXIII.	OTHER THEORIES OF GERMINAL SELECTION	401
XXIV.	THE LAMARCKIAN THEORY	413
XXV.	EVOLUTION TODAY	427
	INDEX	445

TEXTBOOK OF EVOLUTION AND GENETICS

TEXTBOOK OF EVOLUTION AND GENETICS

CHAPTER I

INTRODUCTION

Whatever may be the attitude of the individual toward evolution, memory will tell him that there was a time in his life when he did not think. Experience with others or the words of his parents will show him that he was a living, active organism, carrying on in his small body all of the fundamental life processes which continue in it today, but without consciousness of these processes or of the world about him, or of his individual existence. There is a distinct resemblance between this past oblivion of infancy and the normal condition of the lower animals. We cannot say that they are entirely devoid of the processes of mind which are so highly developed in ourselves; rather it seems that they do the same things in lesser degree, handicapped as they are by inferior brain development and by lack of that convenient means of storage and exchange, language and articulate speech.

If we look back through the long ages of recorded history we cannot fail to note another analogy in the gradually increasing complexity of society, in the development of mankind from savagery to primitive cultures and finally to the great civilizations which have come and gone. Each stage has contributed to the greatness of its successors, each has added to the complexity of human knowledge, each has made man a little more independent of his environment, each has turned his thoughts a little more keenly inward until race consciousness has become an active factor in the shaping of human destiny. Behind this long record we find a few remnants which tell us of the infancy of the human race. Crude drawings on the walls of caverns and the implements which these primitive peoples used disclose something of their limited culture. Bones associated with these eloquent legacies tell us much of the characteristics of the people who left them. Everything points to gradual change, but behind these records—what?

Man has first of all a heritage, in common with all other organisms, a thing without which he cannot exist. He lives surrounded by conditions of various kinds which, in the aggregate, we call his environment. Environment is a second essential; to it the heritage responds within the limits of its possibility. The combination means life. We know from our own experience that life can exist without consciousness. We know too that at some time in the ascending complexity of individual development consciousness dawns, and the individual responds to the world about him not merely as a series of reactions to environmental stimuli, but with an awakening realization of other entities about him, and at last a consciousness of self. Where this point lies in the organic world we cannot say with certainty; it may be that man alone is more than an organic automaton. The light of personal experience clarifies its significance. Have we always, as a species, possessed this quality which must develop in each individual? In view of the records just mentioned this seems unlikely. Back of that crude beginning of our record of man's progress there must have been something. Consciousness must have had a beginning.

Through Beebe's striking powers of description we may share his imaginative conception of this process as he watched the monkeys in his jungle laboratory. "A little monkey climbed down a swaying vine, hand over hand, until his face was close to a quiet pool of sweet water. The day before at evening, he had done the same thing. His mother and his ancestors for generations had done likewise. And always they chattered at the monkey they saw in the water, and finally in anger snatched at him, and their little fingers troubled the water and the monkey vanished. Then they drank eagerly, turned quickly, and clambered swiftly up to rest.

"Today the little monkey began to chatter, then stopped. He moved, and the monkey in the water moved. He brushed away some hairs from his face and the water monkey. Then something happened. He stopped chattering and peered again and again at the face in the water. He put his little paw over his eyes and slowly took it away. Then he forgot his thirst, raised his head and gazed fixedly before him, wrinkling his forehead and remaining very quiet. And the more distant his gaze, the less he seemed to observe, and the deeper became the wrinkles.

" . . . Something introspective had come to pass—a glimpse of the ego—a momentary flash of self-consciousness. The little face in the water was not really another monkey. And the end of this realization was to be man."

The dawn of consciousness alone can have been the beginning of that curiosity which has led man for ages to attempt the explanation of the world about him, and himself. Nothing in the world has been more baffling in this pursuit than the thing called life. Is it some force or quality distinct from all else, or is it merely the product of other forces? Is it divine, or is it an earthly thing? Shall we ever be able to explain it, or must it always remain a great mystery? Whatever may be the answer, Philosophy will continue its attempts to explain and Biology its investigations, and if nothing more accrues, they will at least have clarified our understanding and increased our store of facts.

It is only natural that in other lives, particularly in very different lives, this spirit of curiosity should find a major stimulus. Animals were competitors of primitive man for the bounties of the earth in various ways. They must often have used food which he himself desired, and others must have been ready at any moment to use man himself as food. They must have contributed to his diet early in his existence, and when domesticated they became not only a more important but a more intimate part of his life. They must, as a result of these varied contacts, have impressed themselves upon him as a conspicuous part of his environment. We can imagine a first scientist pioneering in comparative anatomy as he picked the bones of game at his fireside. He might note that both fish and bird have that peculiar jointed axis of bones which we call the spinal column, and might wonder why they should be so different in other ways and so nearly alike in this. He might see the same thing in a rabbit, and the evident resemblance between its legs and the wings of the bird, superficially so different. Or in his chance contacts afield he might wonder why the deer, so different in many ways, should have hair like man and the rabbit, while the bird has feathers and the fish, scales. Out of an infinite accumulation of such observations, leading step by step to greater powers of observation and increasing possibilities for interpretation, has developed the science of Biology, and out of an insatiable desire to explain these relationships of different organisms, all united by the possession of that unknown thing called life and in varying degrees by peculiarities of organization, has come our recognition of that process of nature which we call evolution.

From the first realization of evolution as a natural process by which species are developed from preëxisting species, and from the first sound attempts to explain this process there arose a series of theories which we can still believe in part. It was only a natural outcome of scientific progress that a reaction to this method should take place. Science needs working hypotheses, but sooner or later these must be soundly rooted in fact and the twentieth century has seen a vigorous attempt to discover the underlying principles of all types of development. The most significant field of investigation has been the relationship of individuals of different generations, the process of heredity. The science of genetics has explained many phenomena of heredity. It is still impossible to correlate genetics wholly with other fields of biology and to determine just how the transition from species to species is brought about in evolution but we no longer lack a foundation of soundly organized facts for interpretation.

However willing we may be to refer ultimate causes to faith in God or some mystic force, we cannot fail to admit that in man's knowledge of the living things about him there is much that is within his power to explain on a basis of natural laws. That inquiry into these things need conflict with other fundamental beliefs is a product of the imagination of those who do not, will not, or cannot understand the findings of science; if faith without understanding is beautiful, then faith with understanding is transcendent. We can conclude no better than with the ideas expressed by Erasmus Darwin in his *Zoönomia:* "The world has been evolved, not created; it has arisen little by little from a small beginning and has increased through the activity of the elemental forces embodied in itself, and so has rather grown than suddenly come into being at an almighty word. What a sublime idea of the infinite might of the great Architect! the Cause of all causes, the Father of all fathers, the *Ens entium!* For if we could compare the Infinite it would surely require a greater Infinite to cause the causes of effects than to produce the effects themselves.

"All that happens in the world depends on the forces that prevail in it, and results according to law; but where these forces and their substratum, Matter, come from, we know not, and here we have room for faith."

CHAPTER II

THE HISTORY OF EVOLUTION

It is often difficult to know what is cause and what is effect in past occurrences and it is therefore not easy to decide whether the innate curiosity of developing intelligence first led man to set down records in primitive form and thus stimulated his desire for the accumulation of knowledge, or whether the increase of knowledge led to a conscious desire for some way to record it. In either case facility of written expression increased rapidly with the development of more complex social systems and the earliest civilizations found man able to make permanent records with great accuracy of detail. There is abundant evidence that he took note of the organic world very early in his existence beyond the mere need of supplying himself with food and clothing, but we find nothing like an organized natural science until the Greek and Roman civilizations arose. Several men of those periods are entitled to rank as pioneers in the field of natural history.

The Greek Philosophers. Among the Greek philosophers Anaximander (611–547 B.C.), Empedocles (495–435 B.C.), Democritus (460?–357 B.C.) and Aristotle (384–322 B.C.) and his pupil and associate, Theophrastus (370–286 B.C.), produced works which show a surprising clarity of interpretation for a period when so little was known of the world of nature. In an examination of the beliefs of these men a salient feature is seen to be that striving for an explanation of life and living things which has led gradually up to our modern ideas of evolution. The pioneer work which establishes the early Greeks as the founders of natural history is, indeed, largely lacking in the observation and recording of facts, with the exception of that contributed by Aristotle and Theophrastus; it neglects experimental methods and original investigation, but it strikes at once into the problems which have remained forever since open to investigation. While these men saw but vaguely and expressed themselves fantastically in the light of modern knowledge, we must remember that their investigations and inquiries had no precedent and no foundation in recorded science. This very lack was probably in part responsible for the accuracy of the generalizations which pay such high tribute to their mental powers. "The spirit of the Greeks was vigorous and hopeful. Not pausing to test theories by research, they did not suffer the disappointments and delays which come from our own efforts to wrest truths from Nature. Combined with great freedom and wide range of ideas, independence of thought, and tendencies to rapid generalization, they had genuine gifts of scientific deduction, which enabled them to reach truths, as it were, by inspiration" (Osborn).

Anaximander is conspicuous among these philosophers for his idea of an actual transformation of living organisms from one state into another, particularly from aquatic to terrestrial. He even included man in this theory. Although vague in detail, his work foreshadows our modern idea of the adaptation of organisms.

Empedocles has been called the father of the evolution idea because he first expressed theories to account for the gradual development of different kinds of organisms. These theories were founded on some erroneous and fantastic ideas, but they embody the germ of the evolution conception.

Empedocles believed in the spontaneous origin of living creatures from inorganic matter, but when we consider that this belief was commonly accepted for many centuries thereafter, and was not completely overthrown until late in the nineteenth century, Empedocles' acceptance of it seems less remarkable. His belief that independent parts of organisms arose spontaneously and later became associated to form entire animals seems little short of ridiculous. The thought of heads, bodies and legs wandering about and finally combining at random is contradictory to the simplest biological knowledge of today, but here again, when we remember the centaurs and satyrs of Greek mythology, we realize that there was reason for Empedocles' belief. He had been taught that such anomalous creatures actually existed and it was no more than natural for him to attempt to account for their occurrence along with that of normal animals. He added to this fantastic portion of his theories the belief that some of the random combinations were unable to maintain themselves and so were replaced by more perfect individuals which were able to live and to perpetuate their kind. This view is very close to the idea of competition in nature and the survival of the fittest which persists even today as a logical interpretation of some evolutionary processes.

In spite of the vagueness of his theories, Empedocles therefore dealt with logical ideas of evolution including the gradual development of existing species, the necessity for adaptation, competition among organisms, and the extinction of less perfect creatures which accompanies the persistence of those better fitted for life.

Aristotle. Other Greeks contributed ideas likewise vaguely suggestive of modern scientific beliefs, but Aristotle is generally admitted to be the outstanding thinker of the times. He worked on the same basis as his predecessors, for they had accumulated no dependable facts, but in spite of such limitations he expressed most of the fundamental principles of evolution. Although this phase of his work is of chief interest to us in such a study as this, it is important for a full understanding of Aristotle's place in biological science to note that he did not limit himself to philosophical considerations, but made extensive and in many cases accurate observations of natural phenomena. At least one of his observations, that of parthenogenesis in the honey-bee, is commonly credited to a scientist of the nineteenth century. In the science of botany Theophrastus shares Aristotle's eminence as an accurate and original observer.

Aristotle's ideas in the field of evolution may be summed up as follows:

1. He believed in natural law as the source of evolutionary change.

2. He believed in intelligent design as the ultimate cause of all nature.

3. He did not accept the idea of survival of the fittest.

4. He believed in the development of modern organisms from a primordial soft mass of living substance, essentially as we believe today.

5. His works suggest a phylogenetic series such as we now recognize in living organisms.

6. He recognized rudimentary organs as an evidence of relationship and the unity of groups of related forms.

7. He believed in epigenesis in ontogeny.

8. He recognized fundamental principles of heredity.

9. He believed in prenatal influences and in the inheritance of acquired characters, the former a fallacy and the latter still unproved.

In spite of the inaccuracies of some of these views and certain other erroneous opinions which he held, Aristotle's work eclipses that of all other ancient scientists in this field and he was not surpassed until the beginning of modern scientific methods several centuries later.

Through the Dark Ages. For years after Aristotle's life the contributions which can be said to have any bearing on the problems of organic development and the origin of life have no more than minor biological significance. Pliny (27–79 A.D.) and Galen (131–200 A.D.) are the most conspicuous figures of the few succeeding centuries; the former did little or nothing of sound scientific value, but Galen was a remarkable observer, clear thinker and excellent writer. He was the foremost anatomist of antiquity. Finally the influence of the early Christian church, favoring "traditional knowledge and the special-creation idea in its most literal form", so hindered independent thought that not until the sixteenth century was progress again resumed. It is gratifying to note that even during this dark period three theologians, Gregory of Nyssa (331–396 A.D.), Augustine (353–430 A.D.), and Thomas Aquinas (1225–1274 A.D.) expressed belief in the symbolic nature of the Biblical story of the creation.

Development of Scientific Methods. An inevitable step in the development of true natural science was the departure from unsupported or poorly supported philosophical reasoning and reference to authority, which took place soon after the renewal of scientific thought. During the sixteenth century great strides were made in the development of modern scientific methods, and since then there has been no interruption of progress. Vesalius (1514–1564) in anatomy and Harvey (1578–1667) in physiology are outstanding figures in this period. Each applied to his work sound principles of observation and experiment, and each is known for the accurate contributions to science which resulted from these methods. A little later the microscope was introduced, and investigation of fields hitherto barred from human vision began. Hooke (1635–1703), Malpighi (1628–1694), Swammerdam (1737–1680) and Leeuwenhoek (1632–1723) were among the pioneers in microscopic work, which has been destined to play such a large part in the biological sciences.

Philosophy was not neglected during this period. The names of Bacon (1561–1626) and Kant (1724–1804) especially are cited in connection with the maintenance of the primitive idea of evolution. Their work was destined, however, because of the very nature of purely philosophical limitations, to add nothing more than corollaries to the points so well expressed by the Greeks.

Results of the New Methods. The accumulation of scientific data by observation and experiment could hardly fail to give a different impetus to scientific progress. The old desire to explain life and the relationship of living things was maintained, but new methods of study disclosed such a storehouse of accurate information to be had for the seeking that the observation and recording of material facts came to be, for the time, the prevailing tendency. We find that knowledge of natural facts accumulated rapidly while philosophical interpretation entered a fallow period which lasted, with a few interruptions of importance, for many years. Finally Darwin, at the middle of the nineteenth century, placed the old evolution idea on a basis of sound scientific data, and thus brought it for all time into the realm of biology.

Early Evolutionists. Among the scientists of the eighteenth and nineteenth centuries prior to Darwin, Linnaeus (1707–1778), Buffon (1701–1788), Erasmus Darwin (1731–1802), Lamarck (1744–1829), and Saint-Hilaire (1772–1844), made notable contributions to biology. None of these was destined to bring the theory permanently before the world, but their theories were valuable and show increasing accuracy in the interpretation of natural phenomena.

Linnaeus' chief contribution to biology was the plan of classification, which still prevails, together with the same binomial system of nomenclature now employed. Even his classification of organisms left its impress on that still in use, although it has been almost completely concealed by the corrections and amplification of the intervening years. In spite of the fact that in working out his classification of plants and animals he did much to illustrate their phylogenetic relationships, he did it unknowingly. He believed firmly in special creation as the origin of primary forms, although to this belief he appended a theory of development of the various species from a limited number of such forms.

Buffon "was not a true investigator," although "of a more philosophical mind than many of his contemporaries" (Locy). Buffon believed in the gradual evolution of species, but in spite of the fact that he retained this belief throughout his life he was hesitant in expressing it. His writings are noted for their excellent diction, but on the point of evolution they are vague and obscure. Some writers have attributed this reticence to the weight of ecclesiastical authority for special creation which then obtained, and to this we may add the knowledge that "he was a man of elegance, with an assured position in society." (Locy.) Such standing would hardly be conducive to militant opposition to the church. In spite of the vagueness which he displayed on evolution, there is a general agreement that he was the first to believe in the direct modification of organisms by their environment. He also anticipated Malthus in the idea of struggle for existence as a compensation for overproduction in maintaining the balance of nature, and expressed other opinions which are strongly suggestive of Darwin's theory of Natural Selection.

Erasmus Darwin, the grandfather of Charles, also believed in the inheritance of acquired characters, or environmental effects, but instead of emphasizing the formative power of the environment he recognized the activity of forces within the organism responding to environmental conditions as the basis of change. He, too, recognized the occurrence in nature of a struggle for existence, and carried the idea one step further than Buffon by suggesting its ultimate beneficial results. His works vaguely suggest sexual selection and the idea of protective coloration. Some biologists have speculated on the possible influence of his work on that of Lamarck, but Packard's vigorous defense of the integrity of Lamarck's contributions leads to the conclusion that he did not know of Darwin's writings. It is certain, however, that Erasmus Darwin's work received some contemporary recognition, and since he was a physician and naturalist, it was probably sound enough to deserve even more.

Lamarck (Fig. 1) later and apparently independently developed the ideas of his predecessors to such a degree that he ranks second only to Darwin as the founder of one of the schools of modern evolutionary theory. The available accounts of his life afford an interesting evidence of the adverse conditions under which valuable scientific work may be produced. Lamarck was born in 1744, the eleventh child in a military family. All of his brothers entered the army, so Jean Baptiste was placed in training for the clergy. This was so little to his taste that he followed the army into Germany and in his short period of service displayed "the courage and independence that characterized his later years." He was found physically unfit for a military life and took up the study of medicine in Paris, later becoming a naturalist. He devoted himself for years to the study of botany, earning a scanty living by filling various positions as instructor and curator. During his connection with the Royal Garden in Paris, which was named the *Jardin des Plantes* at his suggestion, he became associated with Cuvier, who was to have such an important influence on his work. When fifty years of age, in 1794, Lamarck turned to the study of invertebrate animals, for which he developed a greatly improved classification. What effect this work may have had on his philosophical conclusions it is difficult to say, but six years after undertaking it he departed from his previous idea of the fixity of species, and in 1809 published the *Philosophie Zoologique* which formulated his theory of evolution (Locy). After the publication of his views on evolution, which he elaborated later, he was

FIG. 1.—Jean Baptiste Lamarck.

strongly opposed by Cuvier. Cuvier's position was superior to that of Lamarck and his influence greater; his scientific conclusions were, however, much less accurate. The resulting unfair disregard of Lamarck's theories, together with poverty and blindness, contributed to the sadness of his declining years, and in 1829 he died, his true greatness for the time unrecognized.

Lamarck's contributions to science include the proposal of the term "biology" and the tree of life, representing the phylogenetic relationships of existing organisms, in addition to his actual theory of evolution. This, when first published in 1809, consisted of two laws, translated as follows:

"*First Law:* In every animal which has not exceeded the term of its development, the more frequent and sustained use of any organ gradually strengthens this organ, develops and enlarges it, and gives it a strength proportioned to the length of time of such use, while the constant lack of use of such an organ imperceptibly weakens it, causing it to become reduced, progressively diminishes its faculties, and ends in its disappearance.

"*Second law:* Everything which nature has caused individuals to acquire or lose by the influence of the circumstances to which their race may be for a long time exposed, and consequently by the influence of the predominant use of such an organ, or by that of the constant lack of use of such part, it preserves by heredity and passes on to the new individuals which descend from it, provided that the changes thus acquired are common to both sexes, or to those which have given origin to these new individuals."

To these he added later the idea that necessity in the organism gives rise to new organs. Other corollaries expressed his belief in various modifying factors, but essentially his theory involves the belief that change springs from within the organism, in response to definite conditions of the environment, and that such changes, once initiated, are transmitted by heredity. The last point has been a frequent subject of dispute, and was further complicated because Lamarck added to these points the assumption that the environment acted directly on plants.

Saint-Hilaire was a contemporary of Lamarck who went back to the belief of Buffon in the direct effect of environment. His chief claim to distinction is that he believed in the occurrence of sudden changes in organisms, giving rise to new species, an idea later developed by deVries.

Charles Darwin (1809–1882), (Fig. 2), is preëminent in the field of evolutionary thought, as is well shown by the common use of the word Darwinism as a synonym of evolution. While this is an erroneous use of the term, his eminence is justified by his works, not because he was the first man to believe in evolution as a natural process, but because he brought to the support of the theory so much evidence, accumulated and prepared with such painstaking care, that he may rightly be termed the first to place it upon an adequate and permanent foundation of scientific fact. Since the appearance of his *Origin of Species* in 1859 there has been no doubt among scientists of the reality of evolution as a process in nature, although Darwin's theory of method has remained, like all other such theories, a subject of dispute. His work is essentially responsible for reforms in all fields of biology

Fig. 2.—Charles Darwin.

which have made possible the modern development of the science.

Darwin's early training included the study of medicine at Edinburgh and preparation for the ministry at Christ's College, Cambridge. During the three years at the latter place he became interested in science, and when H. M. S. *Beagle* was devoted to an expedition from the years 1831 to 1836, he accompanied the survey party as naturalist. His account of this period, under the title *The Voyage of the Beagle,* shows a remarkable capacity for observation of facts of great variety. Much of the time the ship was absent from England was spent in South America, but the brief stop which Darwin was enabled to make at the Galapagos Islands seems to have been a particularly productive part of the trip. The remarkable conditions prevailing in these islands, recently brought to the attention of the world in inimitable style in Beebe's *Galapagos, World's End,* appear to have been a stimulus to his inquiring mind, and later to have furnished him with valuable data in connection with his work on evolution, although he did not begin his first notebook on the development of species until 1837.

After returning to England, Darwin devoted himself to his scientific investigations. Although he was financially well able to do this, he was handicapped for the rest of his life by ill health, and was forced to limit his periods of work to less than two hours each. His first idea of natural selection came as a result of reading Malthus on Population, a work which set forth the part played by overproduction and the consequent struggle for existence in the human race. This was destined to disclose to him the idea of the survival, under similar conditions in the organic world, of those individuals best fitted for life under the existing conditions, and the destruction of those less favored. With the aid of his own extensive knowledge of variation in organisms he was able to formulate the theory which has carried the name natural selection or survival of the fittest. Darwin records that he first set down this theory in June, 1842, and later extended it in 1844, but it was not until 1858 that it was finally made public under circumstances which are a fine example of individual generosity. During the year 1858 Alfred Russel Wallace (1822–1913), as a result of reading the same work which had given Darwin his first idea of natural selection, conceived a theory of the origin of species which was identical with that of his countryman. He communicated his theory to Darwin, who was about to give up his own claim to it when dissuaded by two friends, Hooker and Lyell. The theory thus independently formulated by the two men was presented to the Linnaean Society of London in a joint paper on July 1, 1858, and during the succeeding few months Darwin wrote and published the *Origin of Species* which appeared in 1859. With generosity no less than Darwin's, Wallace recognized the more extensive studies of his fellow scientist on the subject and insisted on relinquishing his own claim to credit. The *Origin of Species* was supposed to be an outline of the subject but in his subsequent work Darwin failed to produce anything which equalled the first in effect. It is interesting as this account is being written to note that the first publication of the work aroused a storm of opposition, much of it similar to that of the present day, against which a vigorous defense was conducted by such men as Thomas Henry Huxley (1825–1895).

After Darwin. The period immediately following Darwin's productive work witnessed much speculative thought on the subject of evolution, but before the close of the nineteenth century scientific activity also showed a fortunate trend toward the accurate examination of the more tangible related subjects. Individual development and individual relationships became an object of careful attention and experimentation. Heredity was investigated by several biologists and the foundations of the modern science of genetics were laid. The statistical method of handling biological data was introduced, out of which biometry has developed.

Herbert Spencer (1820–1903) was the outstanding philosopher among the evolutionists of this period. His work had considerable influence, but in general the hypotheses which he advanced have failed to stand the test of scientific progress.

August Weismann (1834–1914), a German biologist, also contributed notably to the interpretation of facts bearing on evolution. With a thorough knowledge of biological principles as then understood, including some facts of cell structure, he was much better equipped for his work than the earlier scientists and, although his conclusions are now partially disproved, his keen understanding played an important part in the development of modern ideas.

Weismann's work dealt largely with inheritance. It was apparent even before his time that the germ cells were the carriers of hereditary characters, and in dealing with the fundamental problems of evolution it was inevitable that he should enter this field. His idea of the distinctness of the germ cells from the body has come down to the present and is even now a prominent factor in evolutionary thought. It furnished the basis for other theories which were necessary to harmonize known facts with his idea of the germinal origin of characters. None of his theories are now regarded as adequate explanations of the process of evolution.

Sir Francis Galton (1822–1911), a cousin of Charles Darwin, carried on extensive studies of human heredity and published three books, of which the two best known, *Hereditary Genius* and *Natural Inheritance*, appeared in 1869 and 1889 respectively. He is regarded as the founder of biometry, for the nature of his material made necessary some statistical treatment. The complexity of human heredity is so great and its inadaptability to experimental methods so complete that Galton could not approach the results of his contemporary, Mendel, but his work is even now of great value.

Johann Gregor Mendel (1822–1884) (Fig. 3) must be given the credit for laying the foundation of our modern knowledge of heredity. He was an Austrian monk, and a botanist. In his monastery garden at Brünn he experimented with inheritance in garden peas and formulated from his results the laws of inheritance that bear his name. By using peas of several varieties, hybridizing the various kinds, and rearing them through several generations he learned definitely how characters may behave in heredity, and in 1866 published his conclusions in the proceedings of the natural history society of Brünn. This epoch-making paper was lost to the scientific world until the beginning of the twentieth century. Its rediscovery at that time found a number of biologists ready to accept and verify the conclusions which it expressed, and progress in the study of heredity has since been rapid.

Modern Evolution. During the twentieth century many famous names have been linked with progress in our knowledge of evolution. Darwin's and Lamarck's theories have come to be the basis for two leading schools of thought on the subject, and there is an abundance of literature which deals with their extension and verification. Darwin himself, in the later years of his work, indicated his belief that natural selection was not a sufficient explanation for evolution, but that there was also much evidence for the action of environment as a formative influence. The opinions of later writers have carried on varying degrees of controversy to a final recognition of the unproved state of both theories. In addition new theories have shown us that there are probably many different processes of evolutionary change, as might be expected in view of the great complexity of living organisms. Natural selection and the inheritance of acquired characters must take their places with such theories as mutations and kinetogenesis among the numerous probable processes.

Through the development of the science of genetics accurate data have been accumulated on the mechanism of transmission of characters which must be involved in the origin of species as well as in the origin of individuals. The facts available are not yet wholly correlated with other fields of biology but they are sufficient to furnish a sound foundation for future progress. Best of all they encourage the broad thinking which alone can arrive at great truths. Neither the philosophical nor the purely materialistic aspects of evolution seem complete in themselves.

FIG. 3.—Johann Gregor Mendel. (From Locy's *Biology and Its Makers*, with the permission of Henry Holt and Company.)

Late years show an increased emphasis on the philosophical aspects of the problem, and purposive evolution is now popular as a modification of the older and more mechanistic theories. The chief problem of modern evolution, however we approach it, is always, how it occurs. The fact of its occurrence is accepted by all schools of thought and the biologist can scarcely avoid the conviction that there is still much to be explained on the basis of known facts before recourse must be had to purely philosophical assumptions.

That inherent curiosity which prompted man in the beginning to investigate the conditions of life still persists, and with a vast store of accurate knowledge, improved equipment and methods, and probably a gradual increase in his own mental capacity, he may one day solve the problem whose pursuit has already met with such a gratifying degree of success. All of this work must deal with processes, for with Darwin's contribution man became convinced that the relationships which he had striven for centuries to explain were the result of orderly natural development—of evolution.

Summary. The history of biology shows that ideas of the relationship and evolutionary development of organisms are by no means of recent origin. The ancient Greeks foreshadowed many of our modern discoveries and Lamarck, at the beginning of the nineteenth century, formulated a valuable theory of evolution. Between the time of Lamarck and Darwin, numerous contributions appeared on the subject of evolution, but it remained for Darwin to express the theory which gained a permanent place for evolution in biological science. This he did in such a masterly way that his eminence is well deserved. Even though his theories of evolutionary method are no longer regarded as an adequate explanation of the way in which species are formed, they are still accepted as an accurate explanation of natural processes which play a part in evolution. Since Darwin's time a major tendency has been the examination of natural phenomena by the exact methods of observation and experiment. Theoretical contributions have been made, but genetics and other branches of biology have become the most important fields of progress in our knowledge of evolution. The fact of evolution is now generally admitted but there is still much to be known of its processes.

REFERENCES

PACKARD, A. S., *Lamarck*, 1901.
OSBORN, H. F., *From the Greeks to Darwin*, 1905.
LOCY, W. A., *Biology and Its Makers*, 1910.
LULL, R. S., *Organic Evolution*, 1917.
NEWMAN, H. H., *Readings in Evolution, Genetics and Eugenics*, 1921.

CHAPTER III

THE RELATIONSHIP OF EXISTING ORGANISMS

1. CLASSIFICATION

In any primary study of the organic world such as we have assumed to be the starting point of evolutionary thought, the things which an individual might see about him or the larger aggregate which might be collected through the efforts of many observers in different places must necessarily have been the whole source of facts. The accumulated knowledge of modern science has extended this field and added to it records of extinct organisms whose fossilized remains are the material of palaeontology. Together with the information worked out by geologists concerning the relative ages of the rock deposits in which fossils occur, palaeontology gives us in many cases an accurate idea of the past histories of existing plants and animals, and shows within certain limits of accuracy from what forms they spring and through what changes they have proceeded to their present state. The periods covered by this natural record are often unbelievably vast. They show us conclusively that all of our written records together are no more than a page out of the history of the world of nature, and that in the field of evolution, all of our observations of living things are of the present. The few centuries during which we have been making and setting down exact scientific observations are so insignificant in relation to all time that they are but as a moment, and all of the records so slowly and laboriously accumulated are little more than a fairly complete catalogue of the things present in the world at any moment.

Species. In the examination of living things in this brief span of human experience a resemblance is noted at once between certain individual organisms. We see birds in the trees and call them robins or bluebirds or crows. The individuals thus grouped together have rather definite features in common which enable us to associate them easily in the smallest groups commonly used in scientific classification, the species. In spite of the fact that such species as those mentioned are easily recognized and sharply delimited, a closer scrutiny of individual characteristics shows that in many cases it is extremely difficult to say exactly what characters define the species and how they are separated from each other. We have never been at loss for examples of species, yet there has never been unanimity of opinion among scientists as to what species really are.

The belief has been expressed that there is no real group in nature, but that only individuals are real entities and the groups into which we gather them for our scientific records nothing more than conveniences. In contrast the other extreme has been urged from time to time, that as some individuals resemble each other more than they do any other organisms, they must constitute a natural group with definite limits, even though they may vary within these limits. As is usually the case, the opinion has gradually developed that there is sound value in both interpretations. It is now supposed that there are such things as natural groups which may aptly

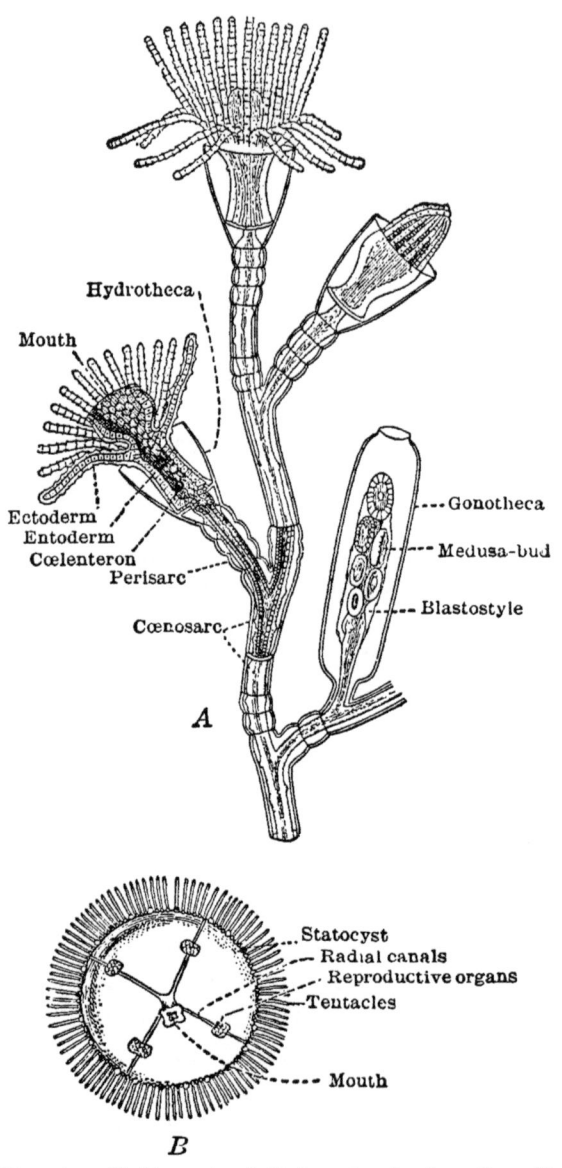

FIG. 4.—*Obelia.* A, stalk bearing hydranths; B, medusoid. (From Parker and Haswell.)

be called species, but that many of our named species blend into each other so gradually that it is difficult or impossible to place certain individuals accurately. These may not be species, and often an abundance of material shows that such is the case. A

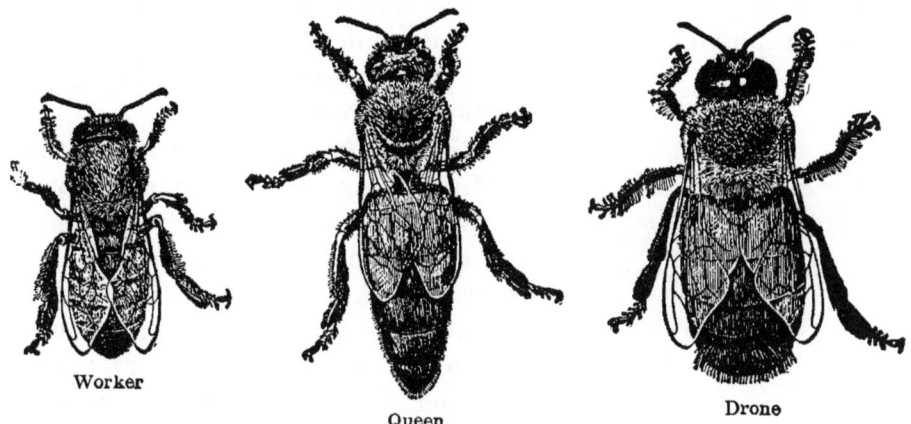

Fig. 5.—Honey-bees. (From Hegner.)

scientist may work with specimens from widely separated regions and find them different, while the later acquisition of specimens from intermediate regions shows that there is a gradual transition from one to the other, or he may be confronted with great variation in a single locality, with blood brothers bearing little resemblance to each other. The only possible conclusion of practical value is that species are not all in the same state; that while many show marked uniformity of characters, others are apparently unstable

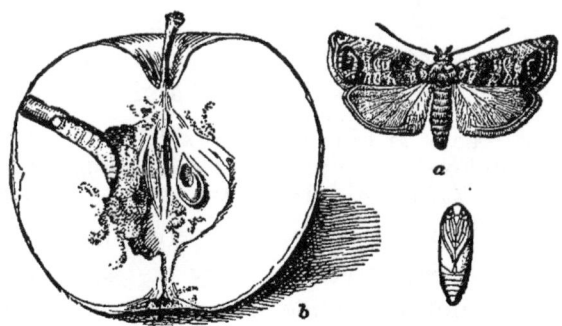

Fig. 6.—The Codling Moth *Carpocapsa pomonella*. *a*, adult; *b*, larva in an apple; *c*, pupa or chrysalis. (From *Farmer's Bulletin* 283, U. S. Dept. Agriculture.)

and in all probability even now undergoing change.

Subspecific Forms. The lack of fixed characters as a universal basis for species is further complicated by the occurrence in many of them of very different types of individuals which nevertheless bear a definite and intimate relationship to each other. This relationship may be in the form of constant association, as in the many colonial animals (Fig. 4) and the social forms (Fig. 5), or it may be in a succession such as the alternation of different reproductive types, and the succession of stages in metamorphosis (Fig. 6). In all of these cases, the ability of the various forms to produce each other is abundant evidence of specific unity, but mere lack of information has often led to the separation of subspecific forms.

Sexual Differences. Sex is a common and sometimes a conspicuous example of intraspecific difference (Fig. 7). In addition to those differences which are essential to complementary reproductive functions, others of an apparently unrelated character often appear. These are called secondary sexual characters. The long tail feathers of male turkeys, chickens, peacocks and pheasants, bright colors in the male sex of many species of birds, the mane of the lion, and other characters are well known examples. Sensory organs in many male insects and the horns of bucks and rams are more evidently useful to the animals. Even in the human race the sexes differ fundamentally, for accumulations of subcutaneous adipose tissue give the body of the female characteristic roundness of outline, while the growth of whiskers and of the vocal cords brings about equally characteristic male development. Among the invertebrates sexual dimorphism sometimes involves the general structure and appearance, as in the common promethea moth, in which color, pattern and shape of the wings differ, as well as the form of the antennae and development of sense organs. The sexes of some of the parasitic worms are even more diverse. While this is true of both flat- and roundworms it is probably nowhere more conspicuous than in *Schistosoma haematobium,* a blood fluke of the eastern Mediterranean region (Fig. 8).

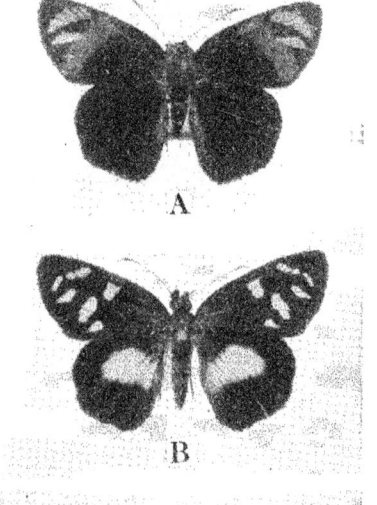

FIG. 7.—*Entheus peleus* Linn. A, male; B, female. The male is dark brown with an orange-red band and transparent orange-yellow spots. The female is brown with white markings.

EXISTING ORGANISMS—CLASSIFICATION

Colonial Forms. Such forms are highly developed in the Coelenterata (Fig. 4), among them such species as the Portuguese man-of-war, which lives in free swimming colonies made up of numerous modified individuals which carry on limited activities for the common good. Many of the simpler hydroids display three forms simultaneously, viz., polyps, asexual reproductive individuals and sexual medusoids similar to small jellyfishes, which later detach themselves from the colony. All of these forms are connected structurally, yet in degree their differentiation is not unlike that of the castes of social insects. In the honey bee colony there are three forms, the functional sexes or queen and drones, and the workers, which are imperfectly developed females with certain modifications peculiar to their own caste. Among the ants division of labor is accompanied by the development of distinct forms by the modification in various directions of the three fundamental types. That these differences of individuals are closely associated with division of labor in all cases is obvious, and the two are generally supposed to have developed together.

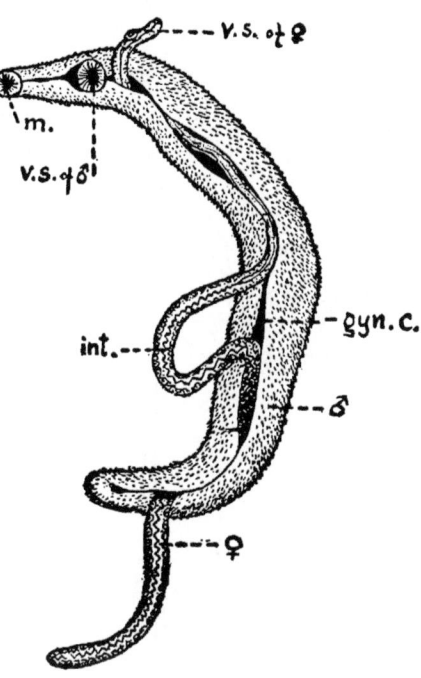

Fig. 8.—Blood fluke, *Schistosoma haematobium*. Male (♂) carrying female (♀) in ventral groove; int, intestine; gyn. c., ventral groove or gynecophoric canal; m, mouth; v. s., ventral sucker. x 8. (Reprinted by permission from *Animal Parasites and Human Disease* by Asa C. Chandler, published by John Wiley and Sons, Inc.)

Alternation of Generations. Still another type of polymorphism is alternation of generations such as that found in many plants and some of the lower animals. In *Obelia* (Fig. 4), for example, a member of the phylum Coelenterata, asexual individuals produce the medusoids by budding, while the whole colony is nourished by the polyps. The medusoids swim away from the colony, mature their germ cells, and by this process of sexual repro-

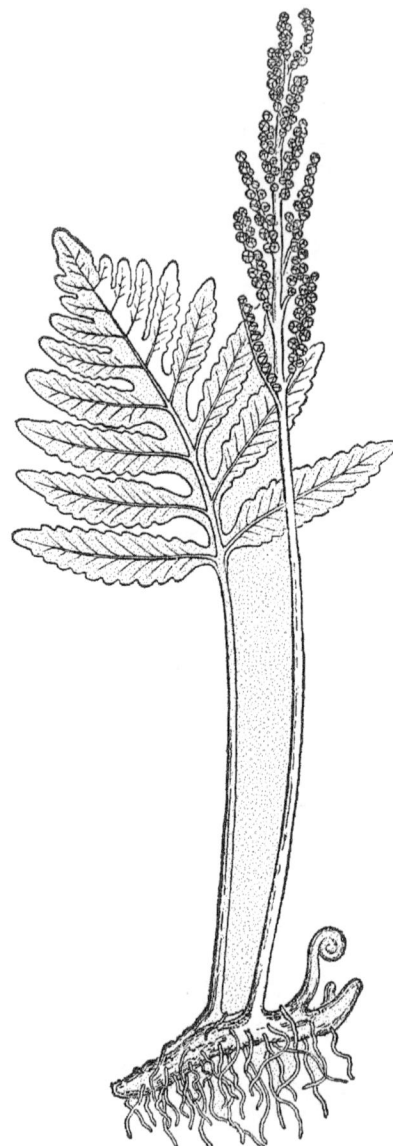

duction give rise to individuals which develop into new colonies. Ferns are asexual plants which produce many spores, small reproductive bodies which are able to develop under favorable conditions into plants of entirely different appearance. The fern of common parlance is the sporophyte (Fig. 9), and the plant produced from its spores the gametophyte or prothallus (Fig. 10). The latter, like the medusoids of *Obelia*, produces germ cells which fuse to give rise by sexual reproduction to a new sporophyte.

Metamorphosis resembles this process only in that the different forms appear in succession; all forms are a part of a single genera-

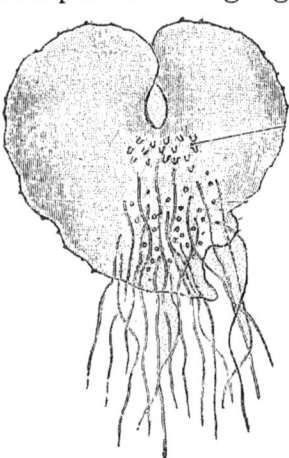

Fig. 9.—The sensitive fern, *Onoclea sensibilis*, showing a vegetative leaf (left) and a spore-bearing leaf (right). (From Woodruff, after Bergen and Davis.)

Fig. 10.—Gametophyte or prothallus of a fern, *Aspidium filix mas*, from below. (From Strasburger, after Schenck.)

tion. Holometabolous insects show a maximum degree of metamorphosis in their transition from egg to larva, to pupa and thence to adult (Fig. 6). The larva is essentially a growing stage, the pupa a resting stage through which transition from the larval to the very different adult structures is accomplished, and the adult is primarily the reproductive stage.

What Are Species? In the face of such diversity of form within many species, it is impossible to lay down definite criteria for the limitation of this unit of classification. The ability of individuals to produce fertile offspring when mated has been adopted by some scientists, but it has been shown that not only do many closely related species cross, but in some cases they produce fertile offspring which maintain themselves within the range of variation represented by their diverse parents. Criteria of morphology fail when great diversity occurs, and rigid delimitation on this basis covers a latitude in single colonial species greater than the differences between some obviously distinct species. Differences of physiology are difficult to judge, yet species exist between which no other distinctions are known. We are forced to the conclusion already briefly expressed, that there are specific entities in nature, although the conditions of their existence are variable. Since the species is the unit whose occurrence evolution proposes to explain, this very instability is significant. Those species which are variable, and apparently undergoing change, seem about to give rise to several different species, while those which are fixed within relatively narrow limits seem to be more definitely established. By referring to the past we see that still other species have come and gone, apparently after passing through a period of senility characterized by inability to adapt themselves to changing conditions. In this, again, palaeontology clarifies our understanding of the condition of modern species, although the modern species alone show that distinctness of kind is relative.

Major Groups. Beyond those relationships which enable us to group individuals together as species, we find other points of similarity which indicate broader associations. Our robins and bluebirds, for example, have definite structural characteristics in which both differ from the crow and the hawks, so we call both thrushes, yet the thrushes and hawks are more closely related to each other than to our domestic animals, because they are birds. Step by step these resemblances proceed through groups of increasing extent, each based on more fundamental characters than the one below it, and consequently embracing a wider range of species. The system of classification developed by Linnaeus and still used with modifications employs for this succession of groups the following terms, beginning with the initial subdivision of living things into plant and animal kingdoms and proceeding through those of decreasing extent:

    Kingdom
        Phylum
            Class
                Order
                    Family
                        Genus
                            Species

Thus man is *Homo sapiens*—the species *sapiens* of the genus *Homo*, which belongs to the family Hominidae in the order Primates. The Primates are members of the class Mammalia, in the phylum Chordata of the animal kingdom. To indicate finer distinctions in classification such modifications as suborders and superfamilies are sometimes used. In the example given everything up to the ordinal name indicates man's exalted opinion of himself, for he stands alone! The order, however, acknowledges his association with the apes and monkeys, the class, to all animals that have hair and nourish their young with milk, the phylum to those which have a backbone and to certain other remote relatives, and the kingdom, finally, to all animals.

Mimicry. Superficial resemblance is usually but not always a dependable index of relationship. We have mammals which may be mistaken for fish, beetles and flies which look like wasps, flies that resemble bumble bees, and a variety of lesser resemblances. Such abnormal superficial similarity has been recognized as playing a definite part in the lives of organisms, and the gradually accumulated knowledge represented by modern classification has relegated these types of resemblance to their proper places and expressed the fundamental relationship which they often obscure. The superficial resemblance of one species to another is called mimicry. This is well illustrated by the resemblance of certain harmless species of insects to others which are either unpalatable to bird or animal enemies, or able to defend themselves. The common eastern butterfly, *Basilarchia archippus* (Cram.), while superficially very different in appearance from its congeners, is much like the milkweed butterfly, *Danaus menippe* (Hbn.), which is apparently unpalatable because of the bitterness of its food plant.

Convergence. Resemblance between animals of different groups is usually due to convergence as fundamentally different organisms become adapted to the same conditions of environment. The fishes are aquatic organisms, admirably adapted to the conditions of their environment. Whales, seals and dolphins are fish-like in many ways, but they show their terrestrial origin in that they must breathe air. Their points of resemblance to the fishes are due solely to the fact that life in the water is possible only if certain conditions of form and locomotion can be met, and the physical conditions of the ocean are such as to limit the ways in which they can be met. The tail of the dolphin is somewhat like that of the fish, and serves the same purpose; likewise the wings of insects and of birds are superficially similar, and have the same function, although they are fundamentally different structures. This resemblance of unlike structures which comes about through the adaptation of different organisms to similar conditions is known as analogy, and is a common corollary of convergence.

Divergence. Relationship may also be obscured by the differences in development of closely related animals. Man is a mammal, the squirrel is a mammal, and the seal is a mammal, but man has assumed the erect posture for terrestrial life, the squirrel lives in trees, and the seal is almost wholly aquatic. The result is a complete overshadowing of their fundamental similarity, yet the flippers of the seal and the front paws of the squirrel are the same vertebrate structures as the hands of man. The term adaptive radiation is applied to this divergence of related forms, and the related structures which thus assume superficial differences are said to be homologous.

Accuracy of Classification. Such accuracy is wholly dependent upon the ability of the taxonomist to go beyond the superficial characteristics of the organism and to interpret fundamental conditions. This has been accomplished to a marked degree in bringing our classification to its present state of reasonable perfection. Studies of evolution have been invaluable in this development in that they have brought about a keener realization of relationship between organisms and the progressive nature of such relationship, and now that our classification is in such an excellent state, it exerts a reciprocal influence of great value as an illustration of the results of evolution.

Classification of Plants. The most common classification of plants is based upon four major groups. The third has been divided into three, which are indicated here merely as components of the one division.
 1. Thallophyta. The algae, fungi and lichens.
 2. Bryophyta. Liverworts and mosses.
 3. Pteridophyta. Ferns, horse-tails or scouring rushes, and club mosses.
 4. Spermatophyta. The seed-bearing plants.

Classification of Animals. Some variation occurs in the classification of animals as various writers estimate differently the importance of certain characters. The following outline includes the major subdivisions, or phyla, which are commonly recognized.
 1. Protozoa. Single-celled animals.
 2. Porifera. The sponges.
 3. Coelenterata. The hydroids, jellyfishes, sea anemones and corals.
 4. Ctenophora. Comb-jellies or sea walnuts.
 5. Platyhelminthes. Flatworms: free-living forms and parasitic flukes and tapeworms.
 6. Nemathelminthes. Roundworms, including many parasitic forms found in man as well as free-living forms.
 7. Rotatoria. Wheel animalcules.
 8. Bryozoa. The moss animals.
 9. Brachiopoda. Tongue or lamp shells.
 10. Echinodermata. Starfishes, sea urchins, sea cucumbers.
 11. Annelida. Jointed worms: earthworms and leeches.
 12. Arthropoda. Crustacea, including lobsters, etc. Myriapoda, spiders, insects, etc.
 13. Mollusca. Snails, mussels, squids, etc.
 14. Chordata. Several obscure worm-like animals, the tunicates and salpians, lancelets, round-mouthed eels, fishes, amphibia, reptiles, birds, and mammals.

The tree of life, adapted to modern classification from the original conception of Lamarck, shows graphically the relationship of these phyla and some of their more important subdivisions (Fig. 11).

EXISTING ORGANISMS—CLASSIFICATION

Relationships. *Protoplasm.* A complete dissertation on the relationship of minor groups of animals would involve more details than could be set down in a single volume, but many points of fundamental relationship are visible in organisms which can readily be appreciated in a brief outline. Broadest in scope is the common basis of all life, plant and animal, the substance protoplasm. Protoplasm is made up of three chemical compounds, proteins, carbohydrates and fats, associated with various inorganic compounds such as water and common salt, which are not changed by the body. The fats and carbohydrates are made up of the elements carbon, hydrogen and oxygen, while proteins include these three together with nitrogen, sulphur and in some cases phosphorus and iron. The properties of protoplasm are the same in all organisms. In addition to its definitive chemical composition, it is enabled to continue its existence and activity, and to grow and reproduce, by taking in other substances, changing them chemically through the process of digestion and incorporating them into itself. This process is called intussusception. After the substances have become an integral part of the body, their potential energy is liberated by oxidation and thus activates the organism. The waste products of oxidation are then passed out of the body by the excretory system. The constructive part of the entire interchange between the organism and its environ-

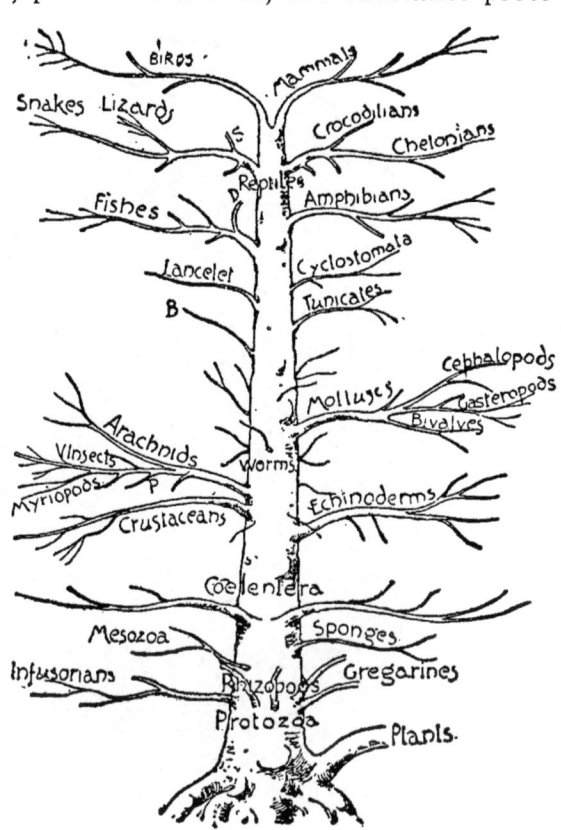

Fig. 11.—Diagram, the tree of life. (From *Outlines of Zoology*, by J. Arthur Thomson, with the permission of D. Appleton & Company.)

ment is called anabolism, the destructive part katabolism, and the whole process metabolism; this is one of the most striking characteristics of living matter. Another distinctive quality is the power to reproduce itself in the organized form which is characteristic of the various species, and finally this remarkable substance has properties which enable it to receive stimuli of light,

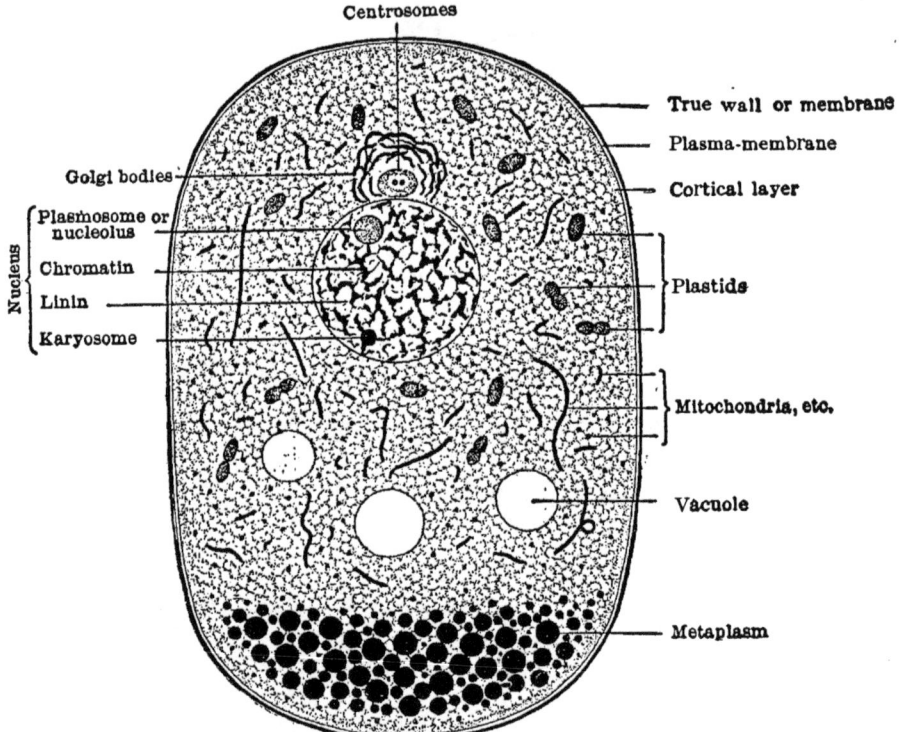

FIG. 12.—General diagram of a cell. (From Woodruff, after Wilson.)

heat, contact, sound waves and chemical substances in its environment, to conduct these stimuli to various parts of the body which it forms, and to respond to them in various ways with the result that the organism fits into, or is adapted to its environment.

The Cell. Protoplasm as we see it in living organisms is found in only one form, the cell (Fig. 12). No animal or plant exists which is simpler than this unit, and all more complex forms are built up of many such units. Cells may undergo great differentiation of form as they become specialized for various tasks, and so we find in the body of man the flat, horny cells of the cuticle, tall cells with waving cilia lining the trachea, long and contractile cells in the muscles, nerve cells with long fibers which coördinate the various organs, and many more (Fig. 13). Wherever found and

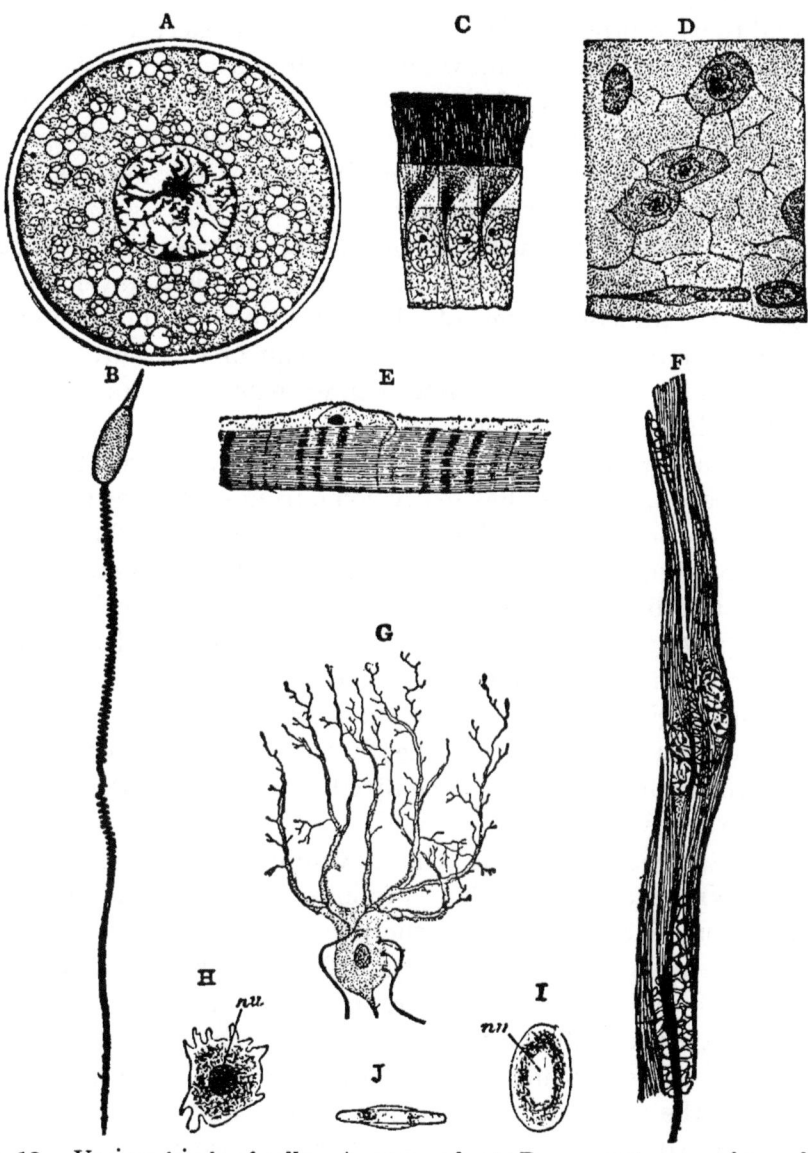

Fig. 13.—Various kinds of cells. A, ovum of cat; B, spermatozoon of a snake; C, ciliated epithelium from the digestive tract of a mollusc; D, cartilage of a squid; E, voluntary or striated muscle fiber from an insect; F, involuntary or smooth muscle fibers from the bladder of a calf; G, nerve cell from the human brain; H, white blood cell of frog; I, red blood cell of frog; J, same, edge view; nu, nucleus. (From Woodruff, after Parker and Haswell, H, I, J; and Dahlgren and Kepner, A–G.)

however specialized, cells are made up of two fundamental parts, the nucleus and cytoplasm, the former embedded in the latter. Each is complex, as will be seen in the diagram, and the two are essential to each other. The nucleus appears to exert a controlling influence over the cytoplasm, while the differentiation of the cytoplasm determines the chief characteristics of the various types of cells. Some cells, such as the red blood corpuscles of most mammals, are without nuclei, but after the loss of the nucleus their lives are short and replacement occurs frequently.

Plan of Animal Structure. *Single-Celled Organisms.* Although it is necessary to call on our knowledge of individual development for accurate interpretation of the conditions found in many organisms, many others show a simple plan of structure; for example, the fact that living matter cannot exist in units less complete than the cell leads at once to the conclusion that single-celled plants and animals are the most simple of all organisms (Fig. 14). For that reason, if all life has really come from such a lowly origin, they must represent the oldest forms now extant. They have existed longer than any other forms, and have had opportunities to become fitted for life under various conditions; hence we find many species differentiated in various ways. Some are parasites in man and other animals. If man is of recent origin, as we suppose, or even if his origin followed that of the other animals, as all admit, this is in itself evidence that the Protozoa have departed from their original condition to become fitted for life in the other bodies; they have evolved.

The Germ Layers. During the development of many multicellular animals a definite plan is followed in which the first step is the repeated subdivision of the original germ cell to form a hollow sphere, in the simplest state. This hollow sphere caves in on one side until the two halves are in contact, thus forming a sac with two layers of cells in its wall, called the gastrula. The inner layer is associated primarily with nutrition and respiration, the outer with the nervous system, sense organs, and protective skin. Lastly, a third layer or mass of cells forms between the two, from which develop the muscular and skeletal systems, circulatory system, excretory system, reproductive system, and many supporting and accessory parts of other structures. These three layers of cells are called the germ layers. A group of similar cells is obviously simpler than a sac with a two-layered wall of which

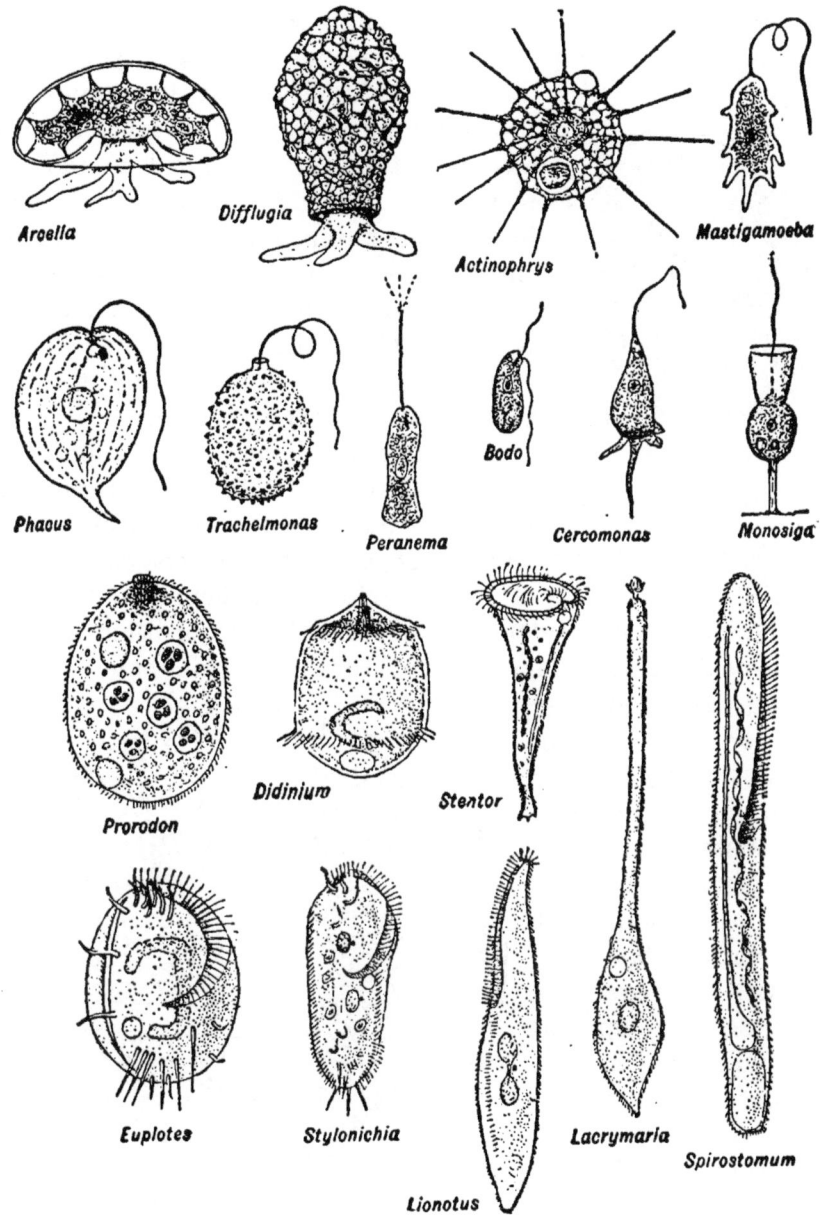

Fig. 14.—Some common fresh water protozoa, greatly magnified. (From Woodruff, after Curtis.)

the cells of the different layers carry on different functions, and this structure is simpler than one in which a third layer is differentiated.

On this basis we are able to group the fourteen phyla. Among the Protozoa most are independent single cells, but some are associated together in groups (Fig. 15), and most of them remain undifferentiated. The next three phyla are fundamentally simple two-layered sacs although they are modified in various ways, and are called diploblastic. One little fresh-water animal, *Hydra*, shows this arrangement very simply (Fig. 16). All other phyla have three fundamental layers, and are called triploblastic. Their relationships are indicated by the development of other structures.

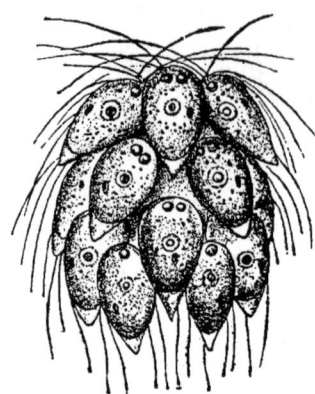

FIG. 15.—A simple colony of unicellular organisms (*Spondylomorum*). (From Hegner, after Oltmanns.)

Triploblasts. *Symmetry.* Complexity of the environments to which these animals are adapted is here reflected in a variety of different tendencies in their structure. Not all of them have followed the same course of development beyond this acquisition of three fundamental germ layers. Thus, the flatworms have merely added to the complexity of the original sac-like digestive cavity, while the rest have developed a tubular structure, opening cephalad at the mouth and caudad at the anus. Yet the flatworms share with the remaining phyla the development of a head in which nervous control is concentrated, and accompanying this, their bodies are formed of two similar halves, flanking the median axis. Below this group, animals are either asymmetrical or made up of parts radiating from a common center, as in *Hydra*. Such a form is known as radially symmetrical, while the bilateral arrangement is called bilateral symmetry.

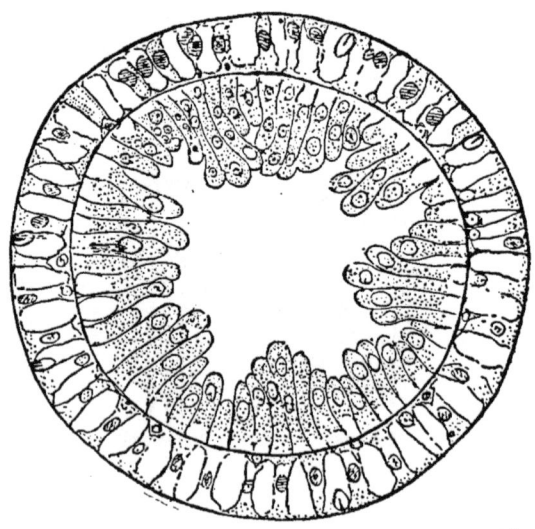

FIG. 16.—*Hydra*. Transverse section, highly magnified. (From Woodruff, after Shipley and McBride.)

The Echinoderms have returned almost completely to the older form, although they begin life as bilaterally symmetrical larvae.

The Body Cavity. Triploblastic animals, even in the lower phyla, develop another characteristic, the body cavity, or coelom, formed by a splitting of the middle layer. This cavity is important in many animals in connection with circulation and excretion, but in the higher phyla is little more than a cradle for the visceral organs. It occurs in a modified form in roundworms, and is found in all succeeding phyla (Fig. 17).

Metameric Structure. The Annelids, particularly the common earthworm, illustrate the development of repeated similar parts, called metameres. These are indicated by the ring-like subdivisions on the outside of the earthworm, but involve also a subdivision of the coelom and the arrangement of parts of various organic systems. This arrangement is modified in many ways, but in the earthworm not to such an extent that the succession of similar organs is obscured. Many segments show a portion of the alimentary tract, a ganglion of the nerve cord, lateral nerves, portions of longitudinal blood vessels, a pair of transverse blood vessels, a pair of nephridia, and other structures. Such a plan is the basis of development in all higher phyla, even in man, although in the adult there is little visible evidence of it save in the spinal column, ribs, and associated muscles and nerves. In the fish, one of the lowest vertebrates, everyone is familiar with the V-shaped bands of muscle which lie along the sides, each representing a metamere.

Appendages. While none of the preceding phyla have paired jointed appendages for walking, grasping and other functions, the Arthropods develop a long series, so characteristic that they have given the name to the order from the Greek ἄρθρον, a joint, and πούς, ποδός, a foot. Metameric structure is conspicuous in the phylum, although modified by the division of the body into three distinct regions, head, thorax and abdomen, and each metamere appears to have been fundamentally capable of producing a single pair of appendages. In the nineteen pairs present on the lobster or crayfish, there is a fine lesson in the possibilities of such structures (Fig. 18). Developed on a common plan of structure, they are formed for swimming, walking, grasping, accessory organs of generation, accessory mouth parts, mandibles, and sensory organs.

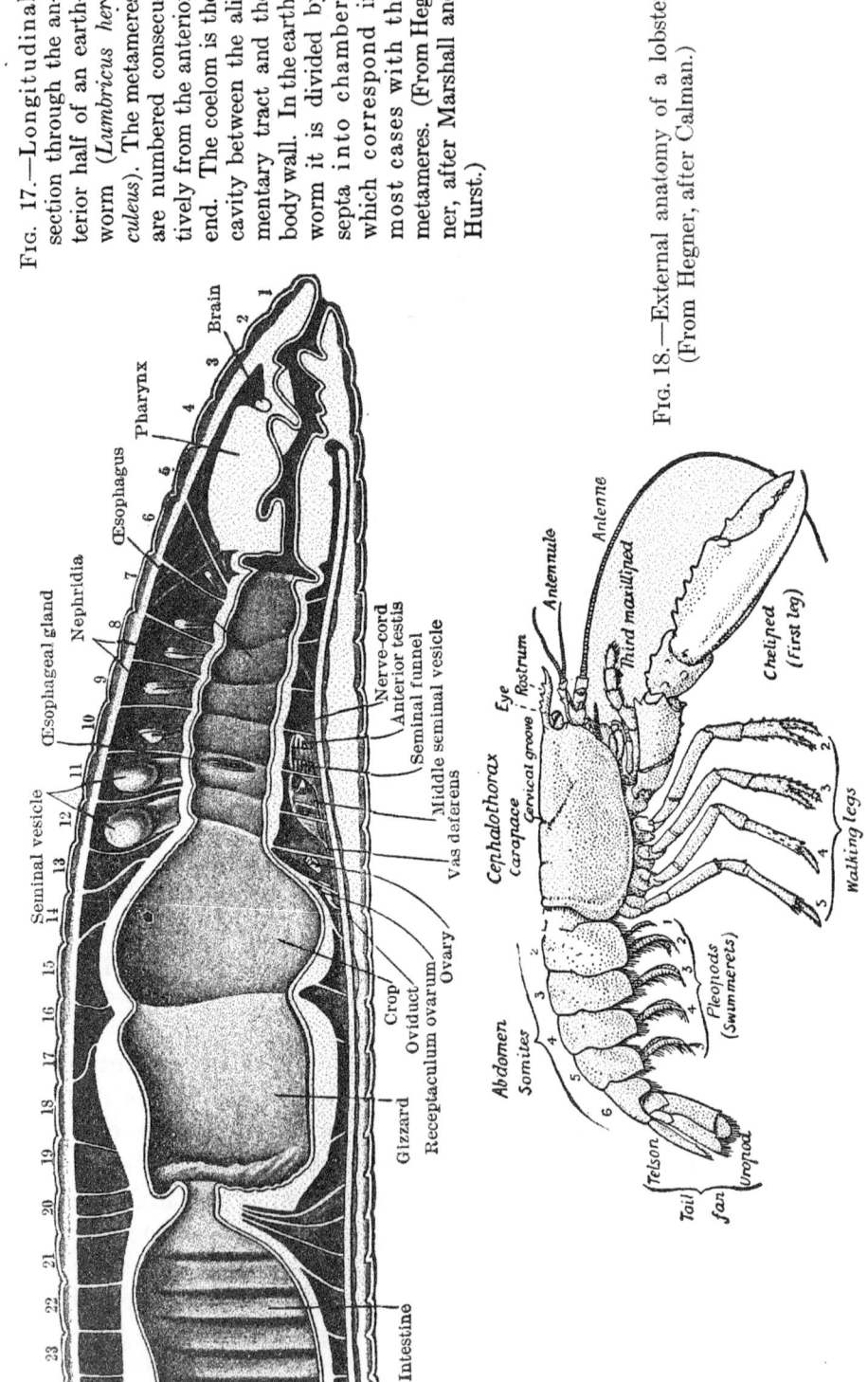

Fig. 17.—Longitudinal section through the anterior half of an earthworm (*Lumbricus herculeus*). The metameres are numbered consecutively from the anterior end. The coelom is the cavity between the alimentary tract and the body wall. In the earthworm it is divided by septa into chambers which correspond in most cases with the metameres. (From Hegner, after Marshall and Hurst.)

Fig. 18.—External anatomy of a lobster. (From Hegner, after Calman.)

The Mollusca. The Mollusca are highly specialized, and different from the other organisms. In the class Cephalopoda, including the squids, cuttle-fish, octopus, and other forms, we see chiefly an illustration of different ways of obtaining the same result, for the blood carries oxygen, but not through the same medium as ours, and the eyes are well developed, but different in fundamental structure from ours.

The Chordata. These organisms have departed so widely from the other phyla that they stand alone. While theories are available to explain their relationship to the remaining phyla, nothing affords more exact information than the facts of general structure which have been mentioned. Their isolation is so marked that the term invertebrate is commonly applied to all other phyla, and that of vertebrate to the better-known members of this one. They are characterized by the possession of an inner stiffening structure, the notochord, in contrast to the exoskeleton, or external stiffening structure developed by the skin of invertebrates; the central nerve cord is dorsal instead of ventral in position; and the pharynx, corresponding to the human throat, is at some stage provided with a series of paired lateral openings, the gill slits or pharyngeal clefts, which communicate with the exterior. The vertebrates make up the greater part of the phylum, and include the more familiar animals of every day experience. Six classes are recognized, as follows:

1. Cyclostomata—Round-mouthed fishes.
2. Pisces—True fishes, with hinged jaws.
3. Amphibia—Newts, salamanders, frogs and toads.
4. Reptilia—Lizards, turtles, crocodiles, snakes.
5. Aves—Birds.
6. Mammalia—Animals which have hair, and which secrete milk for the nourishment of their young.

Relationships among the vertebrates are best emphasized in connection with their embryological development and comparative anatomy, and are of sufficient interest to us to deserve special consideration in later chapters.

The whole great field of classification involves more than 700,000 known species of existing animals and almost 300,000 of plants. One class of Arthropoda, the Insecta, alone includes about 600,000 of these. Detailed knowledge of any limited group, for detailed knowledge is possible only of limited groups in the capacity of one individual, is a constant repetition of relationships of species with species and group with group. It is impossible to go into the minor relationships in such a work as this, but in the major points of similarity mentioned in this chapter we have an outline which greater elaboration merely amplifies.

Summary. The time involved in evolution is so vast that our observations of organisms cover, in proportion, only a moment. Nevertheless our records contain evidence of relationship. These are expressed in the classification of organisms into species and the association of species into successive groups of gradually increasing scope. Within this system relationship is evident because of similarities of structure, of which protoplasm and the cell are common to all organisms. Among the animals a definite plan of structure leads to a secondary grouping of the various phyla. Relationship of the various groups is based on the fact that they always present some expression or modification of the fundamental plan, however different they may be in details.

REFERENCES

MONTGOMERY, T. H., Proc. *Acad. Nat. Sci. Phil.*, 1902.
POWERS, E. B., *Am. Nat. XLIII*, 1909.
COULTER, J. M., BARNES, C. R., and COWLES, H. C., *Textbook of Botany*, 1910.
PARKER, T. J., and HASWELL, W. A., *Textbook of Zoölogy*, 3rd edition, 1922.
THOMSON, J. A., *Outline of Zoölogy*, 6th edition, 1914.
LINDSEY, A. W., Denison U. Bulletin, *Jn. Sci. Lab.* XX, 289–305, 1924.
HEGNER, R. W., *College Zoölogy*, revised edition, 1926.
WOODRUFF, L. L., *Foundations of Biology*, 3rd edition, 1927.

CHAPTER IV

THE RELATIONSHIP OF EXISTING ORGANISMS
(Continued)

2. EMBRYOLOGY OF VERTEBRATES

The relationships of no animals are more striking han those of the vertebrates, and no phase of vertebrate relationship is more fascinating than the similarity of their embryonic development. This field has a twofold bearing on evolution, for it illustrates actual relationships in the formative stage of the individual and at the same time points strongly to the probable succession of changes which has brought about the transition from lowest to highest.

Our knowledge of embryology has contributed so conspicuously to the science of comparative anatomy that it is difficult in many cases to separate the two; indeed, the latter cannot adequately be treated without the inclusion of a considerable body of facts from embryology. For this reason the matter presented here must partake somewhat of both fields. It falls into two categories, (1) the resemblance of embryos of the different classes and (2) the resemblance of embryos to the adults of lower classes.

The Resemblance of Embryos of Different Classes

The individual first becomes an entity when the single cell from which it is to develop is matured, whatever may be the nature of this cell. From this point to the completion of its body it is an embryo. Completion does not necessarily mean birth, for most embryos are complete organisms long before birth. Nor does it mean the realization of all the possibilities of differentiation inherent in the individual, for many of these are not normally expressed until long after birth. It means rather the formation of those organs which constitute a complete individual, whether or not they may later atrophy or undergo further development as life proceeds.

Cleavage. The first step in development has already been mentioned as cleavage, or the splitting of the fertilized ovum into successive generations of cells. Such division always occurs in the vertebrates, but its results may be superficially very different according to the amount of food included in the cell. The minimum is found in the eggs of the lower chordates and many fishes,

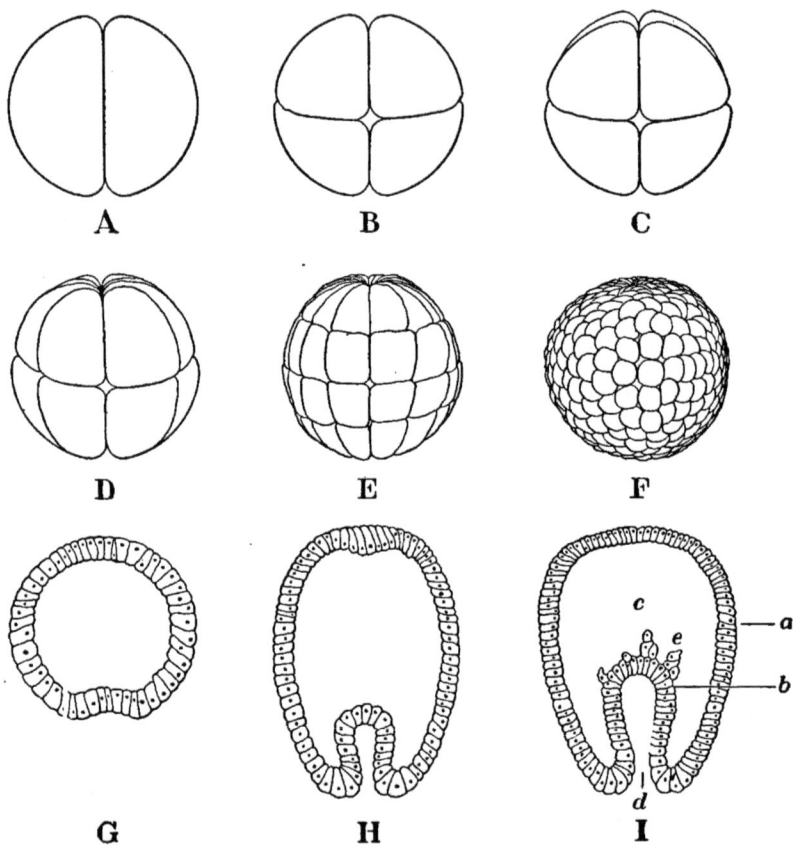

Fig. 19.—Early stages in the development of the egg of a sea urchin. A–F, cleavage and formation of the blastula; G, section of blastula showing the beginning of gastrulation; H–I, early and late gastrula stages. a, ectoderm; b, endoderm; c, blastocoele; d, blastopore, leading into the enteric cavity; e, cells arising from the endoderm, destined to form the mesoderm. (From Woodruff.)

the maximum in the reptiles and birds in whose eggs the entire yolk is the fertilized ovum. In true mammals, whose young are nourished by a placental connection with the mother, such a concentration of food to supply the developing embryo is unnecessary, hence the ovum is small. The stored food or yolk is inert matter which delays the process of subdivision.

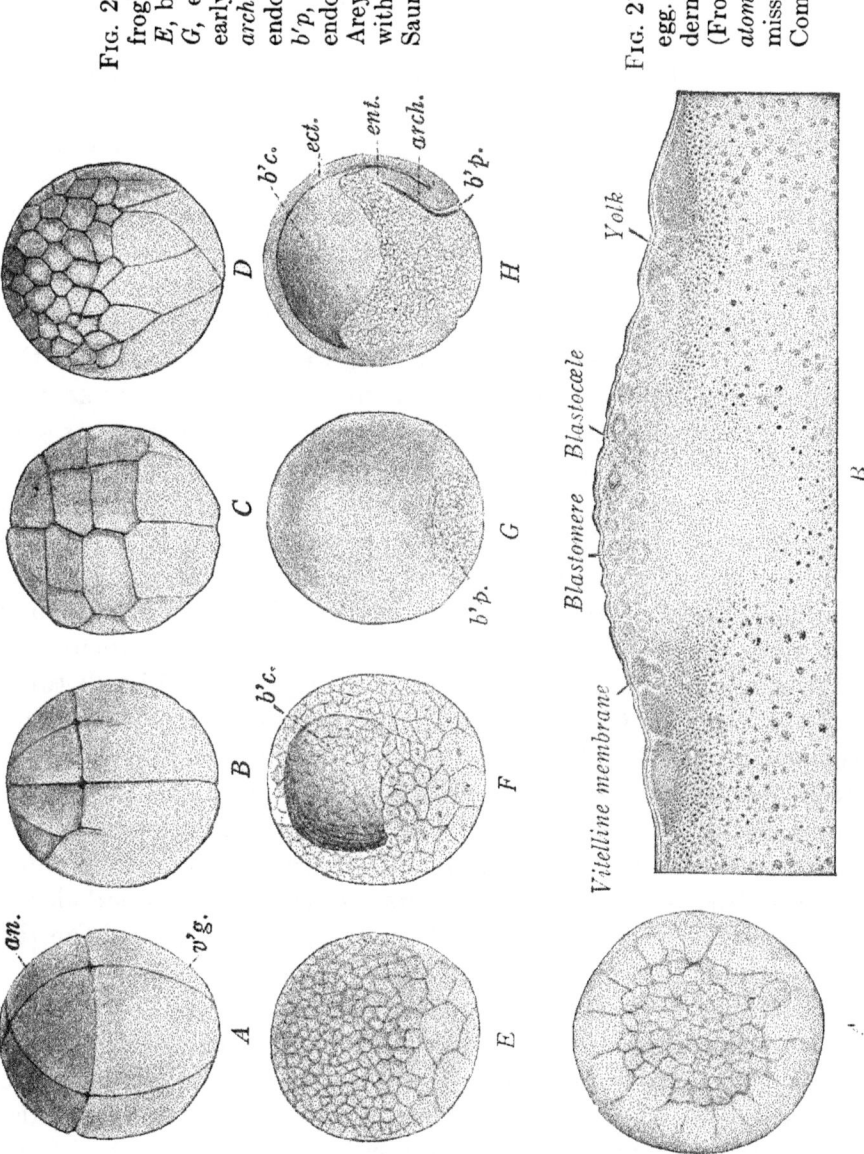

Fig. 20.—The development of the frog's egg. *A–D*, cleavage stages; *E*, blastula; *F*, section of blastula; *G*, early gastrula; *H*, section of early gastrula. *an*, animal cells; *arch*, archenteron, the primitive endodermal cavity; *b'c*, blastocoele; *b'p*, blastopore; *ect*, ectoderm; *ent*, endoderm; *v'g*, vegetal cells. (From Arey's *Developmental Anatomy*, with the permission of the W. B. Saunders Company.)

Fig. 21.—Cleavage of the pigeon's egg. *A*, surface view of blastoderm; *B*, section of blastoderm. (From Arey's *Developmental Anatomy*, after Blount, with the permission of the W. B. Saunders Company.)

Eggs of the first type give rise by a succession of cleavages to a hollow sphere called the blastula, made up of cells very nearly alike in size (Fig. 19). The presence of a moderate amount of yolk gives rise to a slight modification of this form as shown in the blastula of the frog (Fig. 20), while the relatively enormous amount of yolk in the egg of a bird crowds all of the cytoplasm to one pole, with the result that cleavages cannot pass entirely through the sphere. In these eggs the protoplasm alone is cut up into a cap of cells underneath which a small space is equivalent to the cavity of the spherical blastula or blastocoele, so conspicuous in the other forms (Fig. 21).

Gastrulation. Following cleavage the simplest blastulas (Fig. 19) cave in on one side until the concave layer of cells is in contact with the convex side, thus establishing a sac with a two-layered wall by whose formation the blastocoele has been obliterated. The enclosed cavity, or archenteron, opens to the exterior of the embryo by the blastopore. If we compare the series of figures of frogs' eggs (Fig. 20) we note at once that the blastocoele is relatively so small that this process cannot be accomplished. Instead, a crescentic lip evaginates from the thinner part of the blastula, grows down and around the structure until its ends meet and form a rounded opening, and the same result is obtained. The blastocoele is obliterated, while the cavity present is enclosed by a two-layered wall and opens to the exterior of the embryo. That this archenteron is almost completely filled by the remaining yolk mass which bulges from the inner cells is merely incidental to the accumulation of yolk. In the bird's egg the cap of cells turns under at one point, and by proliferation from this point finally becomes double (Fig. 22). The yolk is then capped by two layers of cells in place of one, although it is so bulky that these layers do not enclose it until much later. For homologies here we must note that the blastopore has another definite characteristic: it is bounded by the united outer and inner layers, or ectoderm and endoderm, and this is also characteristic of the point at which the cells on the hen's egg turn under to form the second layer.

The Neurenteric Canal. In order to compare the early development of the mammal it is necessary to pass on to other structures. The mesoderm in all vertebrates arises between the two original layers, and from the three all structures of the body develop. As an early step a region along the mid-dorsal line of the

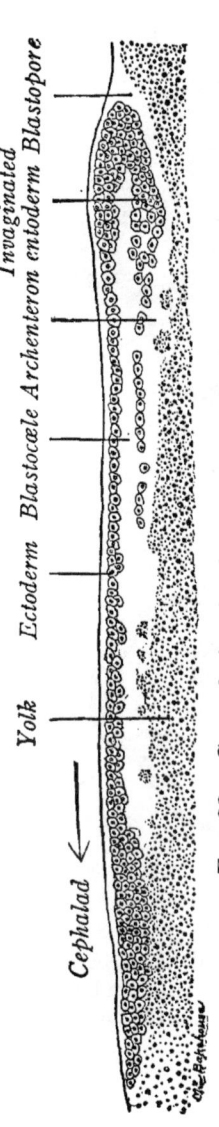

Fig. 22.—Gastrulation in the pigeon, from a longitudinal section of the blastoderm. (From Arey's *Developmental Anatomy*, after Patterson, with the permission of the W. B. Saunders Company.)

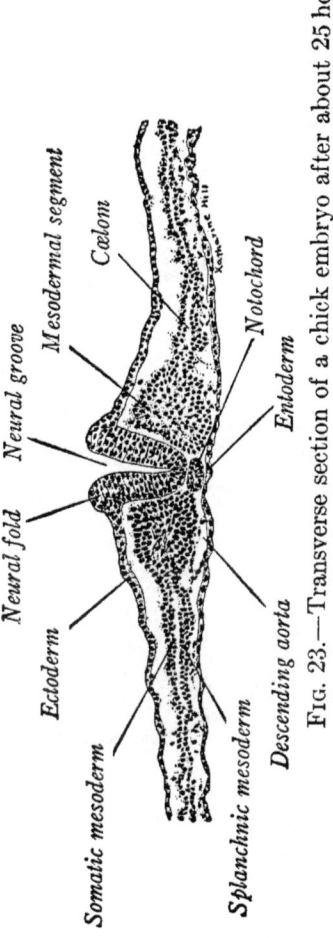

Fig. 23.—Transverse section of a chick embryo after about 25 hours incubation. (From Arey's *Developmental Anatomy*, with the permission of the W. B. Saunders Company.)

body becomes depressed, and flanked by two ridges (Fig. 23). This process of involution finally results in a meeting and fusion of the two ridges, now become folds, so that an inner longitudinal tube of ectoderm is formed, surmounted by a continuous layer of

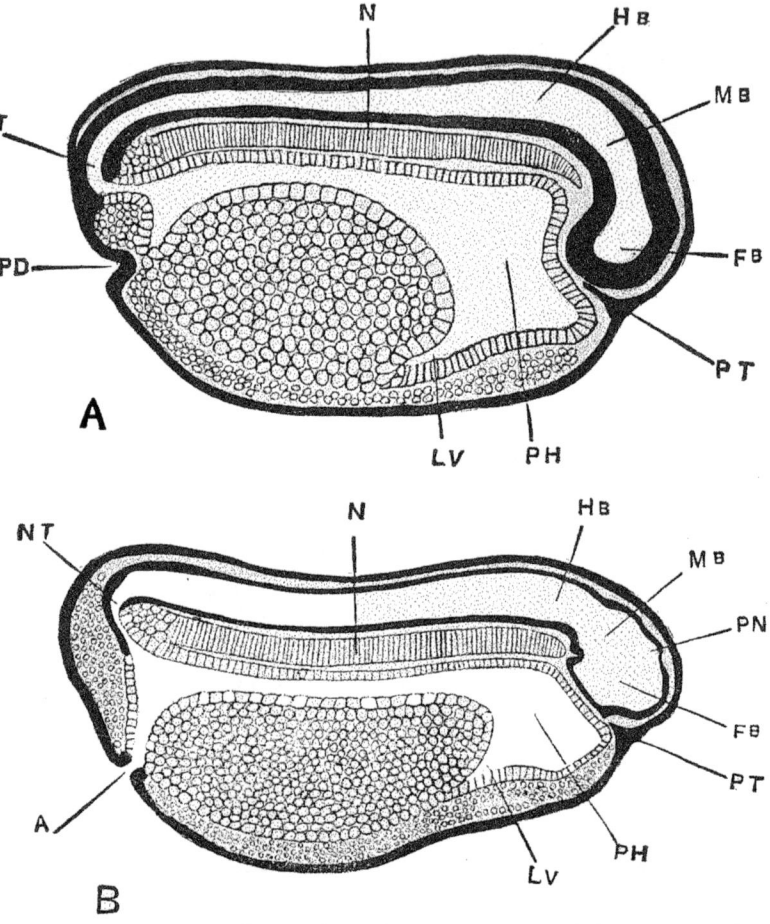

Fig. 24.—Longitudinal sections of developing embryo of frog. A, earlier, B, later stage. N, notochord; Fb, Mb, and Hb, fore-, mid-, and hind-brain; PT, pituitary anlage; PD, proctodaeum; A, anus; PH, enteron; NT, neural tube at the neurenteric canal. (From Holmes.)

superficial ectoderm. As a result of the relation of the folds to the blastopore, the cavity of the tube communicates with that of the archenteron (Fig. 24). The tube is destined to become the central nervous system, and is called the neural tube, hence this continuous passage is called the neurenteric canal. The connection is later broken, the dorsal tube persisting throughout life in the

EXISTING ORGANISMS—EMBRYOLOGY 45

brain and spinal cord, while the ventral part forms anterior and posterior connection with the exterior and gives rise to the alimentary tract and its derivatives.

The Foetal Membranes. Associated with terrestrial life the vertebrates above the Amphibia develop a remarkable series of structures called the foetal membranes, which compensate the differences between the heavy water as a medium in which the lower forms develop and the light, dry air which surrounds the eggs of the reptiles and birds (Fig. 25). Two of these membranes are formed in the hen's egg by the growth of folds from the layers of tissue surrounding the young embryo. The folds consist of an outer layer of ectoderm and an inner layer of mesoderm. By a process of involution, similar to the formation of the neural tube, these folds unite to form

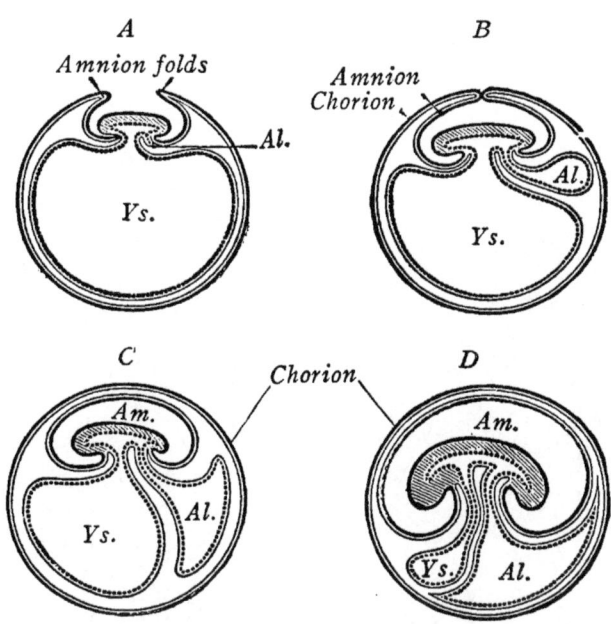

FIG. 25.—Diagram illustrating the development of the foetal membranes Al, allantois; Am, amniotic cavity; Ys, yolk sac. The chorion of this figure is the serosa of the text. (After Gegenbaur in McMurrich; taken from Arey's *Developmental Anatomy* with the permission of the W. B. Saunders Company.)

two layers covering the embryo; the outer consists of ectoderm outwardly and of mesoderm inwardly, and is called the serosa, while in the inner, called the amnion, the order is reversed. The amnion is filled with fluid which protects the delicate tissues of the growing embryo in every way as effectively as the ocean protects the developing embryo of a fish. As the outer edges of the cap of growing tissue extend, they finally meet to enclose the yolk. The serosa is then a hollow sphere, which must be involved in any contact with the outer world. Inside of it another sac is then completed, lined with endoderm

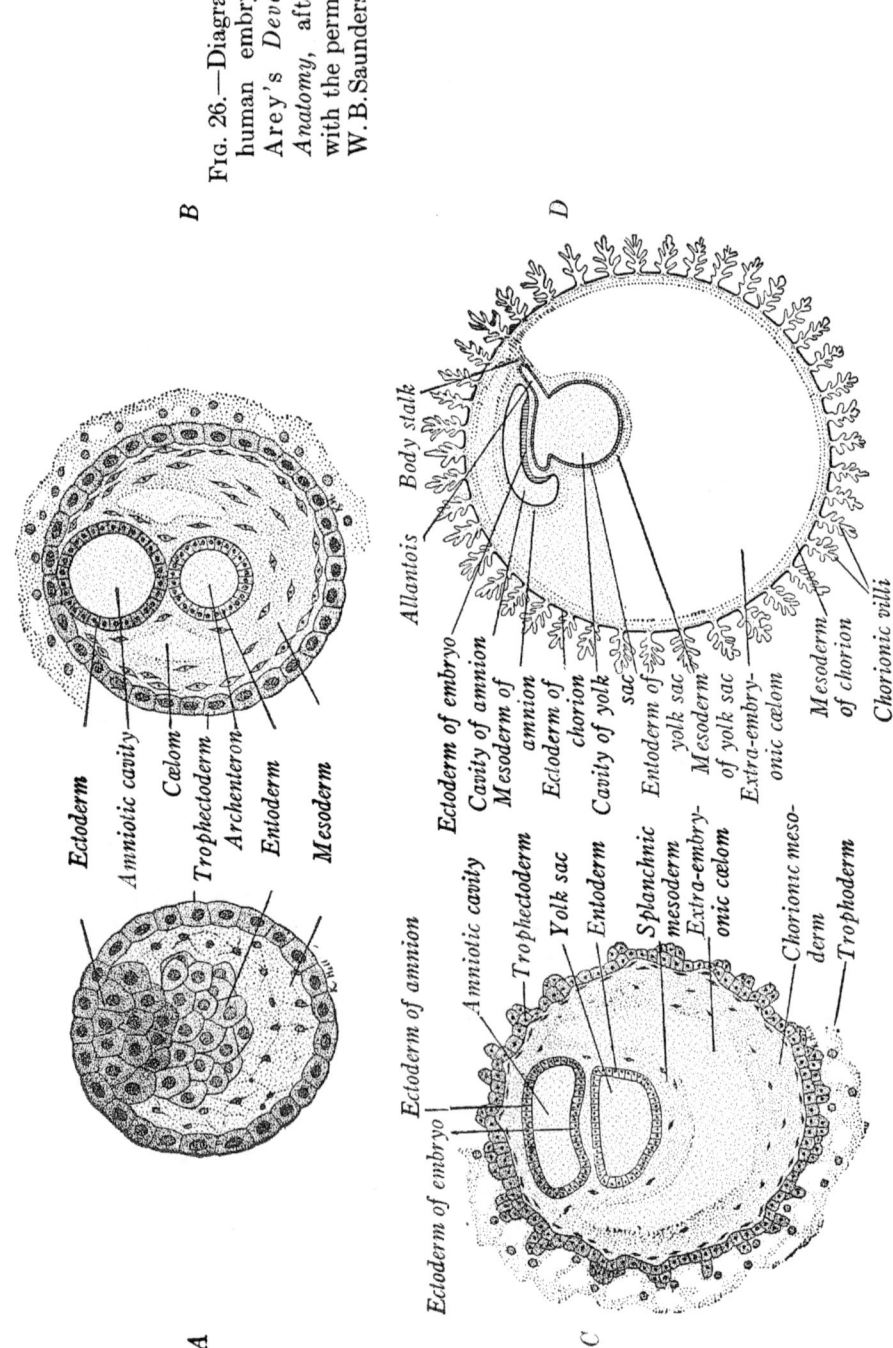

Fig. 26.—Diagrams of early human embryos. (From Arey's *Developmental Anatomy*, after Prentiss, with the permission of the W. B. Saunders Company.)

and covered with mesoderm. The intimate association of this sac with the yolk has given rise to the name yolk sac. Last of all an evagination similar to the yolk sac in structure is formed from the hind gut, expands, and in contact with the serosa provides for respiration in the chick. This is the allantois.

Mammalian Homologies. In the mammalian embryo all of these things are found. The first subdivisions in man produce a hollow sphere which later corresponds to the outer ectodermal layer of the serosa. A mass of cells buds off inside of this, which in turn produces a second mass; these are ectoderm and endoderm respectively. Mesodermal tissue fills in the spaces between all three (Fig. 26A). Finally the masses of ectoderm, endoderm and mesoderm split, forming cavities which increase in size until the original ecto- and endodermal masses are connected by a slender stalk

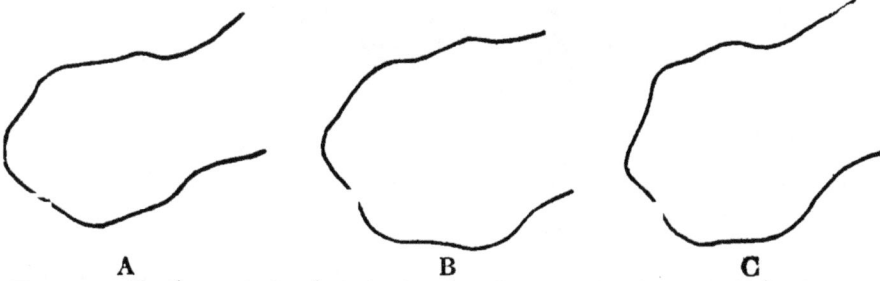

FIG. 27.—Outlines of the limb buds of embryos. A, pig; B, rabbit; C, man. Each shows five rounded prominences which are the first evidence of digits.

with the outer sphere, now obviously similar to the serosa previously described (Fig. 26B, C, D). In the stalk a diverticulum from the hind gut of the embryo runs out toward the serosa, small, it is true, but with exactly the relationships of the allantois, which it is. The embryo develops from the layers of tissue between the cavities of the ectodermal and endodermal masses, and comes to lie in the first of these, which thus corresponds in all particulars with the amnion of the birds and reptiles, and lastly, although the absence of yolk robs it of its original function of absorbing nourishment, the endodermal sac has the connections and structure of the yolk sac. In addition to these homologies we find that in the embryo itself, the neurenteric canal is as well developed as in the lowest chordates.

Organogeny. *General Similarity.* After these early steps in the development of vertebrate embryos, the definite structures which appear show an equally close resemblance. The brain is first a series of three expansions of the neural tube at the anterior end of the body. Limbs first bud out from the body wall as rounded projections, and digits in turn bud from them in those animals which possess such structures (Fig. 27). The mouth and anus form as invaginations of ectoderm, the stomodaeum and proctodaeum, which meet the endoderm and break through to form the continuous tube of the alimentary tract. The nostrils

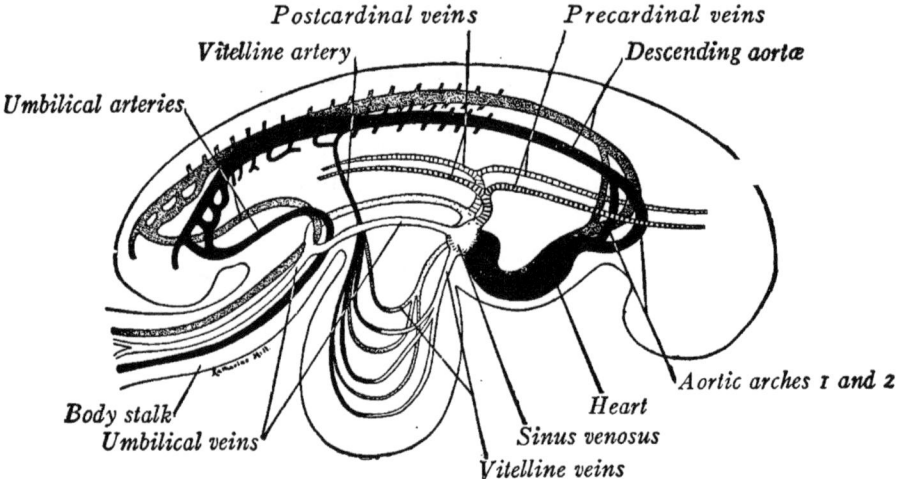

Fig. 28.—Lateral aspect of the circulatory system in a human embryo of 2.6 mm. Diagrammatic. (From Arey's *Developmental Anatomy*, after Felix-Prentiss, with the permission of the W. B. Saunders Company.)

are at first merely depressions of ectoderm which become associated with the central nervous system as organs of smell, while in the terrestrial forms they also play a part in respiration, and for this purpose join the stomodaeum. In terrestrial forms the trachea and its subdivisions in the lungs are first only evaginations from the primitive gut. By later subdivision of the stomodaeum, they are more definitely associated above the amphibia with the olfactory part of the respiratory system. The swim bladder of a fish corresponds in origin to the lungs. Glands, such as the liver and pancreas, in all vertebrates bud out from the alimentary tract.

The Circulatory System. Particularly noteworthy is the circulatory system, for when it once develops tubular arteries and veins and a chambered heart, it is similar in all vertebrate embryos. A pair of veins enters the body from the yolk sac and a pair from the region of the allantois. These join the heart, whence the blood passes forward into a pair of aortae. The aortae curve dorsad and pass along the body, giving off branches which supply both the body and the extraembryonic regions whence the two pairs of veins drain blood. The blood which these branches distribute to the body is collected by a pair of anterior or precardinal veins, and a pair of postcardinals. On each side of the body these vessels unite, to enter the heart as a common trunk. Such a plan is common to all vertebrate embryos early in their development (Fig. 28). Most significant of all is the series of aortic arches formed between the ventral and descending aortae anterior to the

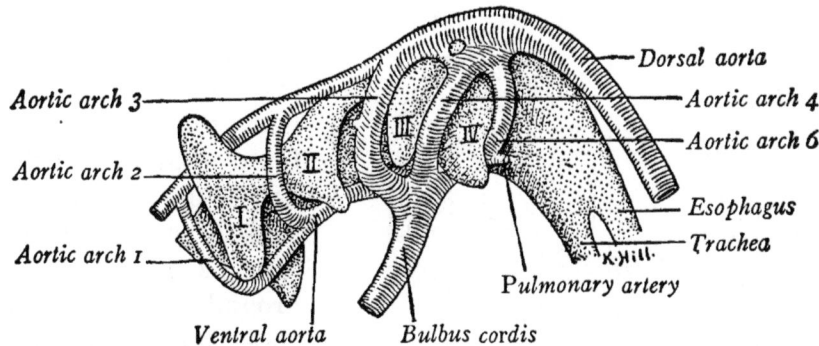

Fig. 29.—Reconstruction of the aortic arches and pharyngeal pouches of a 5 mm. human embryo. The pouches are indicated by Roman numerals. (From. Arey's *Developmental Anatomy*, after Tandler, with the permission of the W. B. Saunders Company.)

heart. The connections by which they are joined originally are the first pair of six. The remaining five pairs form as outgrowths of the two aortae of the same side which meet, forming dorsoventral connections. While these arches are very different in the adults of the several classes, they appear in the embryos with very similar form, and the embryo of man is not excepted (Fig. 29).

The Eye. The development of the eye in vertebrates is initiated by the appearance of an evagination of the lateral walls of the forebrain, the anterior of the three primitive expansions of the neural tube (Fig. 30). The evagination expands at its outer end, and here invaginates in turn to form a double cup, the optic cup, connected to the forebrain by the more slender optic stalk. This cup is destined to form the pigmented and nervous layers of the retina in the adult eye. The lens develops from an invagination of ectoderm opposite to the optic cup, which is later freed from the outer ectoderm as a hollow vesicle. The optic nerve grows back to the brain from the inner layer of the optic cup, and the entire eye is enveloped by the hard sclerotic coat derived from the mesodermal layer. Later on, in all forms which have eyelids, the skin in front of the eye forms two folds which grow until they meet and fuse, later on to separate again as the eyelids. In some animals the lids remain fused until after birth, hence the young are said to be born blind. The skin continues around the edges of these folds and joins the delicate conjunctiva which covers their inner surfaces and extends across the exposed portion of the eyeball. In this process the eyes of all vertebrates correspond in so far as their adult structures correspond.

Accompanying these and many other points of similarity in the parts of vertebrate embryos a striking superficial resemblance has been noted in the early embryos of the five highest classes of

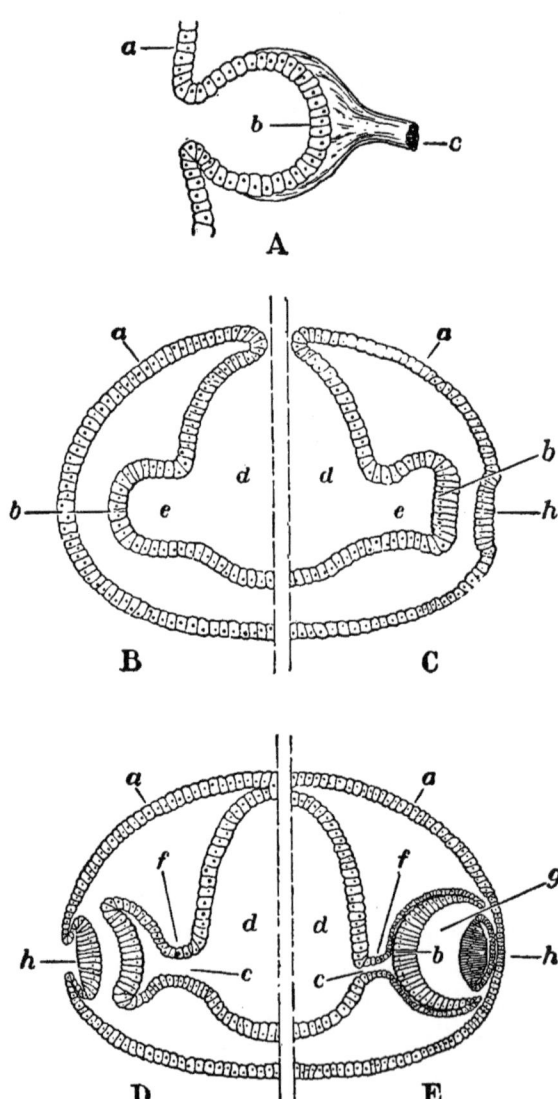

FIG. 30.—Diagrams illustrating the formation of the eye in an invertebrate (A) and a vertebrate (B, C, D, E, successive stages). *a*, ectoderm; *b*, retinal layer; *c*, future position of the optic nerve; *d*, cavity of the brain; *e*, optic vesicle; *f*, optic stalk, later replaced by the optic nerve; *g*, cavity of the optic cup, later the posterior chamber of the eye; *h*, developing lens. (From Woodruff.)

vertebrates. Explanation is inadequate to give an idea of this resemblance, which is shown in Fig. 31. The structures which have been described show that this resemblance is not merely superficial, but is based on fundamental structure.

RESEMBLANCE OF EMBRYOS TO ADULTS OF LOWER FORMS

The resemblance of very early stages of vertebrate development to the lowest animals has already been brought out. All have their origin in a single cell, the fertilized ovum, morphologically equivalent to the single cell which makes up the entire body of a protozoön. The blastula is quite similar to such colonial protozoa as *Volvox* (Fig. 32), although we must recognize that there is some differentiation in the cells of the blastula which is not present in the cells of a *Volvox* colony. The simple gastrula has been compared with coelenterates, such as *Hydra* (Fig. 33) and the modifications of gastrulation brought about by the accumulation of yolk in the ovum result in structures which can readily be homologized with those of simpler forms. Beyond this point we must seek resemblance of vertebrate with vertebrate as their structures appear.

The Notochord. Immediately after gastrulation a structure called the notochord appears in all chordates (Fig. 34). It is a rod of tissue whose origin is similar to that of the mesoderm although their development is not directly associated. The notochord extends through the body longitudinally between the neural tube and the alimentary tract, and just ventral to the former so that it corresponds in position to the main axis of the spinal column. In the cyclostomes it persists throughout life; in the remaining vertebrate classes it is well developed only in the embryo and is either vestigial or absent in the adult. One striking evidence of relationship is the similarity of development of the notochord in reptiles and mammals. Even the birds, which are in general rather closely related to the reptiles, show a marked modification of the process. As the bones of the axial skeleton develop, centra of the vertebrae replace the notochord at least in part.

The Skeleton. *Lower Fishes.* If we examine the most primitive fishes, the sharks, we find that the skeleton consists of cartilage alone. It is divided into axial, appendicular, and visceral parts as in the rest of the vertebrates but the structure of each part is

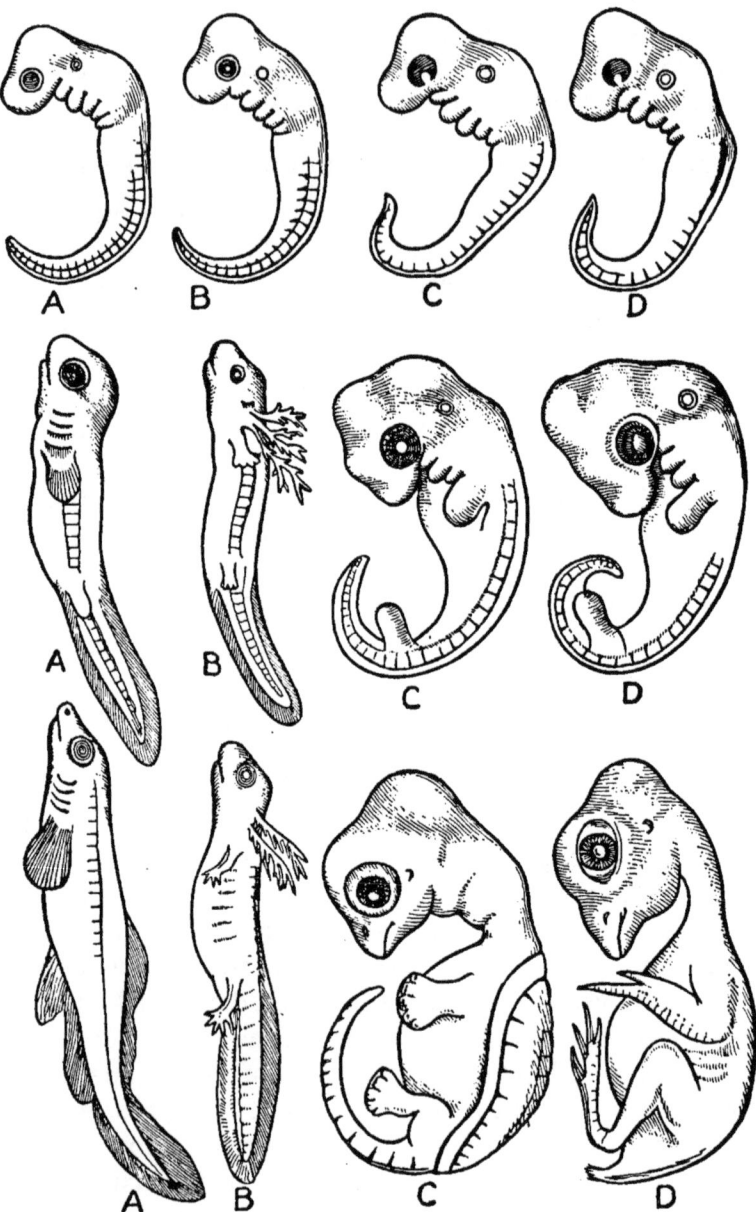

FIG. 31.—Series of vertebrate embryos at three comparable and progressive stages of development. A, fish; B, salamander; C, tortoise; D, chick. (*Continued on next page.*)

EXISTING ORGANISMS—EMBRYOLOGY

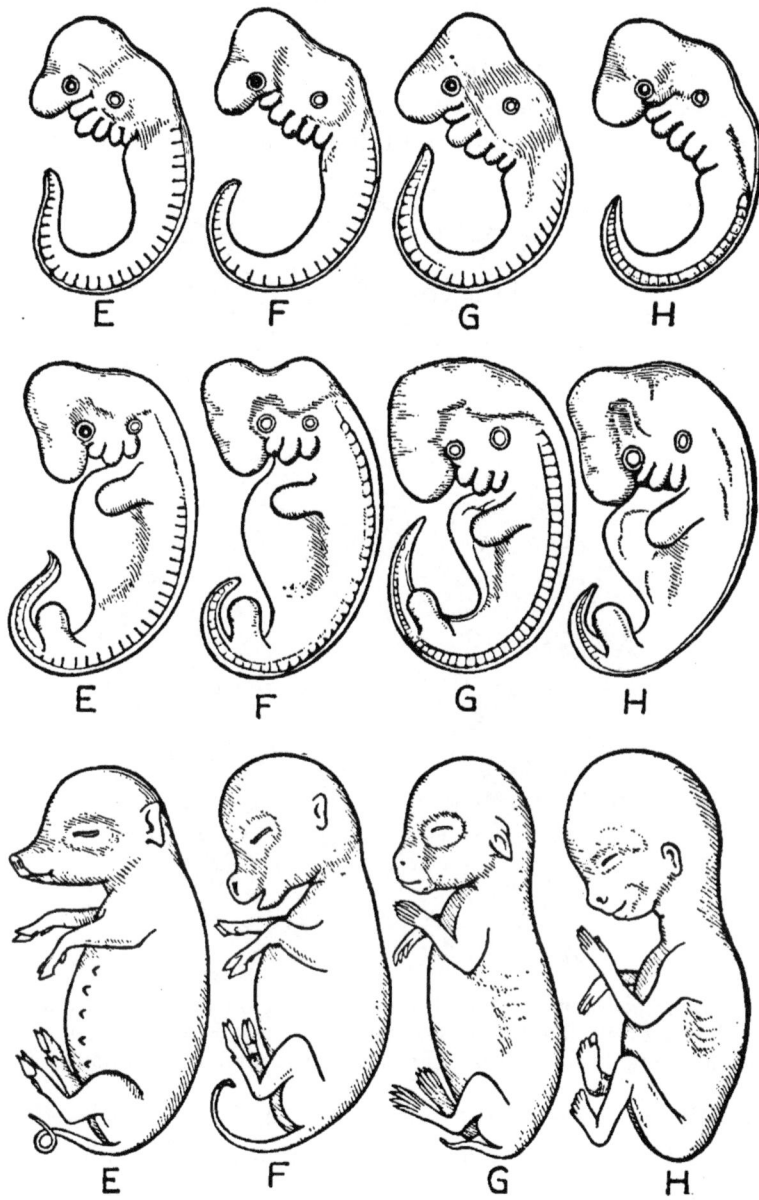

FIG. 31.—(continued). E, hog; F, calf; G, rabbit; H, human. (From Romanes' *Darwin and After Darwin*, after Haeckel, with the permission of the Open Court Publishing Company.)

relatively simple. The head contains no bony shell encasing the brain but does include a mass of cartilage of peculiar form in which the brain rests. This structure is called the chondrocranium (Fig. 35). The vertebral column consists of a series of vertebrae with hollow ends, hour-glass-shaped in longitudinal section, between

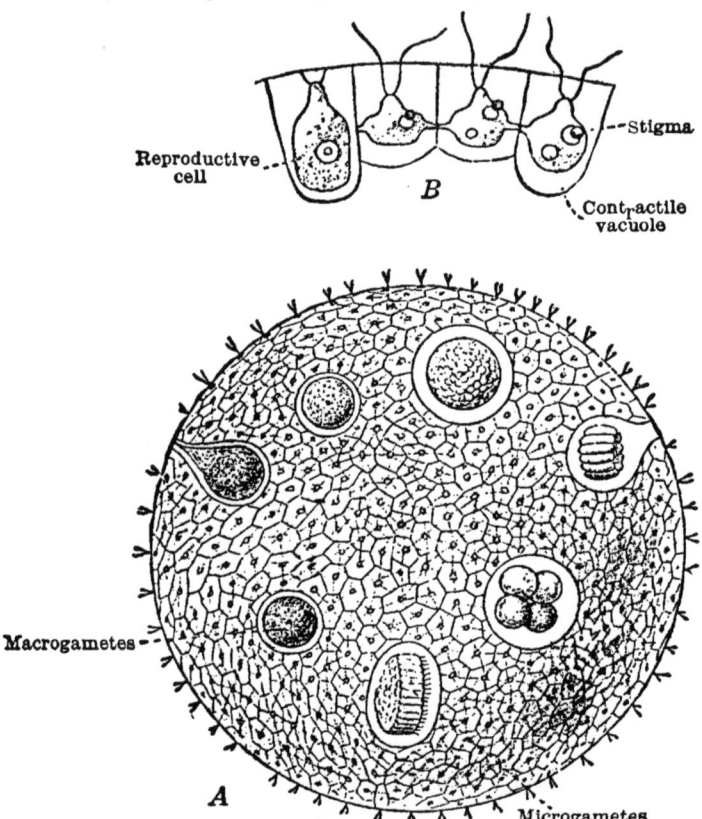

FIG. 32.—*Volvox globator*, a colonial unicellular organism. *A*, a sexually ripe colony showing reproductive cells in various stages; *B*, a portion of the edge of the colony highly magnified. (From Hegner, after Bourne and Kölliker.)

which small remnants of the notochord persist. Associated with the head and pharyngeal region is the visceral skeleton, consisting of the upper jaw (pterygo-quadrate cartilage), lower jaw (Meckel's cartilage) and six pairs of cartilages behind the mouth, the first constituting the hyoid arch and the remainder supporting the gills and called branchial arches. The appendicular skeleton consists of two horseshoe-shaped pieces of cartilage in the pectoral and pelvic regions respectively, each bearing a pair of fins based on cartilage supports.

EXISTING ORGANISMS—EMBRYOLOGY

Higher Fishes. In the higher fishes similar parts appear, but the cartilage is largely replaced by bone. To the chondrocranium, which also becomes bony, are added other bones which form a dorsal shell to enclose the head (Fig. 36). These bones are called dermal bones, and are not formed of cartilage at any stage. The vertebral

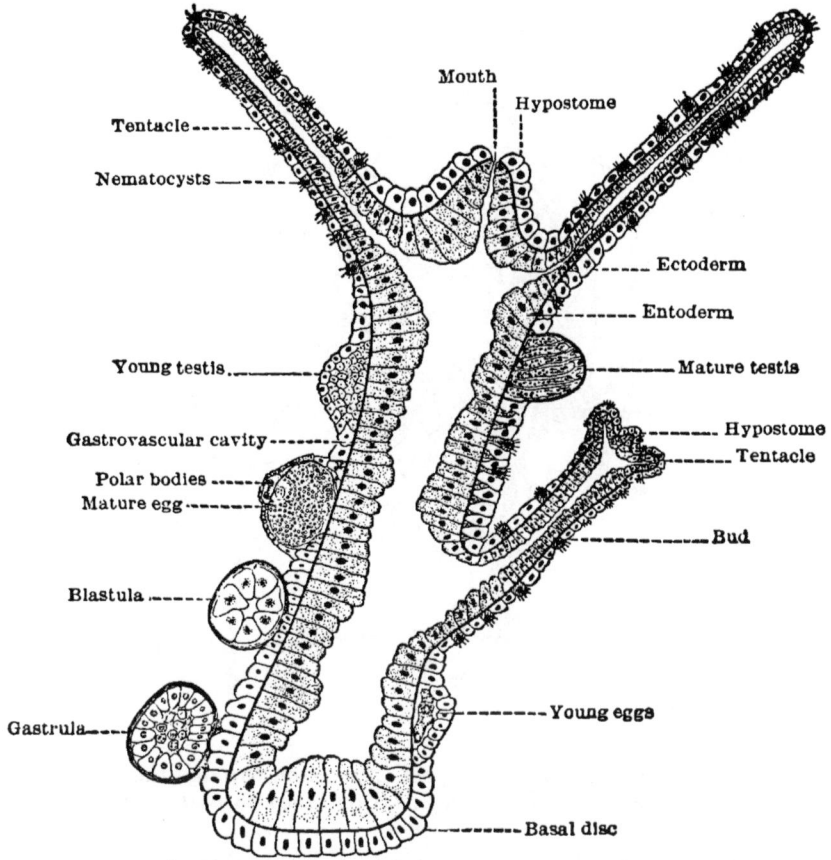

FIG. 33.—A longitudinal section of *Hydra*. Not all of the structures shown occur on one animal at the same time. (From Hegner.)

column may remain much the same in form, although ossified. The jaws, like the skull, are enclosed and strengthened by dermal bones, and similar additions to the pectoral girdle occur. The chief differences are the general tendency to ossification of the cartilaginous structures and the addition of dermal bones.

Other Classes. As the embryo develops in classes above the fishes, the first evidence of skeletal structure after the notochord is the formation of cartilages in various regions. In the head, for example, a mass of cartilage appears similar to the chondrocranium of the dogfish. This mass later ossifies to form most of the occipital, sphenoid, ethmoid and temporal bones, to which are added the other bones of the skull by development directly from membranous regions, without a primary cartilaginous stage (Fig. 37).

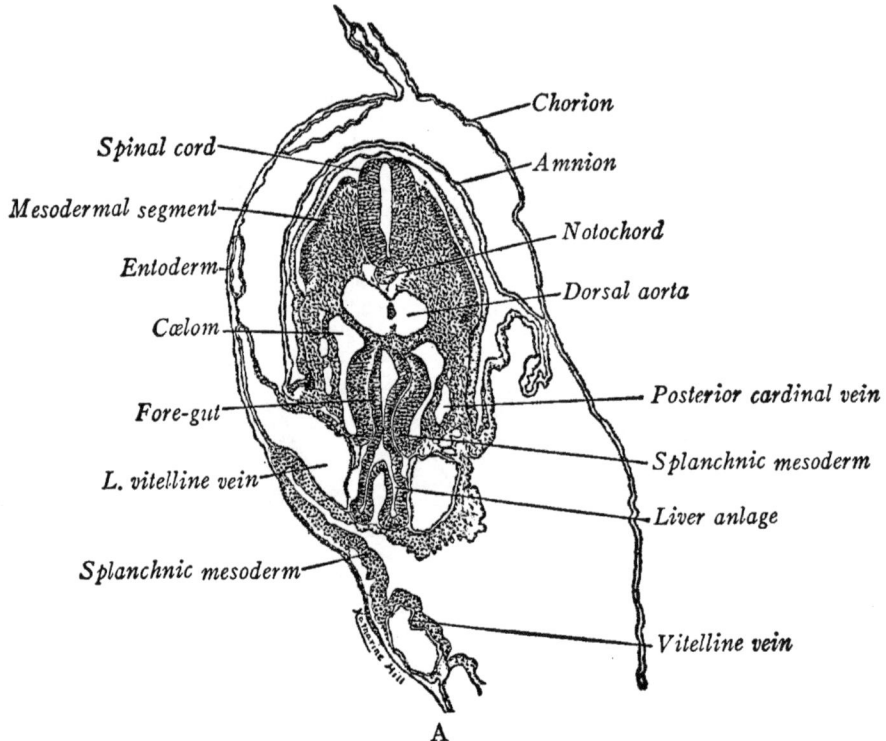

Fig. 34.—Sections of vertebrate embryos. A, section through the liver anlage of a chick embryo of two days. (*Continued on next page.*)

The vertebrae arise first as a series of hour-glass-shaped structures (Fig. 38), surrounding and finally obliterating the notochord as they take on the definitive form of the adult. The visceral skeleton remains largely cartilaginous, and contributes to the formation of the larynx and upper part of the trachea, the bones of the middle ear, and as in the fishes, the hyoid apparatus and lower jaws. The last is a striking duplication of the state illustrated by the various fishes. In early embryos sections of the lower jaw show a rod of cartilage on each side. This is Meckel's cartilage which persists as the sole support of the mandible in the sharks. In slightly older embryos a condensation of mesenchymal

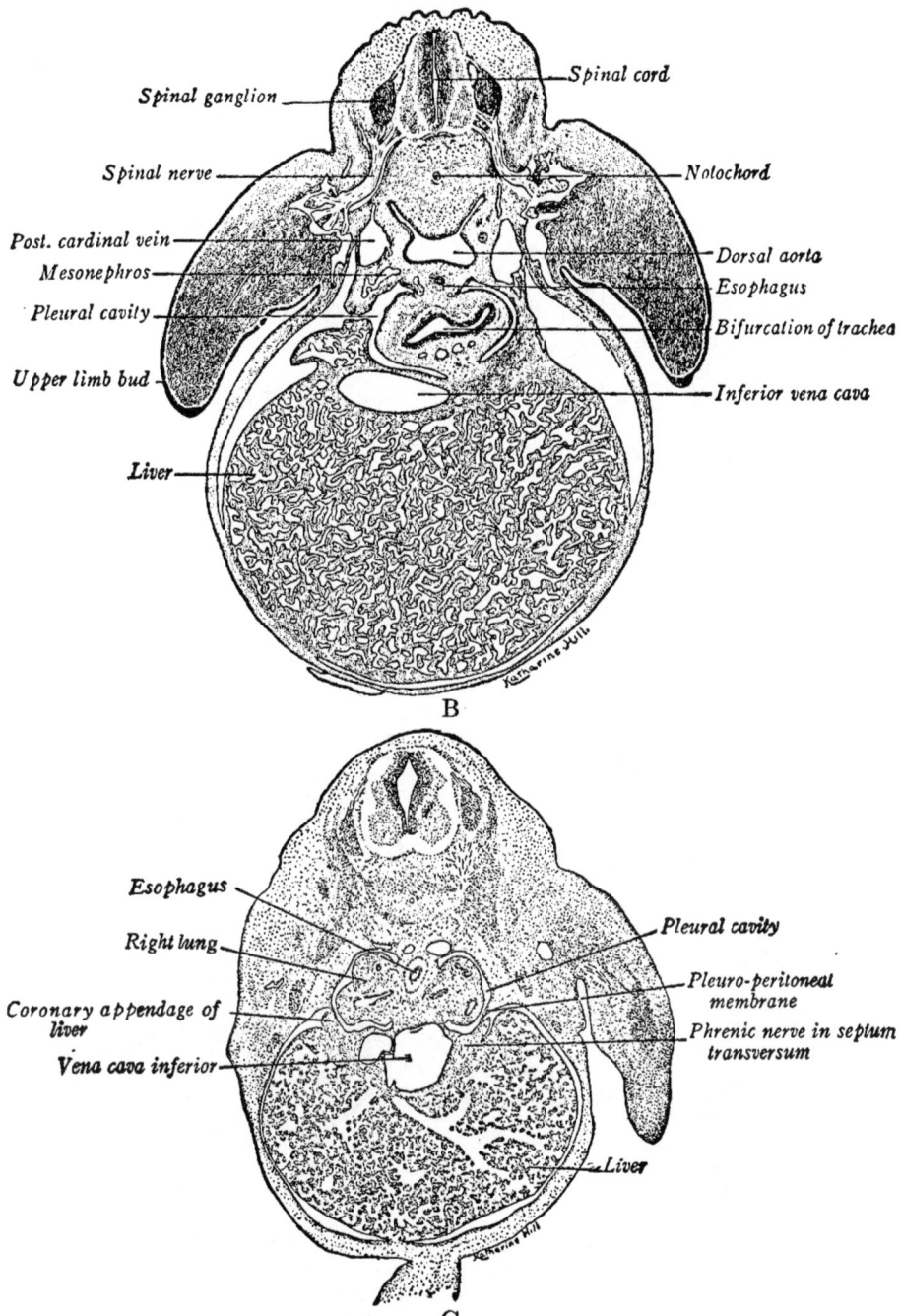

Fig. 34.—*(continued)*. B, section through the liver and anterior limb buds of a 10 mm. pig embryo; C, section through liver and one limb bud of a 10 mm. human embryo (after Prentiss). The notochord is visible in C but is not labelled; compare with B. (From Arey's *Developmental Anatomy*, with the permission of the W. B. Saunders Company.)

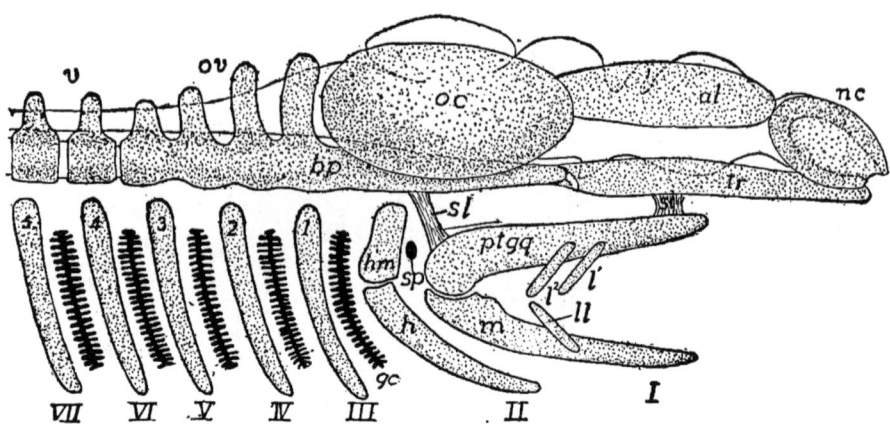

Fig. 35.—Diagram of an early elasmobranch chondrocranium in side view; the brain outlined. *al*, alisphenoid plate; *bp*, basal plate; *gc*, gill clefts; *h*, hyoid; *hm*, hyomandibular; *l*, upper labials; *ll*, lower labials; *m*, Meckel's cartilage; *nc*, nasal capsule; *oc*, otic capsule; *ov*, occipital vertebrae; *ptgq*, pterygoquadrate; *sl*, suspensory ligaments; *sp*, spiracle; *tr*, trabeculae; *v*, vertebrae; I–VII, visceral arches; 1–5, branchial arches. (From Kingsley's *Comparative Anatomy of Vertebrates*, with the permission of P. Blakiston's Son & Co.)

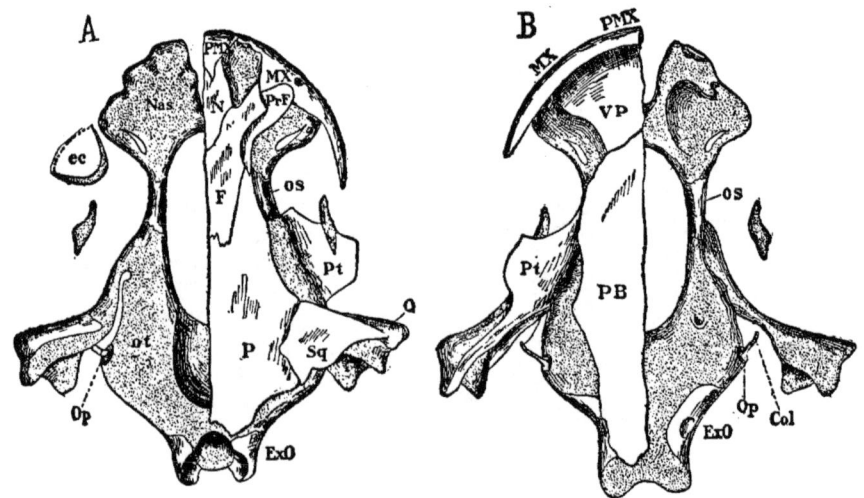

Fig. 36.—Dorsal (A) and ventral (B) views of the skull of *Cryptobranchus alleghemiensis*, a primitive salamander. Dermal bones: PMX, premaxillary; MX, maxillary; N, nasal; F, frontal; PrF, prefrontal; P, parietal; Sq, squamosal; Pt, pterygoid; VP, vomero-palatine; PB, parabasal. Cartilage bones: OS, orbitosphenoid; Q, quadrate; ExO, exoccipital; Op, operculum. Other parts: Col, columella; Nas, nasal capsules; ec, eye capsule; ot, otic capsule. In both figures the dermal bones are removed from the left half. (From Wilder's *History of the Human Body*, with the permission of Mrs. H. H. Wilder and Henry Holt and Company.)

tissue appears near this cartilage, bone spicules form in it, and a definite bony structure makes its appearance. This bone envelops

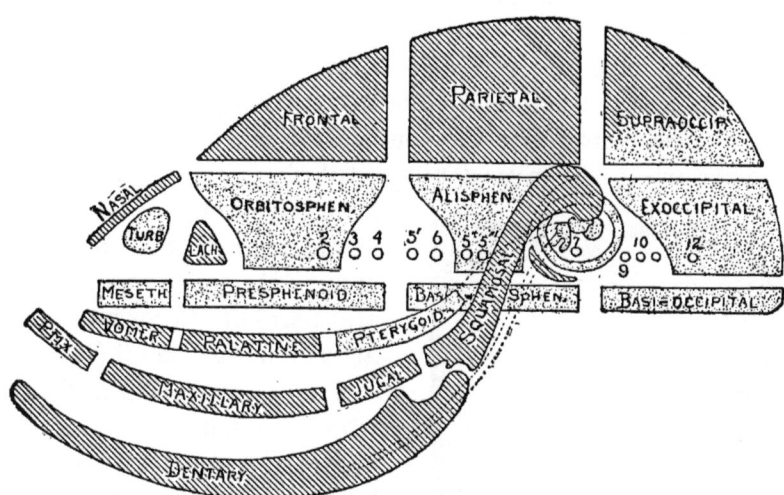

Fig. 37.—Diagram of the bones of the mammalian skull. Cartilage bones dotted, membrane bones lined. 2-12, nerve exits. (From Kingsley's *Comparative Anatomy of Vertebrates*, with the permission of P. Blakiston's Son & Co.)

Meckel's cartilage (Fig. 39), and finally replaces it, but for a time a relationship exists between the two similar to that in the bony fishes. In the appendicular skeleton a similar change takes place during embryonic development, but most of the bones are preformed in cartilage. The clavicle in the pectoral girdle of man is added as a membrane bone. In the pelvic girdle the most striking changes are due to the increased stresses incidental to terrestrial life; one of them is the connection of this girdle with the spinal column by the two dorsal bones, the ilia.

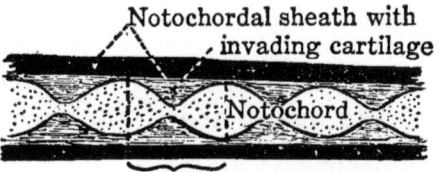

Fig. 38.—Diagram of a longitudinal section through a developing vertebral column to show the invasion of the notochord by the cartilages from which the centra of the vertebrae develop. (From Woodruff, after Walter.)

The Circulatory System. *Fishes.* Although any organic system in the body shows similar evidences of relationship, the circulatory system surpasses all others in its completeness. Returning to the fishes again for the primitive type, we find a heart consisting of two fundamental chambers, but preceded and followed by two others (Fig. 41A, B). The blood enters this heart from the body through several large veins which join the posterior chamber, the sinus venosus. From the sinus venosus it passes into the atrium, thence into the ventricle whose muscular walls contract rhythmically and drive the blood out to the body. As it leaves the ventricle it passes first into the conus arteriosus, which tapers into the truncus arteriosus, a part of the tubular portion of the circulatory system. In some fishes a muscular bulbus follows the conus. From the truncus several pairs of afferent branchial arteries are given off (Fig. 43B) —five in some existing fishes— which pass into the branchial arches and break up into capillaries wherein the blood, laden with wastes, gives up its carbon dioxide and receives a fresh supply of oxygen through the delicate tissues of the gills. It is then collected into other arteries, the efferent branchials, which unite to form the dorsal aorta behind the branchial region. Through this vessel and its branches the pure blood is carried to all parts of the body and distributed to the tissues by another system of capillaries (Fig. 40).

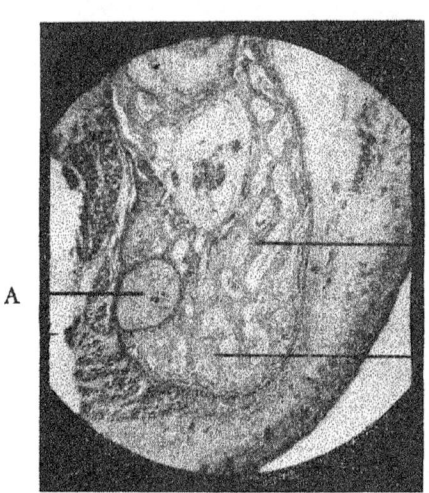

FIG. 39.—Section of the jaw of an embryo kitten. A, Meckel's cartilage; B, bone trabeculae of the developing mandible. From a photomicrograph.

After it has served the tissues the blood is collected again, this time into veins. The veins from the caudal region return their blood to the kidneys, where it is again distributed in capillaries. Such a system is called a portal system. Blood from the alimentary tract is conveyed in a similar manner to the liver, and from these organs other veins convey the blood to the heart, the posterior cardinals that from the kidneys, and the hepatic vein that from the liver. These portal systems are called the renal and hepatic portal systems, respectively (Fig. 40). Blood from the body is also returned to the heart directly through other veins.

In the fish, therefore, we find that the muscular contractions

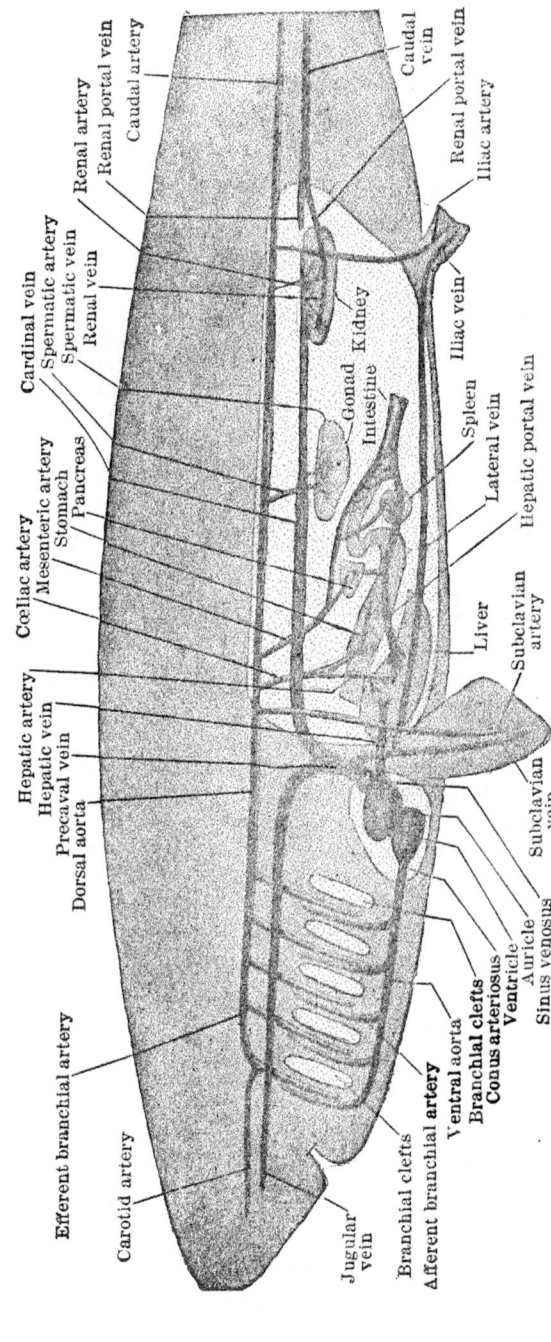

FIG. 40.—Lateral aspect of the vascular system of a dogfish, semi-diagrammatic. (From Hegner, after Parker.)

of one chamber force the blood through a single cycle about the body and back again to the heart, with a maximum of three and a minimum of two capillary systems interpolated in the various divisions of this cycle.

Terrestrial Vertebrates. In all vertebrates above the fishes air breathing is the rule, and in the very lowest of these classes, the Amphibia, another cycle is interpolated. Here we find a three-chambered heart, the atrium divided into two auricles, but

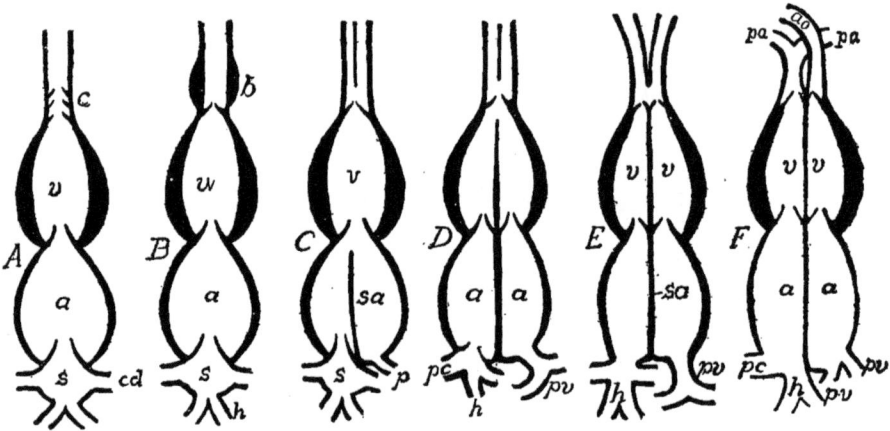

Fig. 41.—Stages in the development of the vertebrate heart, as illustrated by various classes. A, elasmobranch fishes; B, teleost fishes; C, amphibia; D, lower reptiles; E, alligator; F, birds and mammals. a, atrium; ao, aorta; b, bulbus arteriosus; c, conus; cd, duct of Cuvier or common cardinal vein; h, hepatic veins; pa, pulmonary arteries; pc, precaval and postcaval veins; pv, pulmonary veins; s, sinus venosus; sa, interatrial septum; v, ventricles. (From Kingsley's *Comparative Anatomy of Vertebrates*, with the permission of P. Blakiston's Son & Co.)

the ventricle still a single chamber (Fig. 41C). The blood is driven out through the truncus arteriosus into a series of paired vessels, aortic arches, but they are only three in number (Fig. 43C). The posterior pair leads to the lungs and skin, both of which serve as respiratory organs in this class. In the Amphibia which have no gills, no capillary system is interpolated in the aortic arches but the blood is aërated by being forced to the lungs or skin. A limited portion of the body is therefore burdened with a function dealing with all of the blood, and hence a second cycle is established so that blood now courses from right auricle to ventricle to lungs and back to the heart, where it enters the left auricle and is then passed into the ventricle and pumped to the body. This results

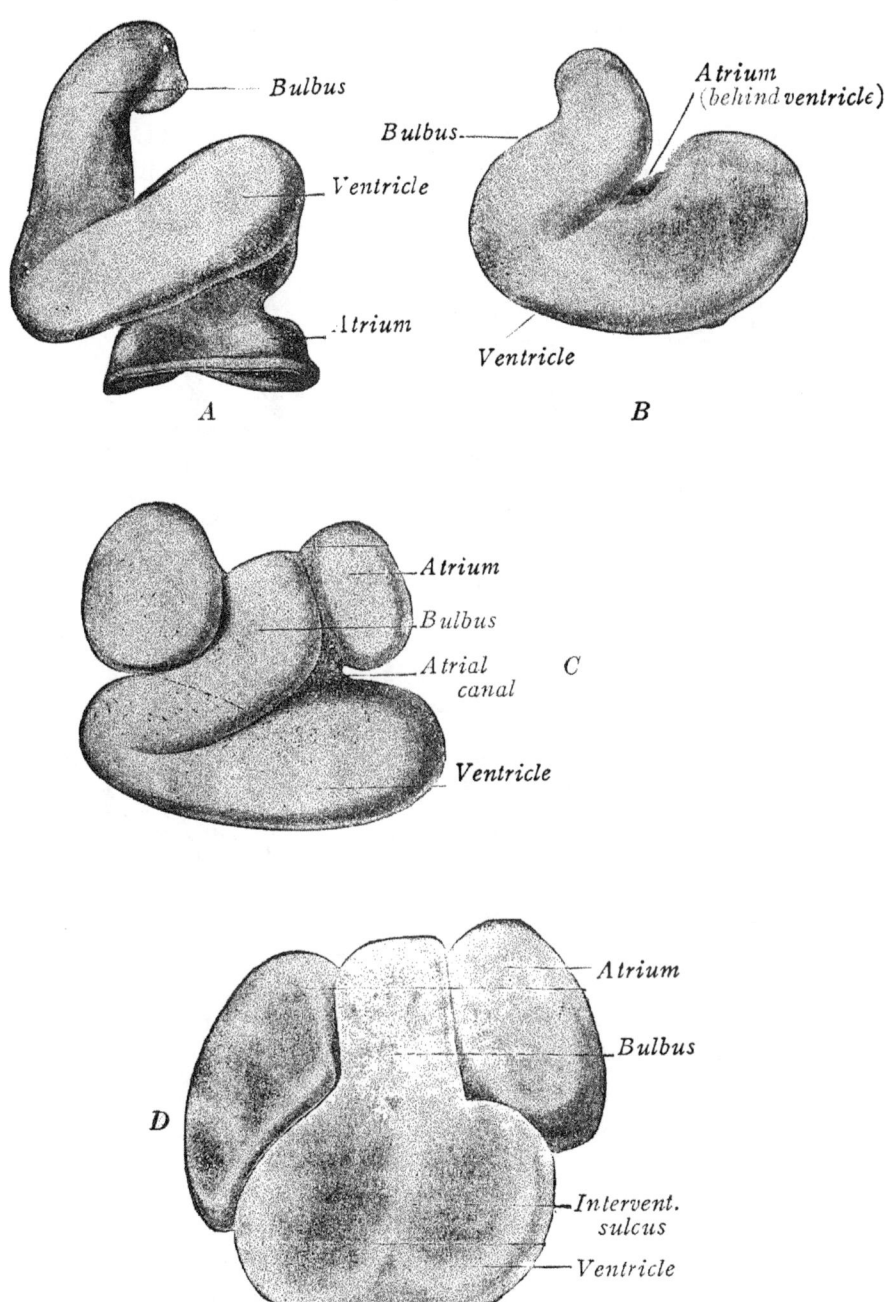

Fig. 42.—External form of the human heart during development. A, from an embryo of 2.15 mm; B, 3 mm; C, 4.3 mm; D, 10 mm. (From Arey's *Developmental Anatomy*, after His, with the permission of the W. B. Saunders Company.)

64 EVOLUTION AND GENETICS

in some mixing of pure and impure blood in the one ventricle, but apparently maintains a fairly effective separation. In the reptiles the two cycles are almost completely separated by the subdivision of the ventricle into right and left chambers (Fig. 41D and E), while the aortic arches are similar to those of the Amphibia (Fig.

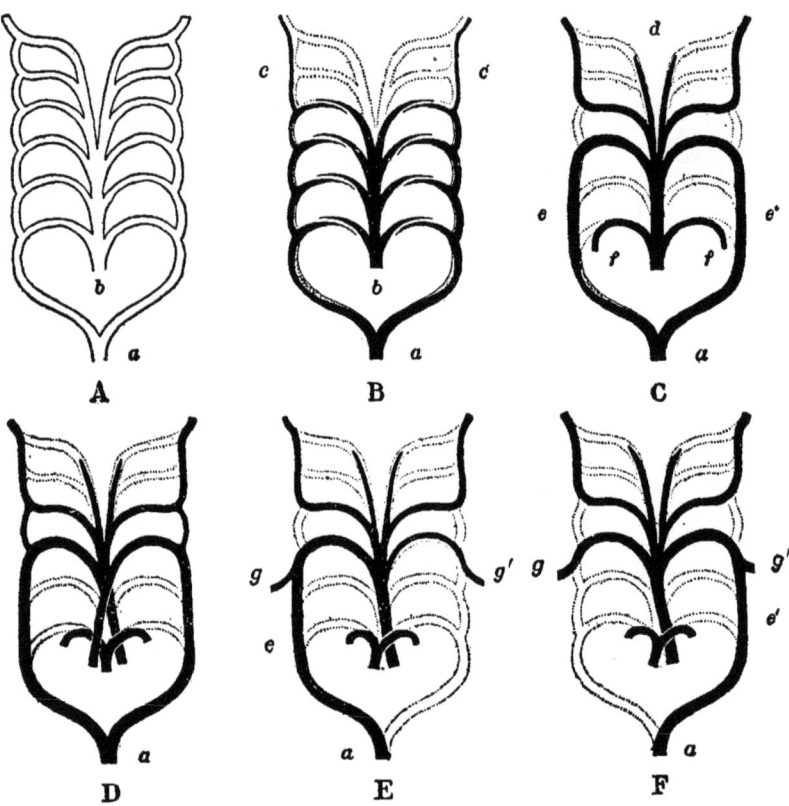

FIG. 43.—Diagram showing the fate of the six pairs of aortic arches in the vertebrate classes. A, the primitive condition; B, fish; C, amphibian (frog); D, reptile; E, bird; F, mammal. *a*, dorsal aorta; *b*, ventral aorta, leading from heart; *c*, internal carotids; *d*, external carotids; *e, e'*, right and left aortic arches; *f*, pulmonary arteries; *g, g'*, subclavian arteries to fore limbs. (From Woodruff.)

43D). In the birds and mammals the two cycles are definitely separated (Fig. 41F), and the great arch which carries blood back to the body is no longer paired. In the birds the left arch of this pair disappears, and in the mammals the right is eliminated, so that in the one class blood is conveyed to the body through a great *aorta dextra*, and in the other through a similar *aorta sinistra* (Fig. 43E, F).

In the embryos of birds and mammals the heart is at first a simple tube, as in such simple chordates as *Amphioxus*. The first differentiation of this tube is its separation into four regions similar to those found in the hearts of adult fishes, viz., the sinus venosus, atrium, ventricle and bulbus arteriosus. Later the first and last are absorbed into the adjacent regions, the auricle is divided by a partition, and finally the ventricle is similarly divided,

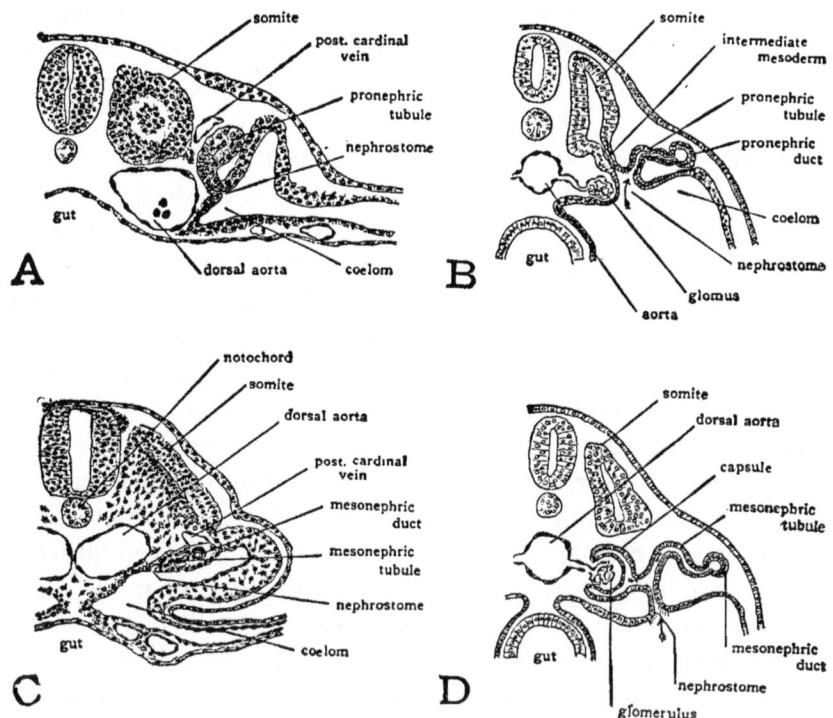

Fig. 44.—The structure of nephric tubules. A, pronephric tubule from a 16-somite chick embryo; B, diagram of functional pronephric tubule; C, primitive mesonephric tubule with rudimentary nephrostome, from a 30-somite chick embryo; D, diagram of functional mesonephric tubule of the primitive type. (From Patten, A after Lillie, B and D after Wiedersheim; with the permission of P. Blakiston's Son & Co.)

so that the organ goes through a transition approximating that represented by the several classes (Fig. 42).

While the heart is developing, a series of aortic arches appear, six pairs in all, which are at first symmetrical and unlike those of the fish chiefly in the lack of capillaries (Fig. 29). As the ventricle divides, the truncus arteriosus splits so that the separation of that portion leading into the last pair of arches, and thus to the lungs, is isolated with the right ventricle, while the remainder carries blood from the left ventricle into the other arches. Of these the fifth pair is never large and soon disappears, along with the first and second. The third pair, and its connections with the first and second, persists to carry blood to the head, while the fourth differs in development in the birds and mammals, but develops into the great aorta in each class. The approximation of the adult structures of lower forms is not limited to these parts of the circulatory system, for the veins also show a gradual transition, but the heart and aortic arches are no less striking, and are more easily understood.

The excretory system develops in vertebrates from paired mesodermal masses, called the nephrotomes, extending from the cervical to the caudal region. In its primitive form it consists of a series of tubules opening into the coelom. By ciliary action these tubules carry wastes from the coelom to the exterior. The presence of a knot of arteries, or glomerulus, projecting into the coelom near each tubule suggests an association with the circulatory system (Fig. 44). Such structures as these arise from the anterior part of the nephrotomes, and make up a pair of bodies called the pronephroi, or anterior kidneys. The association of the excretory tubules with the coelom is soon supplanted by an intimate association with the glomeruli, so that wastes are removed directly from the blood (Fig. 44B). A second pair of kidneys made up of such structures, arise behind the first, and are called the mesonephroi. These become the functional kidneys of adult fishes and amphibia, while the pronephroi occur only in embryos and larvae. At a later stage of development the mesonephroi are supplanted by a third pair of excretory organs, the metanephroi. These are found only in reptiles, birds and mammals, which develop during their ontogeny first rudimentary pronephroi, then functional kidneys of the embryo, mesonephroi, and finally the metanephroi which are to persist throughout life.

Nothing short of a thorough study of embryology can adequately disclose the marvelous resemblances which occur one after the other as the embryos of vertebrates pass through the successive stages of development. In the nervous system, the respiratory system, the development of the pharynx, and in countless other details of structure involving various systems these indications of relationship of the several classes are present. In presenting this brief account the most conspicuous examples have been selected and shorn of all unnecessary details. Such an account is necessarily imperfect, but in this field even scattered facts are striking.

Summary. The relationship of vertebrates is strikingly evident in their embryology. A study of the developmental stages of the various classes shows that animals pass through similar steps up to the point where the divergence of adult structure appears. Even in the initial stages of cleavage and gastrulation modifications occur, but these are incidental to the storage of yolk in the ovum and do not destroy the homologies. After gastrulation such structures as the neurenteric canal and the foetal membranes show other definite relationships. Not only do the embryos resemble each other, but the embryos of higher classes also pass through stages similar to the maximum development of classes below them. The early stages of embryonic development are similar to some of the invertebrates. Such structures as the notochord, the skeleton, and the circulatory and respiratory systems show a gradual transition in adults from the cyclostomes to the mammals which is also followed during ontogeny. The entire development of the individual is a story of relationships which are illustrated by these selected examples.

REFERENCES

NEWMAN, H. H., *Vertebrate Zoölogy*, 1920.
WILDER, H. H., *History of the Human Body*, revised edition, 1923.
MCEWEN, R. S., *A Textbook of Vertebrate Embryology*, 1923.
AREY, L. B., *Developmental Anatomy*, 1924.

CHAPTER V

THE RELATIONSHIP OF EXISTING ORGANISMS
(Continued)

3. COMPARATIVE ANATOMY OF VERTEBRATES

The preceding consideration of embryological relationship has necessarily touched upon some of the salient features of comparative anatomy. The transition of skeletal development, the structure of the heart in the several classes, development of the aortic arches and the venous system may as well be treated in one field as in the other, since they cannot be made clear without reference to both. Comparative anatomy discloses, however, a great many details of homology without reference to development. Many vestigial structures which anatomists have found in man, for example, are shown to have the same relations as functional structures in the bodies of other animals, while the functionally active systems are built up of the same tissues and organs, arranged according to the same plan. While the similarity of functional parts is in itself significant, the occurrence of vestigial organs, particularly those which are found only in occasional individuals, is doubly so. We may content ourselves with necessity as a reason for the presence of useful organs, but obviously useless structures can be explained only on a very different basis.

The Skull. In the skulls of vertebrates from the fishes to the mammals a large number of bones are found, some present in all forms, some in only the lower forms. We have already noted that these are of two different kinds, those originating in cartilage as parts of the chondrocranium and those which develop directly from the embryonic mesoderm. The latter are of particular interest at this point because of the completeness of their history as shown by existing forms. The chondrocranium, or primordial skull, is well developed in the elasmobranch fishes as a supporting structure extending forward from the spinal column beneath the brain, which it does not enclose dorsally or anteriorly except by the development of secondary membranous structures. The same primordial skull develops in forms above the elasmobranchs, but

EXISTING ORGANISMS—ANATOMY

to it is added the series of dermal bones which encloses the brain on all other sides and form the greater part of the skull (Fig. 37).

The Ganoid Stage. The stage in which the dermal bones of the skull are well developed is nicely represented by existing bony fishes, or ganoids, and hence is called by Wilder the ganoid stage. The sturgeons illustrate the origin of these bones as dermal plates, or scutes, which are derived like scales from the corium, the under layer of the skin, and instead of forming an inner bony box, remain an outer bony armor (Fig. 45). Wilder describes these scutes as follows: "The snout, or rostrum, is covered by a series of small *rostral plates*, which extend back as far as the nostrils; back of these openings may be found a pair of *nasals;* behind these again, and between the eyes, is a pair of *frontals*, often accompanied by *prae-* and *post-frontals*. Behind these is a pair of *parietals*, and one or more *supra-occipitals*. On the sides of the head, at about the level of the parietals, are the *squamosals*, and around the eye are several *orbitals*, distinguished as *pre-, supra-, post-orbitals*, etc. The operculum, or gill-flap, which is present in these fishes, is covered and augmented by *supra-, sub-,* and *pre-operculars*."

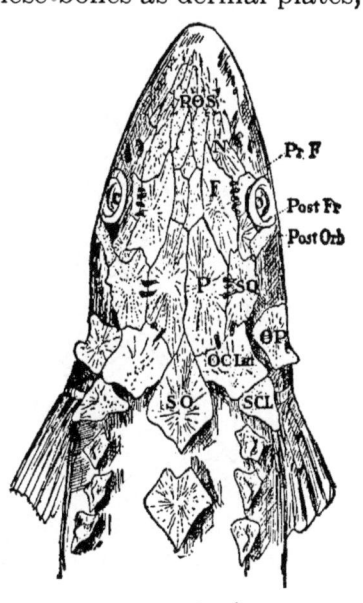

FIG. 45.—Dorsal view of the skull of a sturgeon (*Acipenser*), showing dermal bones. ROS, rostral plates; N, nasal; F, frontal; PrF, pre-frontal; Post Fr, post-frontal; Post Orb, post-orbital; P, parietal; SQ, squamosal; OP, opercular; OCLa, lateral occipital; SO, supra-occipital; SCL, supra-clavicle. (From Wilder's *History of the Human Body*, with the permission of Mrs. H. H. Wilder and Henry Holt and Company.)

The occurrence of bones similar in arrangement to these bony scutes is dependent to some extent, of course, on their functional importance in the various fishes. Terrestrial animals, for example, would not be expected to have bones corresponding to the opercular series, and the shortening of the face and great enlargement of the brain in man must necessarily be accompanied by differences in the skeletal parts involved. Due to these variations in importance, exact duplication in widely different species is not found.

Amphibia. In the skulls of Amphibia the dermal bones are no longer external as in some fishes at any stage of development, but have become definitely incorporated with the cartilage bones as parts of the internal skeleton. In addition the characteristically piscine elements, like the rostrals, the orbitals and those associated with the operculum, have been lost, and the remaining bones

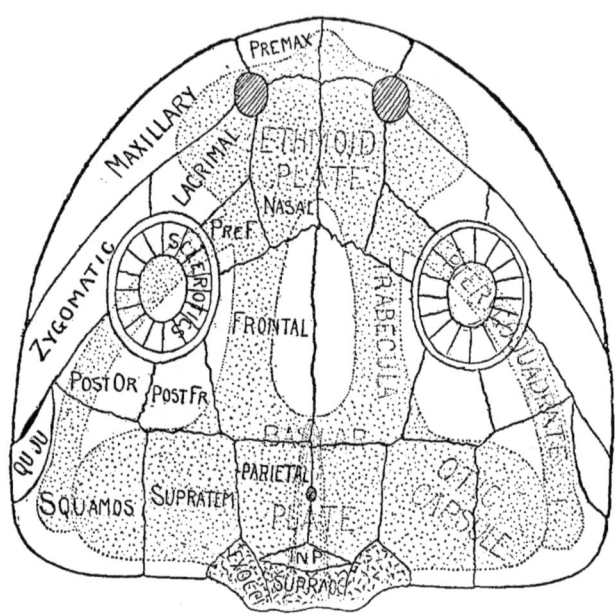

Fig. 46.—Dorsal view of schematic skull, the chondrocranium dotted, membrane bones outlined. premax, premaxilla; pref, prefrontal; postfr, postfrontal; postor, postorbital; squamos, squamosal; quju, quadratojugal; inp, interparietal; exocci, exoccipital; supratem, supratemporal; supraoc, supraoccipital; other names in full. (From Kingsley's *Comparative Anatomy of Vertebrates*, with the permission of P. Blakiston's Son & Co.)

more nearly approximate the number and relationships of the higher terrestrial forms. We find among them a pair of small bones behind the anterior nares, behind which a similar pair of larger bones reach the orbits on the sides. Still another median pair lie behind the orbits. The first correspond in their orientation with the nasals of the fishes, and are called the nasal bones. The next are the frontals, representing these and some of the smaller scutes of the sturgeon and the last are the parietal bones. A pair of prefrontal bones lie behind the nasals and lateral to the frontals, as in the fishes. Caudad this portion of the skull is associated, through the supra-occipital, with the basal or occipital part of the chondrocranium, from which develops the occipital bone. Laterad and ventrad other bones occur, but these are primarily visceral in their associations (Figs. 36 and 46).

Above the Amphibia. In the reptiles, birds and mammals the same essential parts and relationships persist, with some slight

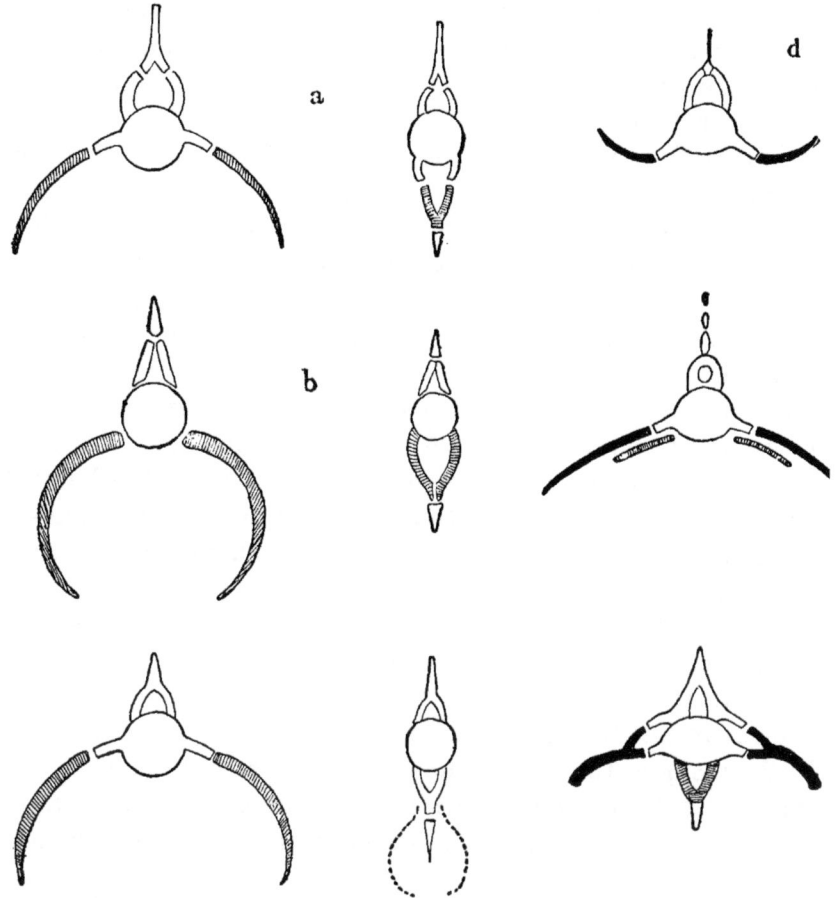

Fig. 47.—Morphology of ribs. a, ganoid fish; b, dipnoid fish; c, teleost fish; d, shark; e, Po*lypterus*, a special case among ganoids; f, urodele amphibian. In the first three the condition in trunk (left) and tail (right) is given. In all figures the "fish rib" is striped and the true rib is black. (From Wilder's *History of the Human* Bo*dy*, after Wiedersheim, with the permission of Mrs. H. H. Wilder and Henry Holt and Company.)

variation in the accessory frontals and in the proportionate sizes of the various bones.

The Human Skull. The skull of man differs in a few conspicuous points, although it is in general similar. The nasals persist as small, paired bones, which may fuse to form a single bone. The frontals also are ordinarily fused to form a single large bone, but in a very small percentage of cases they remain separate. The supraoccipital is usually fused with other elements in the occipital bones, but sometimes persists as a separate bone between the parietals and occipital, as is normally the case in some other mammals. Both frontals and parietals are proportionally much greater in size than in the lower animals because of the large size of the brain, which they enclose.

The Spinal Column. In the remainder of the axial skeleton relationships are quite obvious. Vertebrae in all of the classes above the Cyclostomes consist of a solid centrum from which a dorsal neural arch arises, enclosing and protecting the spinal cord (Fig. 47). This arch is surmounted by a spinous process which furnishes attachments for muscles. Above the fishes the neural arch bears two anterior and two posterior articular processes which aid in preserving a firm articulation of successive vertebrae, and a pair of transverse processes. The centrum may also bear short transverse processes and a ventral arch, called the haemal arch, which is well developed in the fishes and forms a conspicuous appendage of some reptilian vertebrae. The haemal arch terminates in a haemal spine.

Ribs in many fishes are merely the halves of incomplete haemal arches. In some fishes, amphibians and reptiles, ribs of this type are found attached to the same vertebrae that bear other ribs, more dorsal in position (Fig. 47e, f). The former are commonly called fish ribs, and the latter true ribs. True ribs in their typical form have two heads, one of which articulates with the lateral process of the centrum, and the other with that of the neural arch.

The Sacrum. Terrestrial animals have the pelvic girdle attached to the spinal column, and one or several vertebrae are modified for this attachment (Fig. 48). In *Necturus*, a urodele amphibian, only one vertebra is involved, usually the 19th, sometimes the 20th, and rarely the 18th, although oblique attachments are on record in which the pelvic girdle joined the left side of one of these vertebrae and the right side of another. Such vertebrae are called sacral vertebrae, and when more than one is involved a fusion often occurs, resulting in the development of a composite bone, the sacrum. Wilder notes that "this anchylosis of adjacent sacral vertebrae is the most complete in birds and in man, and for the same reason, namely the employment of the hind limbs alone for the support of the body, although in the two cases the number and arrangement of the associated parts differ very considerably." He adds that "variation in the sacral region is not confined to the lower forms, although it is more frequent in these latter (e.g., *Necturus*) and becomes relatively stable in the higher and more specialized classes."

The Visceral Skeleton. In the visceral skeleton and associated membrane bones relationships are in some cases more obscure, but as worked out by comparative anatomists and checked by

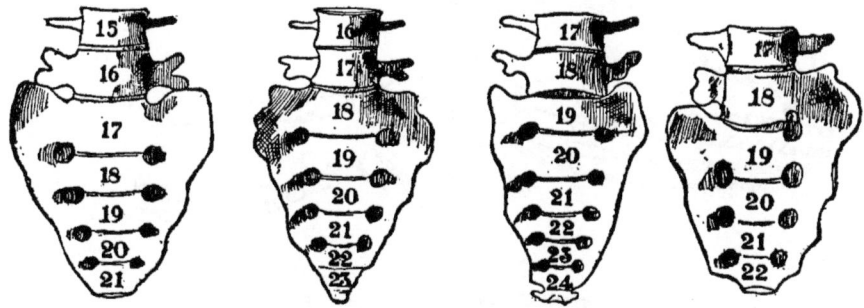

Fig. 48.—Variations in the composition of the human sacrum. (From Wilder's *History of the Human Body*, after Gegenbaur, with the permission of Mrs. H. H. Wilder, and Henry Holt and Company.)

the facts of embryology, they are well established and remarkable (Fig. 49).

The Jaws. The upper and lower jaw cartilages (pterygoquadrate and Meckel's, respectively) have already been mentioned. These belong to the visceral skeleton, and in the elasmobranch fishes are the only skeletal structures about the mouth. In higher groups each is associated with membrane bones which finally supplant it entirely as the skeleton of a jaw. The membrane bones of the upper jaw include the anterior premaxillaries, which lie at the tip of the skull, just below the anterior nares. Behind them are the larger maxillary bones, followed along the outside of each cartilage by a zygomatic (malar or jugal), a quadratojugal and a squamosal bone. On its mesial surface it is associated with an anterior palatine and a posterior pterygoid which form part of the roof of the mouth. The posterior part of the cartilage itself gives rise to the quadrate bone which sometimes intervenes

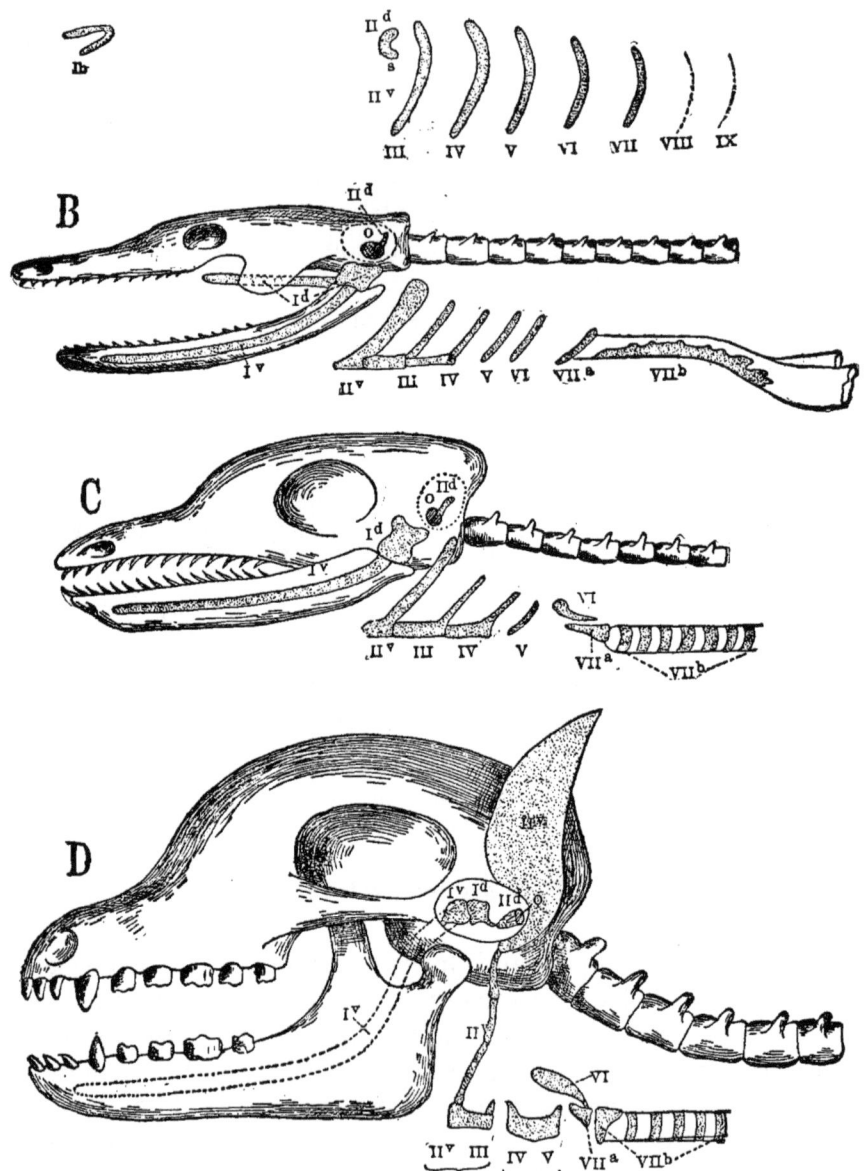

Fig. 49.—Morphology of the visceral skeleton. A, shark; B, amphibian; C, reptile; D, mammal; the successive arches are indicated by Roman numerals; the dorsal and ventral parts of arches I and II are indicated by exponent letters, d and v. lb, labial cartilage; s, spiracular cartilage; o, operculum; VII^a and VII^b are the arytenoid and tracheal cartilages, respectively. (From Wilder's *History of the Human Body*, with the permission of Mrs. H. H. Wilder and Henry Holt and Company.)

between the lower jaw and the skull. The changes in the lower jaw involve in the lower classes the addition of an outer series of bones including the anterior dentale, which usually bears teeth, followed by a splenial and angulare, above which lies the surangulare. The coronoid runs back from the dentale above the surangulare. The posterior end of the cartilage itself may develop into an articular bone by which the jaw is articulated with the quadrate. All of these parts are present in the existing reptiles.

The modification of these bones in the mammals leaves the upper jaw with the same parts, excepting the quadrate bone.

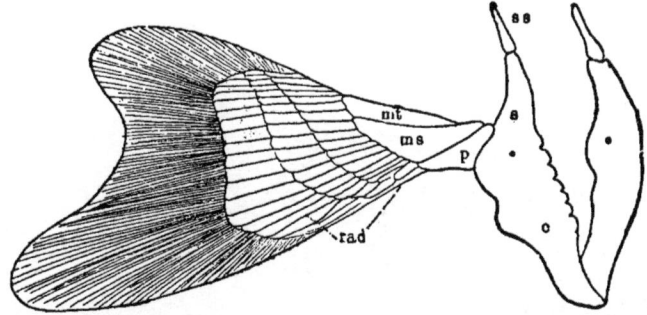

Fig. 50.—Pectoral fin and girdle of dogfish. s, scapula; ss, suprascapula; c, coracoid; p, propterygium; ms, mesopterygium; mt, metapterygium; rad, radials. (From Wilder's *History of the Human Body*, with the permission of Mrs. H. H. Wilder and Henry Holt and Company.)

The lower jaw consists of two mandibles, firmly united in man to form a single bone, consisting of the dentales and possibly the splenial and coronoid bones. Other homologies are obscure, but the angulare is said by Kingsley to be apparently the tympanic bone of the skull. The articulation of the lower jaws with the skull is thus shifted in the mammal to the mandible (dentale?) and squamosal. The articulare and quadrate, freed from this function, are found in the middle ear, the former being certainly the malleus, while the quadrate is possibly the incus.

The remaining parts of the visceral skeleton form the hyoid apparatus and embrace the branchial region in the fishes, furnishing support for the gill arches, while in terrestrial classes their first function continues and they are otherwise represented by the cartilages of the larynx and upper part of the trachea.

The Appendicular Skeleton of Fishes. The appendicular skeleton is simplest in the fishes. Since their bodies are buoyed up at all points by the heavy medium in which they live, the pectoral and pelvic fins and the girdles to which they are attached do not require the strength and rigidity of supporting structures. In the primitive elasmobranchs the pectoral girdle consists of a V-shaped piece of cartilage (Fig. 50) to which is articulated a large basal cartilage of the fins, the mesopterygium. To the mesopterygium are attached two other pieces, the propterygium and metapterygium. These basal cartilages, of which only one is found in the pelvic fins, bear a fan-like series of radial cartilages to which the thin terminal portion of the fin is attached. The V-shaped cartilage obviously consists of the ventral portion that

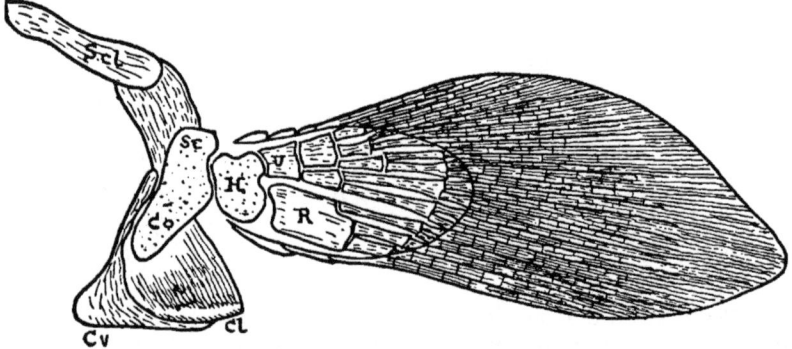

FIG. 51.—Right pectoral fin of *Sauripterus taylori* from the Upper Devonian. cl, cleithrum; co, coracoid; cv, clavicle; sc, scapula; s.cl., supracleithrum; H, humerus; R, radius; U, ulna. (From Lull, after W. K. Gregory.)

runs between the fins and a dorsal portion extending dorsad from each fin. Each dorsal piece is surmounted by another small cartilage. The ventral part is called the coracoid, the dorsal piece the scapula, and the ultimate dorsal cartilage the suprascapula.

Transitional Fins. Homologies of the structures described in the preceding paragraph are not entirely clear, but in one group of bony fishes, the *Crossopterygii*, a significant modification of structure and use of the pectoral fins is found. These lobe-finned ganoids have the strange habit of resting on the bottom of the water in which they live, and supporting themselves by the front fins in a manner distinctly similar to the use of the front legs of the terrestrial organisms.

The fins of some extinct forms of Crossopterygii are well preserved, and in one of these, *Sauripterus taylori*, from the Upper Devonian, the entire skeletal structure of the pectoral girdle and fin is shown (Fig. 51). In this species it is at once evident that the principal bony framework of the fin is not unlike that of the appendages of terrestrial vertebrates. The fin is articulated to the girdle by a single basal bone, to which are attached two other bones, and following these is a series of bones of less regular arrangement. The fringe of the fin is, of course, a structure adapted to aquatic life, and hence plays no part in comparison with terrestrial forms, but the rest of it foreshadows both in structure and use the conditions found in higher animals.

The Appendicular Skeleton of Terrestrial Vertebrates. Above the fishes we have to deal only with terrestrial animals and those

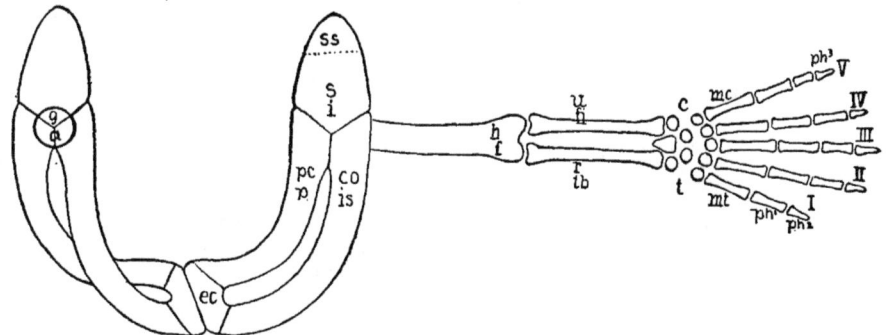

Fig. 52.—Diagram of girdles and appendages from the posterior side; upper letters, fore limb; lower, hind limb. a, acetabulum; c, carpus; co, coracoid; f, femur; fi, fibula; g, glenoid fossa; h, humerus; i, ilium; is, ischium; ss suprascapula; mc, mt, metacarpals and metatarsals; p, pubis; pc, procoracoid; ec, epicoracoid; ph 1-3, phalanges; r, radius; s, scapula; t, tarsus; tb, tibia; u, ulna; I-V, digits. (From Kingsley's *Comparative Anatomy of Vertebrates*, with the permission of P. Blakiston's Son & Co.)

which have become secondarily aquatic, of which there are examples in all classes except the birds. Among birds aquatic habits are always associated with terrestrial and usually also aërial life. Throughout these four classes the girdles still consist of a ventral part extending between the appendages, and lateral parts extending dorsad, which may be attached to the spinal column, as in the pelvic, or supported by muscles as in the pectoral girdle (Fig. 52).

The Pelvic Girdle. The pelvic girdle is regarded as probably more conservative (Fig. 53). It consists in the primitive salamander, *Necturus*, of a broad ventral plate of cartilage, bearing two bones, the ilia, running dorsad. The acetabulum, a cavity in which the hind limb articulates, is located at the junction of the two parts. At the posterior angles of the cartilage plate are two centers of ossification. In higher Amphibia two anterior centers also appear. The bones formed from these four centers are the anterior pubic bones and the posterior ischia, between

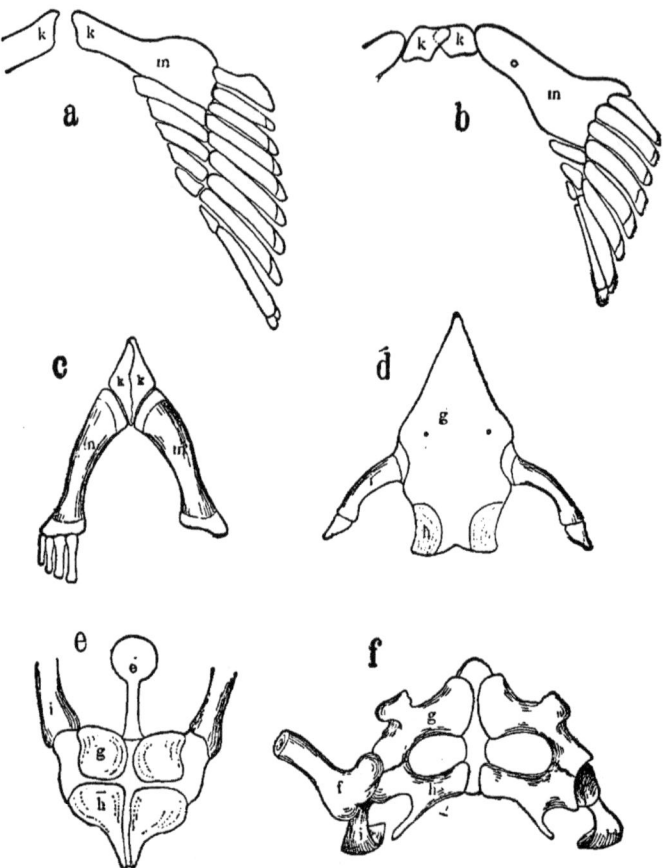

Fig. 53.—Series illustrating a theory of the development of the pelvic girdle. a, sturgeon, *Acipenser;* b, a ganoid fish, *Scaphyrhynchus;* c, a ganoid, *Polypterus;* d, a primitive salamander, *Necturus;* e, a South African frog, *Dactylethra;* f, turtle. In a the part m is formed by a fusion of the anterior rays. The pieces kk, segmented off from m in b, form in c a rhomboidal plate. In d this plate has grown large and bears a pair of ossified ilia, i, and a pair of centers of ossification, the ischia, h. In e two more centers of ossification, the pubes, g, have appeared. f, is a typical pelvic girdle with all its parts. The epipubis, e, is incidental and unimportant. (From Wilder's *History of the Human Body,* with the permission of Mrs. H. H. Wilder and Henry Holt and Company.)

which in reptiles develop the obturator foramina which sometimes join to form a single large opening. While pronounced modifications of these bones occur in higher forms, such as the ventral separation of the ischia in man and of both ischia and pubes in birds, and the broadening of the ilia in man to form the basin-shaped pelvis correlated with his erect posture, the three bones retain characteristic fundamental relations.

The Pectoral Girdle. This girdle is further modified. The scapula remains in Amphibia, surmounted by the cartilaginous supra-scapula, as part of the girdle extending dorsad from the anterior limb. The coracoid develops a pair of distinct bones, corresponding to the ischia of the pelvic girdle, while anterior cartilaginous strips, the procoracoids, separated from these bones by the coracoid foramen, correspond to the pubes. Three membrane bones, the clavicle, interclavicle and cleithrum, are associated with the cartilaginous parts, but the two last are rare above the fishes (Fig. 54). A median element, the interclavicle, is sometimes present between the ends of the clavicles, and epicoracoid cartilages may join the clavicles and coracoids.

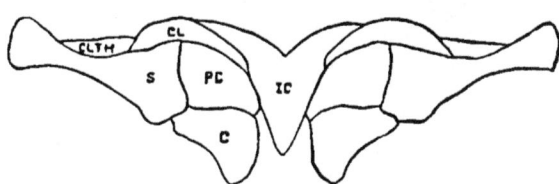

Fig. 54.—Diagram of the shoulder girdle of a primitive reptile, showing the complete series of elements found in the vertebrates above the fishes. Dermal bones: CLTH, cleithrum; CL, clavicle; IC, interclavicle; Cartilaginous elements: PC, procoracoid; C, coracoid; S, scapula. (From Wilder's *History of the Human Body*, with the permission of Mrs. H. H. Wilder and Henry Holt and Company.)

The coracoid and scapula at least take part in the formation of the glenoid cavity, in which the fore limb is articulated. The ventral parts are complicated by association with the median epicoracoid cartilages and with the sternum. Of these bones the dorsal scapula persists as a more or less blade-like bone, the shoulder blade of man. All ventral parts are found in reptiles and both coracoid and clavicle in birds, but only the clavicle in man. In the primitive Prototheria all parts of the primitive girdle are found, and in some of the higher mammals the scapula is the only bone present (Fig. 55).

The Pentadactyl Appendage. The appendages of terrestrial forms, both pectoral and pelvic, are invariably attached to their respective girdles by a single bone, the humerus of the anterior limb and femur of the posterior. To these are articulated two bones, the inner radius and tibia, and the outer ulna and fibula of the anterior and posterior limbs respectively. These are followed by a series of small bones, the carpals and tarsals, which include a proximal pair, a distal transverse row of five, and two median in position. The five distal bones are followed by five long bones, the metacarpals and metatarsals, and these by five series of shorter phalanges which form the skeleton of the digits. This primitive type is called the pentadactyl appendage (Fig. 52).

Specialized Appendages. Modifications are present in such highly specialized structures as the wing of the bird and the foot

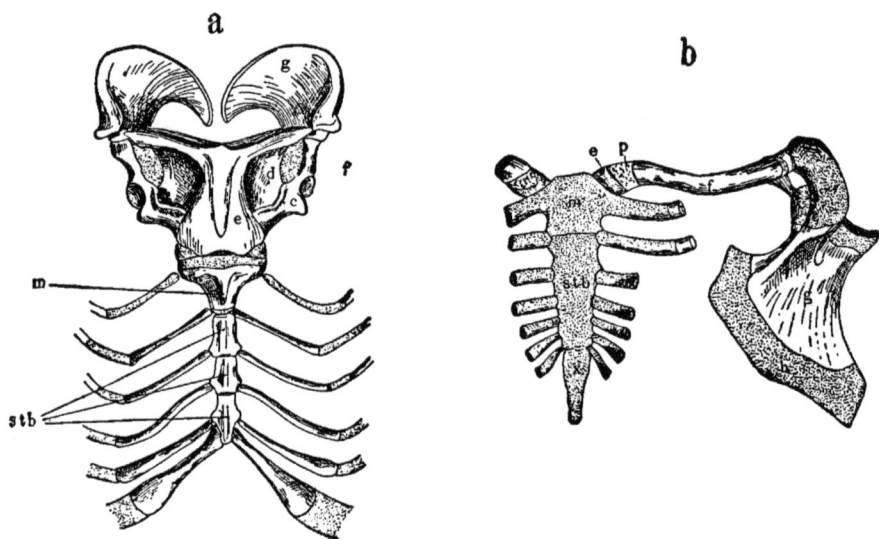

FIG. 55.—Sternum and shoulder girdle of mammals. a, the duck-mole, *Ornithorhynchus*, a primitive oviparous mammal; b, human embryo. c, coracoid; d, epicoracoid; e, episternum; f, clavicle; g, scapula; h, suprascapula; m, manubrium; stb, sternebrae; x, xiphisternum. (From Wilder's *History of the Human Body*, after W. K. Parker; with the permission of Mrs. H. H. Wilder and Henry Holt and Company.)

of the horse, involving the reduction and fusion of parts, and in the marine mammals the phalanges are multiplied for the support of the flippers; but in a large majority of terrestrial vertebrates the fundamental plan of the pentadactyl appendage is distinctly traceable (Fig. 56). It is well represented in a primitive state in the salamander, *Necturus*, and is not highly modified in lizards and *Crocodilia;* in mammals its variations are numerous but often not extreme. Conspicuous examples of modifications in mammals are found in the fore limb of the bat, in which the second, third, fourth and fifth digits are greatly prolonged to support the wing membrane; in the horse, with its single persisting third digit; and in the dolphin, which has some digits prolonged by the multiplica-

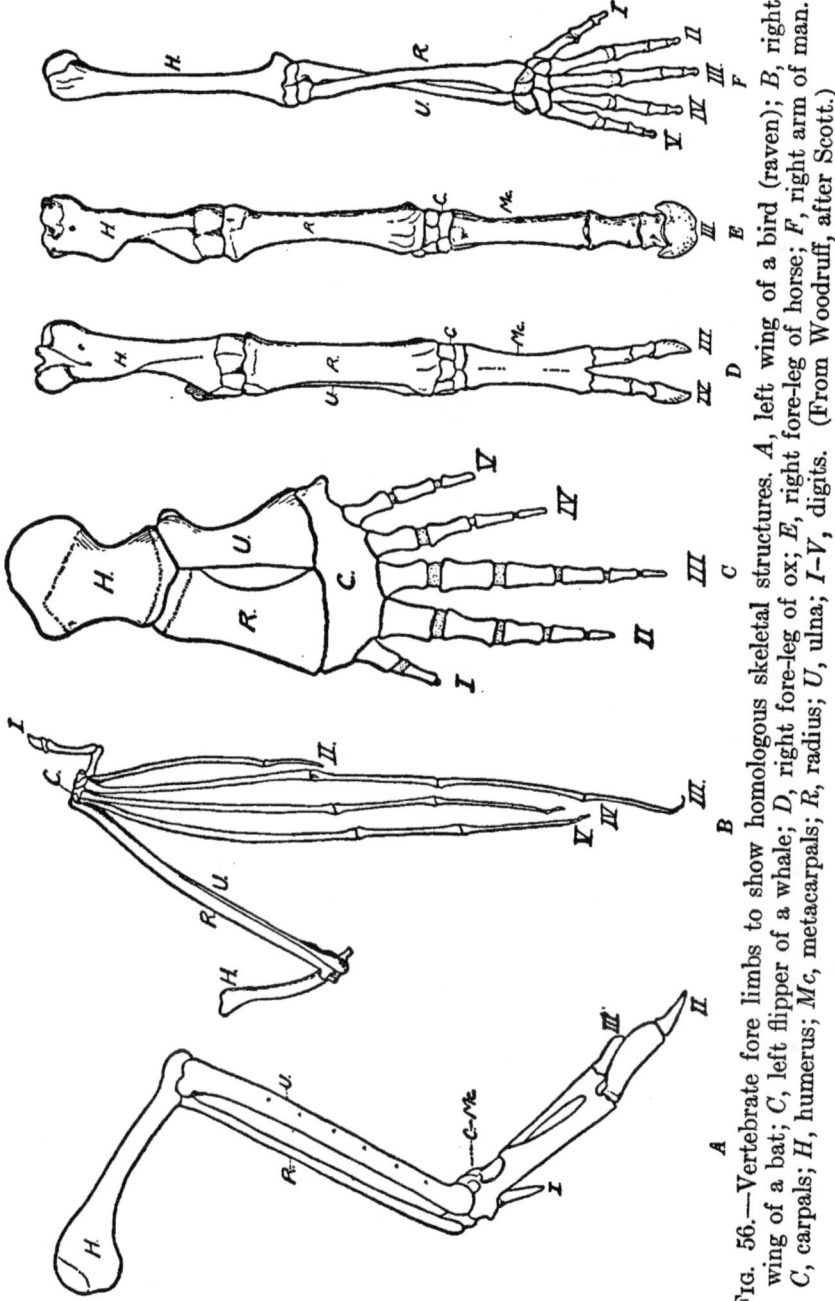

Fig. 56.—Vertebrate fore limbs to show homologous skeletal structures. *A*, left wing of a bird (raven); *B*, right wing of a bat; *C*, left flipper of a whale; *D*, right fore-leg of ox; *E*, right fore-leg of horse; *F*, right arm of man. *C*, carpals; *H*, humerus; *Mc*, metacarpals; *R*, radius; *U*, ulna; *I–V*, digits. (From Woodruff, after Scott.)

tion of phalanges and others shortened, even to a rudimentary state.

The Exoskeleton. The integumentary structures of vertebrates include, in addition to the skin itself, glands, claws, hoofs, nails, horns, teeth, scales, feathers and hair, among which several homologies are evident.

Placoid Scales. If we return again to the elasmobranchs, we find that their skin is studded with minute scales of a type quite

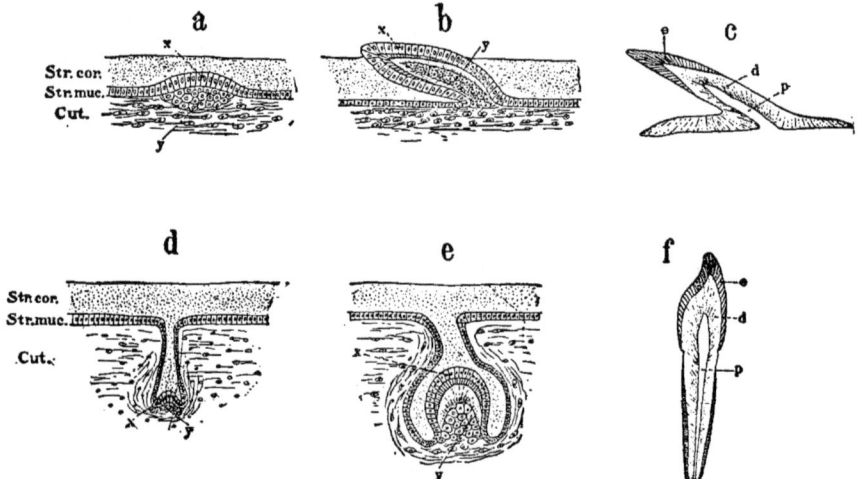

Fig. 57.—Comparison of the development and structure of a placoid scale and a tooth, a, b, and c represent the scale; d, e, and f represent the tooth. The corneous layer of the epidermis is dotted, the germinative layer is represented as a single layer of large cells with nuclei. The dermal layer of the skin is represented as a fibrous layer with scattered cells, x, enamel-producing cells; y, mesodermal papilla; e, enamel; d, dentine; p, pulp cavity. (From Wilder's *History of the Human Body*, with the permission of Mrs. H. H. Wilder and Henry Holt and Company.)

different from those of other fishes. These are called placoid scales, and consist of a flattened base formed of dentine, a fine, compact bony substance, from which projects an oblique cusp. The whole is covered by a layer of enamel which is thickest at the tip of the cusp. A mesodermal papilla projects into the hollow under surface of the scale and furnishes it nourishment. A study of the development of these scales discloses that the dentinal portion is derived from the inner, or dermal layer, of the skin, while the enamel is formed from the enveloping epidermal layer.

Scales and Teeth. The jaws of these fishes are armed with numerous teeth arranged in rows, which bear a conspicuous resemblance to the much smaller placoid scales in form. Unlike the teeth of higher vertebrates, these are superficially attached by a broad base. Their structure is exactly similar to that of the placoid scales, consisting of an inner core of dentine surrounding a mesodermal papilla and covered by a hard enamel layer. In all respects, these teeth are so like the placoid scales that the transition of development is almost obvious. They are larger, but since they are found in a region where transition from skin to buccal epithelium occurs, it is conceivable that even the slight roughness caused by placoid scales might be useful in holding prey and that usefulness might account for the greater development of the teeth.

The teeth of higher vertebrates differ in being deeply seated in sockets in the bones but when we remember that the bones which support them are not derived from primitive cartilages like those of the elasmobranch jaws but are developed from the dermal layer, this relationship is not surprising. During development the teeth are differentiated from portions of the ectoderm which grow into the underlying tissues (Fig. 57).

In structure teeth consist, like the placoid scales and teeth of elasmobranchs, of an inner layer of dentine surrounding a mesodermal core, the pulp, and an outer enamel layer which is thickest on the exposed points. Cement, a substance which covers the root and is conspicuous in the teeth of some animals is added by outer mesenchymal tissue. The two principal parts of the tooth have exactly the same origin as the primitive scales and teeth of the sharks, for the enamel is epidermal and the dentine, dermal. Teeth vary greatly in form according to the habits of animals. Some are pointed for grasping and tearing, others are sharp edged for cutting, and still others have broad rough surfaces for grinding. All three types are present in the human mouth. Rodents have highly developed chisel-like incisors and some snakes have fangs for injecting poison. Such animals as the anteaters have no use for teeth, and they are accordingly reduced or absent.

Scales, Feathers, and Hair. The scales of fishes above the elasmobranchs have no enamel covering and so are wholly dermal in origin; nor are they like the scales of reptiles and birds, which consist of epidermal folds, cornified to some extent and nourished by dermal papillae. The feathers of birds are of exactly the same origin as the scales of this group and the reptiles, though they are much more complex in structure. Scales in the mammals are

84 EVOLUTION AND GENETICS

relatively rare, but are commonly found on the feet and tails of such animals as rodents, and in the armadillo and a few other

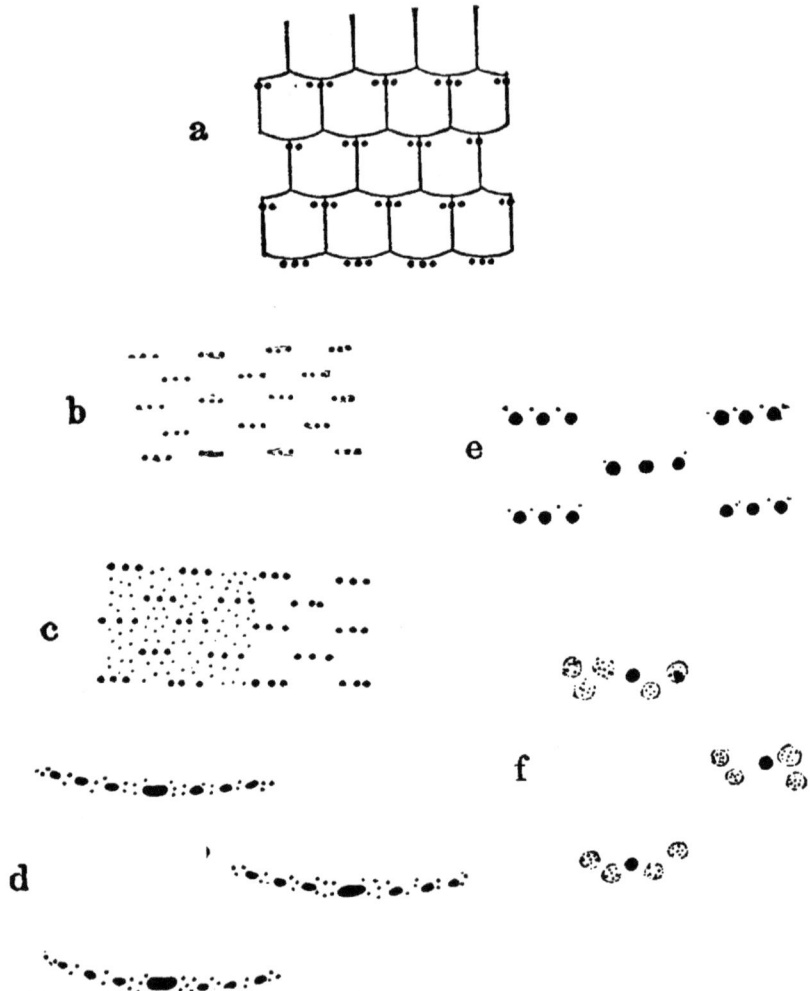

Fig. 58.—Hair pattern in mammals, diagrammatic. a, tail of *Myopotamus*, a South American rodent, with scales and hairs; b, back of *Midas*, a Brazilian monkey; c, back of pig, *Sus vittatus*; d, back of *Coelogenys paca*, a South American rodent; e, back of *Dasyurus viverrinus*, an Australian marsupial; f, back of *Loncheres cristata*, a South American rodent. (From Wilder's *History of the Human Body*, after de Meijere; with the permission of Mrs. H. H. Wilder and Henry Holt and Company.)

mammals the dorsal surface is covered with them. These scales are also similar in form and derivation to those of the birds and reptiles, but are usually less horny. Hair is very different from scales, but the two are associated in some mammals and the pattern of arrangement persists in some that have lost all trace of scales (Fig. 58).

Claws, Hoofs, and Nails. These structures are likewise similar in origin (Fig. 59). The first consist of approximately equal convex dorsal and concave ventral plates in the more primitive birds and reptiles. Mammalian claws have the ventral plate greatly reduced and the tip of the toe covered ventrad by a terminal pad which is scarcely evident in the other groups. In hoofs the dorsal

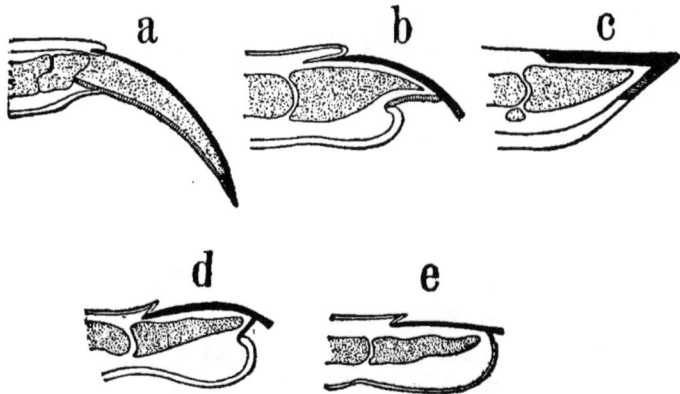

Fig. 59.—Diagrammatic longitudinal sections through digits of various mammals to illustrate the morphology of claws, hoofs and nails. a, *Echidna*, a primitive oviparous mammal; b, a typical clawed mammal (unguiculate); c, horse; d, monkey; e, man. The dorsal plate is in black, ventral plate striped, bones stippled. (From Wilder's *History of the Human Body*, a after Gegenbaur, b-e after Boas; with the permission of Mrs. H. H. Wilder and Henry Holt and Company.)

plate is enlarged and thickened, while the ventral plate is also extensive and horny, though still a rather soft structure. The terminal pad is lacking. In the monkeys nails are found which have a broad dorsal plate extending little if any beyond the tip of the digit; the ventral plate is reduced to a transverse strip beneath the tip of the dorsal plate. In man this reduction is carried still further and the ventral plate is vestigial. The terminal pad is evident in the primates only by the persistent friction ridges with definite pattern which occur on the tips of the digits.

Vestigial Structures in Man. For complete homologies of the remaining systems, reference must necessarily be made to embryological development, involving greater detail than can be included here, although the resulting comparisons are as conclusive as those

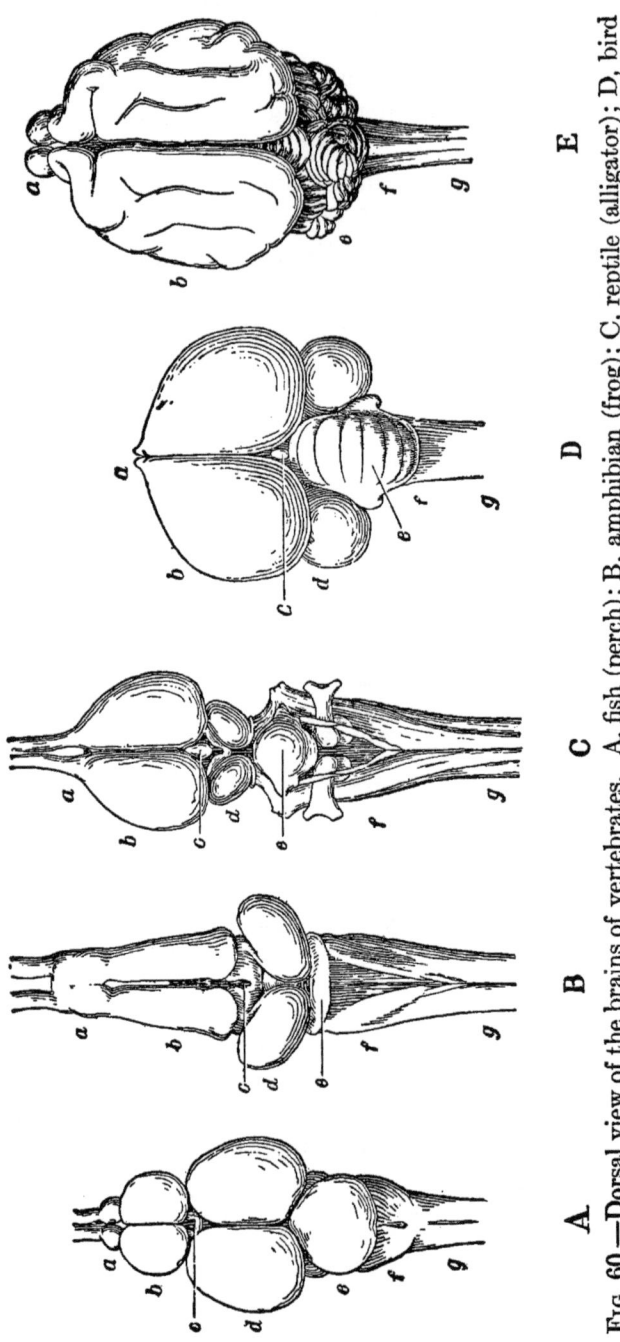

Fig. 60.—Dorsal view of the brains of vertebrates. A, fish (perch); B, amphibian (frog); C, reptile (alligator); D, bird (pigeon); E, mammal (cat). a, olfactory lobes; b, cerebral hemispheres; c, pineal body; d, optic lobes; e, cerebellum; f, medulla; g, spinal cord. (From Woodruff.)

already considered. The circulatory system is among the best to demonstrate relationships, and has already been considered under embryology. The nervous system is very evidently formed by the modification of the same primitive parts in the various classes, as is shown in the figure comparing the brain structure of vertebrates (Fig. 60). In all systems there are evidences of structures having persisted beyond their period of usefulness in the form of vestiges. These are especially interesting in man.

Supernumerary Mammae. Wiedersheim records numerous instances of the occurrence of supernumerary mammary glands. These glands develop embryonically in a ventrolateral milk line, of which all traces are obliterated except the normal adult mammae. In such animals as dogs, cats and pigs which bear several young at one time, these are several in number and lie in two rows indicating the position of the embryonic milk line, while in animals whose young are usually limited to one or two at a birth a more extreme localization limits the functional adult mammae to a pectoral pair (Primates, etc.) or a pelvic pair or group (domestic animals). In rare cases individuals, both male and female, have in addition to the two normal pectoral teats of man a series of additional rudimentary teats in a row, as in lower mammals. In females supernumerary mammae may be functional or mere vestiges, while in males, since the teats are normally vestigial, the same is true of others which may appear. Wiedersheim cites data, dealing largely with soldiers, which show a surprisingly high percentage of polymasty. One observer whose records he uses recorded this condition in more than 5 per cent of cases.

Persistent Hair. This peculiarity has also been recorded in many individuals, such as Jeftichjeff, the "Russian Dog-Man" and Julia Pastrana, a woman whose face was almost completely covered by hair. These cases are probably due to the persistence of the lanugo, a coat of hair which covers the body of the embryo but is normally shed before birth and replaced by the restricted hair of the adult. More significant evidences of relationship of man with the lower animals are found in the resemblance of the hair tracts on his body to those of quadrupeds.

The Tail. While the tail in man is normally reduced to a series of fused vertebrae, the *os coccyx*, completely enclosed within the body, it too may develop occasionally as an external appendage. Wiedersheim cautions against interpreting all such appendages as true tails, but he records two cases in which they contained vertebrae and were very probably vestigial remnants of the tails normally found in quadrupeds.

The Third Eyelid. The eye is the seat of another vestigial structure which normally persists. This is the thin pink fold at the inner corner called the *plica semilunaris*, supposed to be a remnant of the third eyelid of birds and amphibia. The full development of the structure can easily be seen by watching an owl close its eyes. As the upper and lower lids approach each other, the third lid, a filmy, grayish membrane, moves across the eyeball behind them.

Vestigial Muscles. In a few parts of the human body muscles persist which are usually useless and only rarely functional, among them the muscles which move the ears and scalp. These are usually vestigial, but a few individuals retain the ability to contract them at will.

The Appendix. One of the most familiar vestiges is the vermiform appendix. Essentially the same in structure as the intestine, it varies in different individuals from a tubular diverticulum to a solid structure, and in length from three quarters of an inch to nine inches. It is attached to the caecum, a blind end of the large intestine which extends beyond the union of that tract and the small intestine. Both caecum and vermiform process are found in animals below man, and in some species, particularly herbivorous animals, the caecum often attains relatively enormous dimensions and is a correspondingly important organ. According to Wiedersheim it may be longer than the entire body, although in Carnivora and several other orders it is reduced as greatly as in man.

Wisdom Teeth. The third molars, or wisdom teeth, are likewise vestigial. Although the total number of thirty-two teeth is less than the primitive number, we are losing four more, as is apparent from the variable development of the wisdom teeth and their usual ineffectiveness.

The Significance of Vestigial Structures. In these and many other structures a condition is visible in man which can be interpreted only as the persistence of parts for which he has lost or is losing all need. Throughout the various organic systems of the body more or less conspicuous examples of this type of resemblance may be found, and in all systems a very evident similarity of minute structure, arrangement of tissue, and general plan of gross morphology. The resulting adaptations often appear to be but a poor makeshift. They are effective, since their effectiveness is necessary to existence of the species, but they seem crude in comparison to organs of other animals which are used in the same way. Enough examples can be observed among the many species of animals of very different ways of attaining the same end effectively, such as the wings of birds and insects, to make possible only one interpretation of the existing conditions of resemblance among vertebrates, and among invertebrates as well. We can conclude only that the fundamental similarity of their often different structures in any system is evidence of definite relationship; that the dolphin's flippers and the human hand have evidently homologous structure not because such structure is the only possible foundation for a swimming organ and a prehensile appendage but because their ancestors were related in possessing just such a foundation as the pentadactyl appendage, of which they have made different uses.

Summary. The anatomy of adult animals of the vertebrate classes shows many points of similarity. All parts of the skeleton are evidently based on the same plan of structure. The differences which appear are easily correlated with special habits. Exoskeletal structures of different kinds also show fundamental similarity. Relationship of other systems is closely linked with embryology, but the presence of vestigial organs is significant, especially in man. Since these organs are useless or nearly so the only possible explanation of their presence is that they are the remains of once useful organs which the animal has not yet entirely lost.

REFERENCES

ROMANES, G. J., *Darwin and After Darwin*, 1892.
WIEDERSHEIM, R., *The Structure of Man* (translated by Bernard), 1895.
———, *Vergleichende Anatomie der Wirbeltiere*, 7th edition, 1909.
REYNOLDS, S. H., *The Vertebrate Skeleton*, 2nd edition, 1913.
WALTER, H. E., *The Human Skeleton*, 1918.
CUNNINGHAM'S *Textbook of Anatomy*, 5th edition, 1921.
NEWMAN, H. H., *Readings in Evolution, Genetics and Eugenics*, 1921.
WILDER, H. H., *History of the Human Body*, revised edition, 1923.
KINGSLEY, J. S., *Comparative Anatomy of Vertebrates*, 3rd edition, 1926.

CHAPTER VI

THE RELATIONSHIP OF EXISTING ORGANISMS
(Continued)

4. PHYSIOLOGY

We have so far dealt primarily with the structure of living things but the action of the parts composing an individual organism, the processes which are even more evidently essentials of life, are no less significant in evolution. The structure of organisms is a more tangible evidence of relationship but it is important to remember that it is only one of the three essential factors in the existence of living things. While environment, response, and hereditary structure are distinct, we cannot have life or individual lives without the intimate correlation of all three. The response of the organism or its functional activity constitutes the subject matter of physiology.

Fundamental Physiological Properties. Among the characteristics of living matter we find that a certain few belong to this category. While chemical composition and definite form are distinctive, the qualities of irritability and contractility are no less so, yet they are of an entirely different nature since their existence depends both on the things which the organism has received from its ancestors and upon the conditions under which it lives. These physiological properties include irritability, conductivity, contractility, metabolism and reproduction. They may be known equally well as the adaptive properties since it is through them that the organism is adjusted to its environment. With the exception of reproduction alone all are involved in the correlation of the individual and its environment, and the excepted property is, of course, essential to the adaptation of species.

It is in this group of characters that the kingdoms of organisms are chiefly distinct from each other, although there are species which are both plant and animal even in physiological processes.

In the single-celled organisms, where we can observe the processes of life at their simplest, it is obvious that all of these things are active in any cell, be it plant or animal. Such cells as the little green alga, *Sphaerella*, and the Protozoön, *Paramecium*, are in many respects the same. Each moves about actively, an evidence of its inherent contractility. Each, if it comes in contact with some object or substance in the water, indicates that it has received a stimulus from that object or substance. The indication may be in the form of a movement involving the end of the body opposite to that which received the stimulus, hence we know that some impulse has passed through the organism, an evidence of irritability and conductivity, and in addition of response through contractility, which produces motion. Finally, if we watch the organisms long enough, and conditions in the environment are favorable, we note that by the acquisition of substances from the environment the individual grows and reproduces itself in some way. Most of these properties can be equally well observed in a complex animal, such as man, but irritability, conductivity, and contraetility are so little emphasized in the higher plants as to be totally obscured except in rare cases and in normally inconspicuous phenomena.

Plant and Animal Metabolism. In these two simple one-celled organisms, however, we see the fundamental manifestation of life processes, but even here there is a strange difference. The animal is practically colorless, but the body of the plant is green. The animal, as it swims about in the water, is constantly occupied in sweeping into its gullet minute organisms which are massed together and passed into the cytoplasm of its body as food vacuoles, which in a few minutes are changed into a part of the animal itself. The plant does none of this. We know through the investigations of scientists that the two differences are closely related. The green substance in the plant is called chlorophyll, meaning literally the green of leaves, and its presence is the basis for an entirely different metabolism, the foundation of all life.

Photosynthesis. Chlorophyll acts in the plant as a catalytic agent, a substance by means of which a chemical action goes on, although the catalyst is not changed in the process. Plants which possess it, that is green plants, secure from their environment the inorganic substances, carbon dioxide, water and certain salts such as nitrates and phosphates, which may be called their food. In the presence of sunlight as a source of energy, the plant combines the first two, carbon dioxide and water, through the agency of its chlorophyll, to form sugars and starches. A different arrangement of component elements produces fats, and the addition of elements derived from inorganic salts forms proteins. While the complexity of the processes involved is infinitely greater than this, the essential result is the formation of these three fundamental compounds of living matter through synthesis of inorganic substances and the addition of energy derived from sunlight, the whole depending on the plant's possession of chlorophyll.

The plant's activity is thus chiefly constructive from the point of view of living things. Careful study has shown that this is not its entire activity, however, for while it uses carbon dioxide in the synthesis of carbohydrates, and liberates oxygen as a result, exactly the converse of the process which liberates energy by oxidation in the animal body, it also oxidizes some of the substances which it has elaborated, for the liberation of energy. The skunk cabbage, familiar to anyone who has studied nature in the eastern part of the United States, furnishes an excellent example. It blooms in swampy places as early as February, sometimes before the ice has entirely disappeared, and its release of energy in these frigid surroundings is so great that a temperature difference of six degrees has been recorded between the inside of its curious spathe and the cold outside air.

Animal Metabolism. The animal is in general a much more dynamic organism, for it takes the substances synthesized by the plants, with their abundant potential energy, and carries on its own activities solely by releasing this energy by oxidation, after breaking down the substances into their simpler components and resynthesizing these into the similar compounds of its own body. The animal is thus dependent either directly or indirectly upon the plant for its existence. It is wholly unable to build up the substances that it needs from inorganic materials.

In four of the five physiological properties then, as well as in morphological characteristics, the plants and animals are definitely related. In the fifth, metabolism, we see that there is fundamental difference due to the ability of green plants to carry on photosynthesis. However there is also a degree of similarity that is even more striking in parasitic plants like the fungi, which are as devoid of ability to utilize inorganic compounds, as dependent on the green plants, as are the animals.

Plant-Animals. In a few organisms both of these powers are resident. These are called plants and animals, and as a matter of fact are both. The genus *Euglena* includes a number of species not unlike *Sphaerella* to the extent that they are green with chlorophyll and move about by means of a slender flagellum. It has been found that cultures of *Euglena* kept away from light lose their green color, and after a few generations become more definitely animals in appearance through the lack of chlorophyll. Such colorless *Euglenae* can be kept alive by the addition of soluble organic foods to the water in which they live, and since they are kept away from sunlight, they carry on, obviously, the metabolism of animals. Their production of chlorophyll is, however, merely interrupted, for they again become green if brought into the light, and carry on the metabolism of plants. They are definite connecting links between the otherwise different kingdoms.

Other Types of Metabolism. Although the method by which both animals and plants liberate energy is the one most widely prevalent in the organic world, it is not the only possible means. Some of the bacteria carry on a fundamentally different process of metabolism in which energy is secured by the oxidation of very different substances. *Beggiotoa* and *Thiothrix*, the sulphur bacteria, for example, oxidize hydrogen sulphide and store up sulphur as a by-product; this is later oxidized and excreted as sulphuric acid. At least one of the iron bacteria, *Spirophyllum ferrugineum*, is equally dependent upon ferrous carbonate as a source of energy. Still other plants, the yeasts and anaerobic bacteria, do not need any free oxygen to liberate energy, but accomplish this result by modification of food substances in a way illustrated by ordinary fermentation. In this process yeasts break down sugars, forming alcohol and carbon dioxide and releasing the energy which the organisms require.

Specialization. It is of little or no use to consider the similarity of details of physiological processes, for it is difficult to compare the functions of different organs in animals which are not closely related, and in similar organs similarity of function would be a natural consequence. However, some generalizations are possible in connection with the fundamental physiological properties of matter and their distribution in the complex organism. The ectoderm of Metazoa, since it retains direct contact with the environment, might be expected to display a greater development of the quality of irritability and this is true even in the highest phylum, where it is the source of the highly developed nervous system. The function of conductivity is naturally retained in some degree by all cells, but it too is highly developed in the specialized ectoderm of complex nervous systems. Contractility, the means of accomplishing immediate response to environmental stimuli, likewise becomes an ectodermal function in the Coelenterata, but with the development of the third germ layer, this function shifts almost completely to the third layer, the mesoderm.

Inasmuch as the body form in the Coelenterates almost completely removes the endoderm from environmental contacts and protects it by the ectoderm from all necessity for direct response for protection or the securing of food, it is not to be expected that this layer would retain the same qualities as the ectoderm. The additional fact that it lines a cavity of sufficient size to contain particles of food too large for ingestion by single cells, suggests its logical association with digestion, and we find that the initial steps of metabolism are henceforth functions of endodermal tissue. The development of large and complex glands such as the liver and pancreas, involving quantities of mesodermal tissue, still involves the endoderm as the source of the epithelium—the glandular portion—of these organs.

The Endocrine Glands. One evidence of similarity in the functions of similar organs is sufficiently striking to be worthy of comment, inasmuch as valuable therapeutic results have been obtained on this basis. The several ductless or endocrine glands of vertebrates produce secretions known as hormones which exert specific correlating influences on the body through their power to activate or inhibit the development and functions of various parts. These glands include the pituitary body, thyroid, thymus, gonads and various other parts. Because of their importance in the human body they have been made the subject of extensive study in lower animals, and have been found to exert the same influence upon individuals of the species producing them and upon others, even of different classes.

The secretion of the thyroid, whose effect is evident through modification in the various types of goiter and in the congenital insufficiency of cretinism, is now extracted from domestic animals for therapeutic use. Its effect on the normal course of physical development is exerted not only on man, a member of the same class, the mammals, but also on tadpoles and larval salamanders. Tadpoles to which the extract (thyroxin) is administered undergo metamorphosis before attaining the normal growth, while the Axolotl, a salamander which normally remains aquatic, may be caused to develop into a terrestrial animal through the usual amphibian metamorphosis by the administration of the same substance. Likewise, the symptoms of cretinism may be corrected and normal development induced in infants by supplying thyroid extract secured from domestic animals.

The part played by insulin, a hormone secreted by the islands of Langerhans in the pancreas, has recently been given great publicity because of its isolation and the discovery of its effect on the metabolism of carbohydrates. Insufficiency of this secretion results in the disease diabetes, whose effects are now minimized or completely abated by the administration of the commercially extracted secretion of other animals.

These two are familiar examples, but are duplicated by the behaviour of all the other endocrine glands. The interspecific effectiveness of hormones is so dependable that a considerable number are now commercially available, all extracted from the glands of domestic animals, and are used extensively in endocrine therapy. The one limitation is not their lack of potency, but our incomplete knowledge of their action.

Blood Tests. One evidence of physiological relationship has received much attention in the literature of evolution, namely Nuttall's famous precipitin tests for blood. The physiological basis for these tests is treated in Nuttall's book from which we may draw.

Immune Sera. Immunizing properties of blood serum removed from animals which had developed natural immunity from such diseases as diphtheria and tetanus were noted late in the nineteenth century. Ehrlich then experimented with the toxic substances ricin and abrin. He found that animals treated with increasing doses of these poisons developed increased tolerance, or immunity, and that the blood sera of these animals neutralized the poisons *in vitro*, and, of course, when injected into other individuals rendered them immune from the effects of the poisons. He proved also that a serum capable of neutralizing ricin had no effect upon abrin and vice versa. Ehrlich concluded that definite compounds were produced in the blood in response to the poisons. These he called antitoxins, or antibodies. As Nuttall points out "we now know that normal serum contains a number of antibodies having similar actions to those artificially produced as a result of immunization with this or that substance, we know of normal agglutinins, haemolysins, bacteriolysins, antitoxins, antiferments, etc., all of which go to prove the correctness of Ehrlich's views in this respect."

The theory of antibody formation and structure is complex in its details. For our purpose it is sufficient to note that the various types of antibodies are antitoxins, antiferments, cytotoxins of various kinds, agglutinins and precipitins, each named for the substance whose presence in the serum gives rise to it, or for its action upon that substance.

Precipitins. Thus the precipitins have the power to form a precipitate when mixed with the substance which has produced them. Little is known of their nature, but the precipitates produced by their reaction with proteins of blood sera show characteristics of proteins in several recorded cases.

Preparation of Precipitins. Nuttall's procedure involved the use of rabbits, chiefly, as the source of antisera (i.e., sera containing antibodies). Injections of the blood or sterilized blood serum of other animals were made at intervals of several days until three to twenty had been administered. After the completion of this treatment six to fifteen days were allowed to elapse, the rabbit was then killed and bled, and the blood serum extracted and preserved with the necessary precautions to maintain its sterility. In this way antisera were developed for the blood of a number of species of animals, as well as for other substances, such as cows' milk and egg albumen.

Nuttall states that "we have sufficient evidence to show that precipitins are not formed in the serum of closely related animals." He cites the experiments of Bordet and Hamburger, in which precipitins were not produced when rabbit serum was injected into guinea-pigs. Nolf likewise found it impossible to produce antisera in pigeons treated with fowl serum. This indicated a sufficient similarity in the bloods of the animals concerned to account for tolerance of the one species for the blood of the other, an evidence of blood relationship which harmonizes with the fact that the animals are related in a morphological way.

Precipitin Tests. In using the antisera thus prepared, samples of sera were collected in two ways, viz., fluid and dry, although the latter method proved to be the more practicable. Dilutions or solutions in salt solution were prepared in test tubes, and a drop or two of antiserum introduced. The result in the case of related sera and antisera is the formation of a precipitate; unrelated sera do not react when the antiserum is added. Variability in the results is explained by Nuttall's statements that "where a powerful antiserum is added to its homologous blood dilution, the reaction is almost instantaneous, in other cases it takes place more slowly. In the case of a strong antiserum, the reaction takes place as a rule rapidly in related bloods, more slowly in distantly related bloods. The rate at which the reaction takes place may depend also upon the concentration of the blood dilution, the more concentrated dilutions, within limits, reacting earlier than higher dilutions. A weak antiserum will act more slowly than a powerful one." Thus we might expect, all other factors being equal, that the reaction would correspond to the nearness of relationship as determined on the basis of classification, and that in cases of doubtful relationship precipitin tests might furnish a valuable corollary to the usual evidences.

A few examples from Nuttall's extensive tables of results are given in the table on page 98. In the horizontal line are listed the antisera, in the vertical column, the blood tested. Only easily interpreted examples are given, and the symbols have been modified from the original to indicate great reaction, marked reaction, moderate reaction, slight reaction, and no reaction, in order as follows: 1, 2, 3, 4, 0. Blank spaces indicate the lack of a test.

Even in the few cases here presented, it is obvious that the most pronounced reaction is usually obtained with so-called homologous sera, and that animals of different species react in a degree similar to the degree of relationship determined in other ways. This is, of course, subject to error, like all pioneer procedure in science, but in the first three cases two show a maximum reaction of human blood with human antisera, one a slightly greater reaction with the antiserum of an anthropoid ape, all some reaction with antisera of other primates and with some antisera of more distantly related mammals, but none with antisera of the other classes. In cases 5 and 6 the bloods of the orang and chimpanzee show a maximum reaction with antisera of man and the anthropoids, but none with other classes. However, in cases 12, 13, 14, and 15, the bloods of various species of reptiles and birds show some reaction with antisera of these two classes but with no others.

	1. Man (blood from cut)	2. Man (blood from cut)	3. Man (blood from cut)	4. Orang	5. Chimpanzee	6. Donkey	7. Dog	8. Dog	9. Fox	10. Cat	11. Hedgehog	12. Birds (several species)	13. Domestic Duck	14. Green Turtle	15. Alligator
Man	1	1	3	1	1	4	0	2	0	0	0	0	0	0	0
Chimpanzee			2	1	1										
Orang		4	4	2	2										
Monkey		4	4	4	4	0	0	0		0		0		0	0
Hedgehog		0	0	4	4	0	0	0		0	1				
Cat	0	0		4	0		0	0		4		0	0	0	0
Hyena		4	4	3	4	4	2	3		4	4				
Dog		4		0	0		1	1	1	0	3	0	0	0	0
Seal	0	4		0	4		0	4		0	0				
Pig		4	4	2	0	3		0	0	4		0	0	0	0
Llama	0	0		0	0		0	0		0	0				
Hog Deer	0	0		0	4		0	0		0		0	0		0
Mexican Deer	0	4		4	0			4		0		0	0	0	0
Antelope	0	4		0	0		4	0		0		0	0	0	0
Sheep		0	0	0	0		4	0	0	0	0	0	0	0	0
Ox		4	0	0	0	0		0	0	0	0	0	0	0	0
Horse	0	0		0	0	1	0	0	0	0	0			0	0
Donkey						1									
Zebra						2									
Wallaby	0	0		0	0			0		0		0		0	
Fowl		0	0	0	0		0	0		0		4	1	0	0
Ostrich		0		0	0		0	0				4	1	4	4
Fowl Egg		0	0	0	0		0		0			0	4	4	4
Turtle		0		0	0		0		0			0	4	1	1
Alligator				0	0		0					0	4	4	4
Frog				0	0		0					0		0	0
Lobster				0	0		0					0		0	0

Quantitative studies of the precipitates produced in these reactions serve as a measure of finer degrees of relationship, but all tests are subject to modification by various conditions. The following excerpts from Nuttall's work are a partial consideration of these factors.

He writes: "Uhlenhuth agrees with me in finding *that the zoölogical relationships between animals are best demonstrated by means of powerful antisera*. He judged from reactions with such antisera that the ox is not so closely allied to the sheep, as the sheep is to the goat. He found that weak anti-sheep serum produced no reaction in ox blood. In my paper of 21, XI. 1901, I wrote '*The more powerful the antiserum the greater is its sphere of action upon the bloods of related species*'. For instance, a weak anti-human serum produced no reaction with the blood of the *Hapalidae*, whereas a powerful antiserum did produce a reaction and proved what I may be permitted to call the '*blood relationship*' in the absence of a better expression. . . . I also noted that reactions took place 'to a lesser extent in the bloods of allied animals, than in the homologous blood.'" In the following paragraph he shows that the serum may react with a remotely related antiserum if allowed to stand for some time, but he notes the facts that have been brought out here, "that anti-mammalian sera only produce these later reactions in mammalian bloods, anti-avian sera similarly in avian sera alone."

Reproduction. Finally, the reproductive processes of organisms afford unsurpassed evidence of relationship. Many cells retain the property throughout their lives; others are formed as end products in the highly specialized organisms and normally lose it. Some cells of the latter type are capable, however, under the stimulus of abnormal conditions in the organism, of reproducing during the process of regeneration of tissues or parts even though reproduction is not a constant function. A degree of specialization which completely eliminates this function is relatively rare.

Amitosis. In its simplest state cell reproduction apparently consists of a simple elongation, constriction, and separation of the parent cell, resulting in the production of two daughter cells. This process is known as amitosis. It occurs in certain parts of metazoa, as for example the well-known laboratory illustration, the ovary of the cricket, apparently in highly specialized cells.

100 EVOLUTION AND GENETICS

Fission. Single-celled organisms often reproduce by a superficially similar process called fission which also produces two similar individuals at the expense of parental loss of individuality, but this process foreshadows a more complex type of reproduction called mitosis. It is said by some authorities to be a true mitosis.

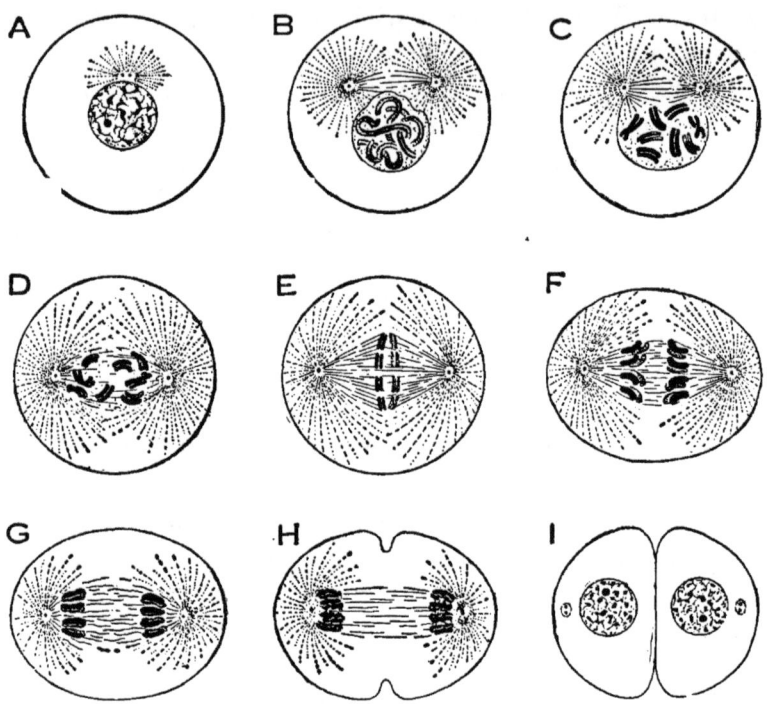

Fig. 61.—Typical stages of mitosis in which the chromosome number is assumed to be eight. A, beginning of prophase: chromatin in a reticular form, centrosome divided and astral fibers formed about it; B, early prophase: chromatin in spireme, centrosomes moving apart and spindle forming between them; C and D, later prophase: chromosomes forming and remainder of nucleus breaking down; E, metaphase: the chromosomes arranged in the equator of the spindle and split longitudinally; F, G, anaphase: the chromosomes migrating toward the centrosomes; H, telophase: a gradual return to the original state of the nuclear constituents and centrosomes, accompanied by constriction of the cytoplasm; I, the two cells formed at the completion of mitosis. (From Woodruff.)

Mitosis. This process involves primarily the chromatin of the nucleus (Fig. 61). During mitotic division of animal cells the centrosome divides, the two halves move away from each other, and about each a series of radiating fibers or apparently fibrous structures appears. Between the two a series of connecting fibers develops. The entire structure is not unlike the field of a bar magnet, with the two centrosomes representing the poles and the fibers lines of force. For obvious reasons the radiating fibers are called astral and the others spindle fibers. During the formation of the spindle the chromatin becomes condensed into a compact thread called the spireme, the nuclear membrane breaks down, and all nuclear structures except the spireme become a part of the cytoplasm. The spireme breaks up into a number of parts called chromosomes, and these bodies migrate into the equator of the spindle. The number of chromosomes varies in different species; there are four in some worms, forty-eight in man, and even more in some animals. This much of the process is called the prophase. During the next stage, the metaphase, each chromosome splits longitudinally into apparently equal halves. Following this the halves migrate toward the nearest aster in the anaphase, and the final telophase includes the constriction and splitting of the cytoplasm into two parts, each including one centrosome and one set of chromosomes, and the reconstruction of the nuclei as in the original parent cell.

Reproduction of Individuals. In its simplest state the reproduction of individuals is no more than cell reproduction. In the lower Metazoa, however, it passes beyond this simplicity since it must involve many specialized cells, but even here a process of fission or budding is common which is little more than the reproduction of numerous cells of the parent to form a new, similar individual. In all Metazoan phyla, an additional specialization occurs, either associated with the process of fission or budding or the sole method of reproduction. In it the reproductive power of certain cells of the body is emphasized to the extent that they are developed solely for the purpose of producing new individuals, so that the complete differentiation of the species from a single cell occurs anew with every generation. To this is added gametic or sexual reproduction, involving the union of two independent germ cells. The modification of cells accompanying sexual reproduction is a further addition to the intricacies of mitosis, but it has such an important bearing on other phases of our subject that it will be taken up in connection with the laws of genetics.

Accessory Reproductive Functions. With the establishment of sexual reproduction the germ cells, or gametes, are formed from ectoderm or endoderm in diploblastic animals, but in the triploblasts they are consistently mesodermal as far as can be determined.

It is certain that the gonads, the organs in which they are developed, originate from that layer. The gonads are commonly formed in the wall of the body cavity, and the gametes discharged into that cavity. Openings in some animals connect the coelom with the exterior apparently solely for the escape of the germ cells, but the excretory tubules or nephridia with the same relations provide another convenient means of egress, which results in a common association of reproductive and excretory systems.

The formation of two types of gametes, the ova and spermatozoa, demands different provisions for their escape and union, but the association of nephridia and gonads remains obvious throughout the higher phyla.

Reproduction of Vertebrates. In the vertebrate classes the transition of reproductive functions illustrates first in the cyclostomes, fishes and amphibia the simpler stages of sexual reproduction. The gametes of most species are discharged from the body and united in the water, where reproduction occurs in all three classes, with a very few exceptions among the amphibia. Terrestrial reproduction in the Sauropsida, including the reptiles and birds, demands additional modifications which are chiefly expressed in the foetal membranes treated under embryological relationship. Internal fertilization of the ovum is, however, a functional modification of equal importance, since the fluid medium is essential for the union of the germ cells.

The addition to the ovum in oviparous species of enough food to carry the young animal through its development to a point where it is capable of functioning as a more or less independent terrestrial organism is accomplished by the assumption of the function of secreting these substances by the tissues associated with the passage of the egg from the body.

The structural and functional modifications associated with the development of large eggs containing much food are not changed in the lowest mammals, but here a new function, with accompanying modification of structures, makes its appearance; viz., the secretion of milk for the nourishment of the young. These mammals are the lowly oviparous Prototheria, the duck mole and spiny anteater of Australia. The next division of the mammals, the Metatheria, includes such species as the kangaroos and opossums. In these the highly developed ducts of the female gonads assume a new function, that of providing nourishment through the circulation to the developing embryo, instead of secreting it as an addition to the egg. Here development is carried to a state of partial perfection, and the young animal is then carried in a pouch on its mother's abdomen and nourished with milk until able to shift for itself. Finally, the Eutheria, or true mammals, illustrate an elaboration of the same process. The connection of embryo with adult is so intimate that it is looked upon as essentially a parasitic relationship, and development proceeds not only to morphological completeness, but through a considerable degree of growth before parent and offspring are separated by birth.

Throughout the multitude of functions of the living organism of which these are only a few illustrations, occur the fundamental relationships which have here been emphasized. From simplest to most complex they are no more than the five fundamental functions, yet in these and in the complexities of their distribution among the parts of the specialized organism we see one more evidence of common relationship of all living things, and of lesser relationships of various kinds among the different groups.

Structure, development, function, and the classification which we have derived from the study of these and other things, all point to the truth which is so well established in modern science, that all living things are related. This much cannot be logically doubted; it is the function of evolution to explain their relationship.

Summary. The relationship of organisms is as well shown by their physiological processes as by their structure and development. Plants in general have one fundamental process of metabolism, animals another. Some organisms carry on both types and others, the colorless plants, form a connecting link between the green plants and animals. Exceptions to these normal processes make it all the more evident that similarity means fundamental relationship. The activity of the endocrine glands is a striking evidence of relationship among animals. Blood tests, based on the immunizing reaction of the blood of an animal to that of another species, have been worked out so extensively that they not only illustrate the fact but also the varying degrees of relationship. The process of cell reproduction, mitosis, links all organisms, and in the reproduction of individuals we find indications of less extensive relationship within phyla. The vertebrates, for example, show a gradual transition of reproductive functions.

REFERENCES

NUTTALL, G. H. F., *Blood Immunity and Blood Relationship*, 1904.
LOEB, J., *Studies in General Physiology*, 1905.
HOWELL, W. H., *Textbook of Physiology*, 9th edition, 1920.
DERCUM, F. X., *Biology of Internal Secretions*, 1923.
JORDAN, E. O., *General Bacteriology*, 8th edition, 1924.
ROBERTSON, T. B., *The Chemical Basis of Growth and Senescence*, 1926.
WOODRUFF, L. L., *Foundations of Biology*, 3rd edition, 1927.

CHAPTER VII

EVIDENCES OF EVOLUTION

1. EXISTING ORGANISMS

Relationship. The facts set forth in the four preceding chapters are such as to leave no logical doubt of the relationship of the various kinds of organisms now extant. It is obvious that there is some degree of resemblance between any species which we may choose, no matter how remote they may seem, even though extreme cases force us back to the cell as a common basis of structure and protoplasm as the one living substance. Other things may be fundamentally different, but these at least connect all living creatures.

The entire body of scientific facts is merely an elaboration and extension of such obvious things as we have assumed to be man's first observation of organic relationship. While some organisms are very obviously related, others are less evidently so, but through his gradual accumulation and interchange of knowledge, man has arrived at an understanding of organic unity. Science makes clear the fact that relationship among organisms is universal.

The Significance of Relationship. The interpretation of such facts beyond this point is somewhat different. To the scientific mind they can have only one meaning, but it is evident that there are others who find it possible to believe in the independent origin of these related organisms. When men were struggling to explain such natural phenomena as are in accord with evolution, even the contemporary origin of new organisms from inorganic sources did not seem absurd to them. We are now in possession of abundant evidence of the non-existence of spontaneous generation as a source of organisms, but the dim records of the past are less easily demonstrated by empirical methods. The nature of organic relationship must therefore be based on logical interpretation of the available facts. Whether or not the processes which have come to be accepted on this basis will some day be demonstrated in the laboratory, it remains for the future to prove. Many things are accepted with less definite proof than that now available for evolution.

Community of Origin. When relationship is mentioned, the immediate thought aroused is of similarity. Further analysis shows that we cannot have similarity, i.e., relationship, without some degree of community of origin. Thus among inanimate objects we speak of sedimentary rocks, igneous rocks, of bricks, of porcelain, of automobiles and radio sets. In any of these categories similarity is evident, although it is not merely likeness which has given us similarity, but the fact that in the one category all things have a common origin. Sedimentary rocks must be laid down by water and igneous rocks must be cooled from a molten state. Bricks are produced by burning clay, and porcelain by a similar process from a different kind of clay. Automobiles of many kinds are the various developments of a single idea, and all radios have been produced by the elaboration of the original mechanism of wireless transmission.

Within the same category we find that the relationship of things is again directly proportional to community of origin. There are various sedimentary rocks. Limestones differ from sandstones and both from shales. Automobiles are the same to a certain degree, but only those produced by the same maker are even approximately identical.

Forces and Materials. This analogy is faulty in more than one way, but it serves to emphasize a fact easily overlooked, viz., that similarity is due to a common origin. It is conceivable, of course, that similar things should be independently produced, but significant that they rarely are. Moreover when similar things are independently produced, either by man or nature, we can be certain that at least the same forces or materials or both have entered into their production. These factors may be widely disseminated if independent from the product, but in living organisms we see that they are concentrated wholly within organic matter, so that this alone is a demonstrable source of living things. It is an old biological principle that all life comes from preëxisting life.

Relationship of Individual Organisms. Such a relationship is even more evidently a matter of origin. Organisms are brother and sister because they are produced by the same parents, or cousins because their parents were so related. The more remote this common source, the more distant the relationship, until the limitations of human records lose sight of it altogether and we look upon our friends and associates as wholly unrelated. Supposedly unrelated individuals marry. The union of any more closely related than cousins is frowned upon, yet after all these are only degrees of relationship.

The relationship of these apparently unrelated individuals serves as an excellent illustration of the course of development of a population. Assuming that one individual is produced by unrelated parents, and that his parents are derived from equally distinct lines, only a few generations back we find his ancestors multiplied to a ridiculous point. The number of ancestors doubles with each generation in geometrical progression. Allowing twenty-five years to each generation, a reasonable average for the period covered by recorded history, we find that seventy-seven generations have passed during the Christian era. By carrying the ancestry of our one individual back through one-half of that period—2^{38}—we reach the stupendous total of 274,877,906,944 ancestors. The present population of the world is approximately 1,748,000,000 and it has increased constantly. Or on the basis of the closest union, that of cousin marriages, assuming even the impossible constant of siblings marrying cousin siblings, we find that the total ancestry of one hundred unrelated individuals vastly exceeds the total population of the world fifty generations back. The absurdity of these results is ample evidence that all members of a given species are related in some degree.

By carrying out such computations of ancestry and comparing them with the increasing population of the world, it is obvious that the individuals of any species have sprung from a very limited number within the period of recorded history. It is neither difficult nor illogical to carry the idea back to an initial unit, an individual or pair. Both biologically and through other sources this view becomes available.

Relationship of Species. As we apply this analysis of relationship to the species which make up the organic world, we find that the direct connection of one with another which is evident in individual relationship is not apparent. The period covered by human records is so brief that it does not afford an opportunity to view the transition of one to another. That this transition may yet be seen is strongly suggested by the production of distinct forms by mutation, a process which will be considered in detail in a later chapter, but even this has not resulted in the undisputed origin of species under human observation.

Since relationship as indicated by structural and functional similarity is evidence of development through similar processes from similar things in all cases which we can examine, however, it is indicative of a similar origin of related species. Through the concentration of all factors in the production of living things in organic matter alone, we may logically conclude that robins and bluebirds, or butterflies and moths, are related because they came from common beginnings, and that birds and Lepidoptera are likewise related in more fundamental particulars because of a more remote derivation from a common source. This view is logically tenable, but in the evaluation of details it is likely to be confusing if not substantiated by extensive knowledge.

Examination of the several evidences of relationship supplies the needed support for details of evolution. The field is so vast that complete analysis is not to be expected, but as many examples are available as the individual may care to seek.

In such species as are found in the genus *Euxoa*, of the moths, the most intimate degree of relationship is apparent. Some have been named from one region, others from another, and material from intermediate regions has later resulted in their union. Those individuals which occur in the Rocky Mountain region have no direct association with those which fly in the Mississippi valley. They are related in structure, pattern, color and habits. Why? Careful evaluation of the facts leaves only one possible conclusion. They must have been derived from the same source, and since they are individuals of the same species, such derivation is easily understood.

In connection with this case, one step leads us to that of obviously different species belonging to the same genus. Suppose that the two widely separated lots just mentioned should have proved to be actually different, no matter how much material from intervening regions might have been secured. They are no more independent in fact than the extremes of the one species, yet we are unable to demonstrate any connection between them other than a certain similarity. Why should they display this similarity? Again the only possible answer is because of a similar origin. Through the characters of the genus, the two are the same. They must then, have had a common origin a little more remote than that of the forms of the one variable species, and so, step by step, the varying degrees of relationship made evident by our classification are evidences of more and more remote common sources.

Ontogenetic Succession. For an illustration of the succession of changes which may have passed in the development of existing forms, it is possible to look to the actual record, which is now available in sufficient extent to be here treated independently. But the things to which man first had to turn in his attempts to explain the relationship of living creatures are entirely within the organism.

In studying the common characters of the vertebrates we have noted the succession of forms characteristic of the several phyla. That these forms may be looked upon as a chronological succession, and not merely a succession in degrees of complexity is made evident by the combination of anatomical and embryological facts. In the skeleton of vertebrates, for example, the occurrence in some fishes of a cartilaginous cranium alone, in others of such a cranium partly ossified together with external bony plates with characteristic arrangement, and in the higher classes of a skull which embryology shows to be made up of a similar cartilaginous portion in the beginning, from which certain bones are derived by ossification and to which others are added by development directly from mesodermal tissue, is indicative of relationship in chronological succession. The same applies to other bones of the body. This gradual succession of stages during embryological development which correspond to those represented by adults of the several classes is clear evidence that the higher forms have come from lower in this phylum. We see in the few points mentioned that the formation of cartilage is not an essential step in the formation of bone, so the transition of some bones can mean only that they still pass through the stages which they have followed in the past.

The circulatory system, especially in the development of the aortic arches, is a similar case. While they are symmetrically paired in the fishes, amphibia and reptiles, although reduced in number in the last two, they are still further reduced and asymmetrically developed in the birds and mammals. Superficially the resemblance is slight, but when we consider the embryological succession of parts, and see even in the human embryo the development of six symmetrical pairs of arches, of which some are later resorbed to bring about the adult condition, we can conclude only that the six are there because they were the original source of those that persist. Nature is not in the habit of producing useless structures, and the appearance of unnecessary structures such as these, which are so very similar to the adult arches of the fishes, points strongly to the derivation of the higher forms from fish-like ancestors.

The Recapitulation Theory. The same interpretation can be applied to any of the evidences of relationship previously brought out. Pharyngeal clefts, Meckel's cartilage, vestigial structures, all such apparently useless things, can be interpreted only as remnants of an ancestral condition, and since these things resemble functional parts of existing organisms, those organisms may logically be interpreted as near that ancestral condition.

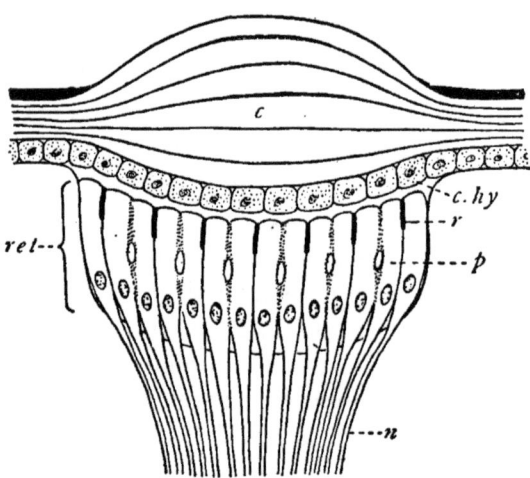

Fig. 62.—Diagram showing the structure of a primary ocellus. *c*, cornea; *c.hy.*, corneal hypodermis; *ret.*, retina; *n*, ocellar nerve; *p*, accessory pigment cell; *r*, rhabdom. (From Comstock's *Introduction to Entomology*, with the permission of the Comstock Publishing Company.)

This resemblance of embryonic stages to a succession of adults of different classes has given rise to the recapitulation theory, the belief that in its individual development an organism repeats the steps of phylogenetic development which gave rise to its kind. The repetition is undoubtedly modified in many cases according to the conditions of existence of the various species, but there is every reason to believe that it is in general true. In the transition from a single cell to a small group, to the hollow spherical blastula, the sac-like gastrula, and on through such details of development as have been mentioned, it is highly probable that in embryological development, or ontogeny, we have before us a partial record of past changes.

The Significance of Homologies and Analogies. Those who find special creation an adequate explanation of diverse living things see nothing more in these facts than the will of the Creator to produce such organisms as now exist. The activating force, in other words, is regarded as independent of the necessary materials, a belief which is not borne out by observed facts. If this were the foundation of life, there is every reason to suppose that every creature would be given the best possible equipment for its mode of life. Instead, organisms often have structures which show definite resemblance to those of other species living under very different conditions. They cannot logically be supposed to have been made from unspecialized raw materials.

Fins and Flippers. Such homologies, very common among living things, do not indicate that the same end cannot be met by different organs, for analogous structures are also fairly common. The whale has flippers which show definite homology with the fore limbs of terrestrial vertebrates, although they are more like the fins of fishes in function. Its respiration is carried on by the same organs as those used by terrestrial species. It would obviously be better fitted for purely aquatic

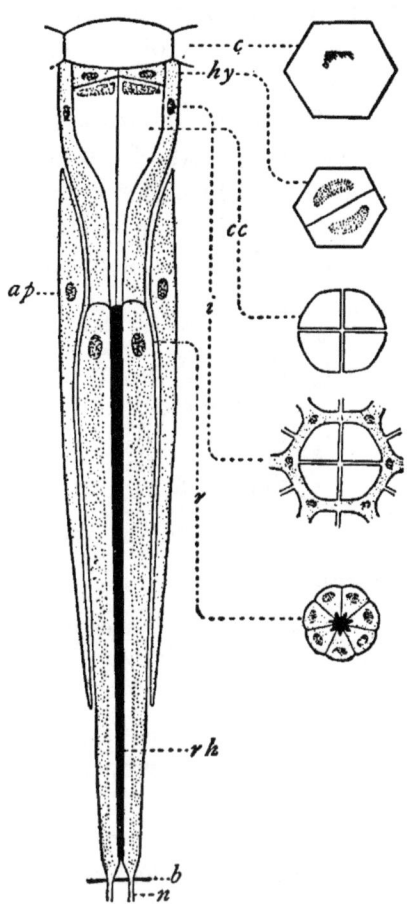

Fig. 63.—An ommatidium of *Machilis*. c, cornea; hy, corneal hypodermis; cc, crystalline cone cells; i, iris pigment cells; r, retinula; rh, rhabdom; b, basement membrane; n, nerve; ap, accessory pigment cell. (From Comstock's *Introduction to Entomology*, with the permission of the Comstock Publishing Company.)

life if it could breathe without rising to the surface, but this it is unable to do. On the other hand its need for such an organ as the tail of the fish is met by a very similar structure which is merely analogous.

Eyes. In the eyes of Arthropoda, Mollusca and vertebrates we find a remarkable example of this kind. All are special sense organs for the reception of light stimuli, and in their highest development, for the formation of visual images, yet they are very different structures.

Insect eyes are of two types, simple and compound; the former may be composed of numerous visual cells grouped beneath a transparent lenticular cornea, developed from the hypodermis (Fig. 62). In the compound eye similar visual cells form units with accessory cells and a separate cornea. These units are called ommatidia (Fig. 63) and are associated in large numbers in the most highly developed eyes. Their action is explained by Müller's theory of mosaic vision. According to this theory each ommatidium records a point of light, not a complete image. The result of numerous points of light recorded by reflection of rays from different parts of an object is an erect mosaic image. This image would depend for its resemblance to the original on the number of ommatidia in the eye, and the resulting completeness of reproduction of details.

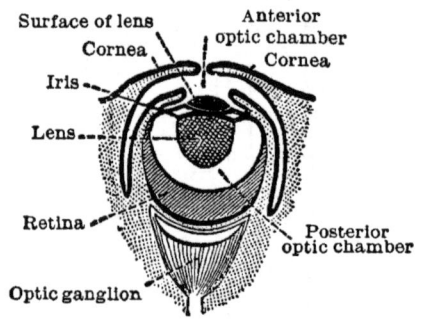

Fig. 64.—Diagrammatic section of the eye of a squid, *Loligo*. (From Hegner, after Grenacher.)

Eyes of Molluscs and Vertebrates. Molluscan eyes as developed in the *Cephalopoda* and vertebrate eyes are very different in optical function from the insect eye. Each is provided with a lens which forms on the retina a complete image of any object within the field of vision. This is, of course, an inverted image. In visual function the two eyes are similar. In structure and origin, however, they are different (Figs. 64 and 65).

Both cephalopod and vertebrate eyes have an outer cornea, behind which is an anterior space or chamber. Between this and a larger posterior chamber lies the lens. In front of the lens the iris governs the size and shape of the pupil, and at the back of the posterior chamber the light-sensitive retina is located. In the vertebrates, however, the posterior chamber of the eye, and consequently the retina, are derived from the first brain vesicle, while in the Cephalopoda they develop directly from the outer ectodermal layer. In the former the lens is derived independently from the outer ectoderm, but in the latter it comes from the optic vesicle. The outer chamber of the eye of the squid is never completely closed. In the vertebrates it is never open to the exterior, but forms by a splitting of mesenchyme outside of the lens. The iris in the vertebrate eye is inside of the sclerotic layer, while in the

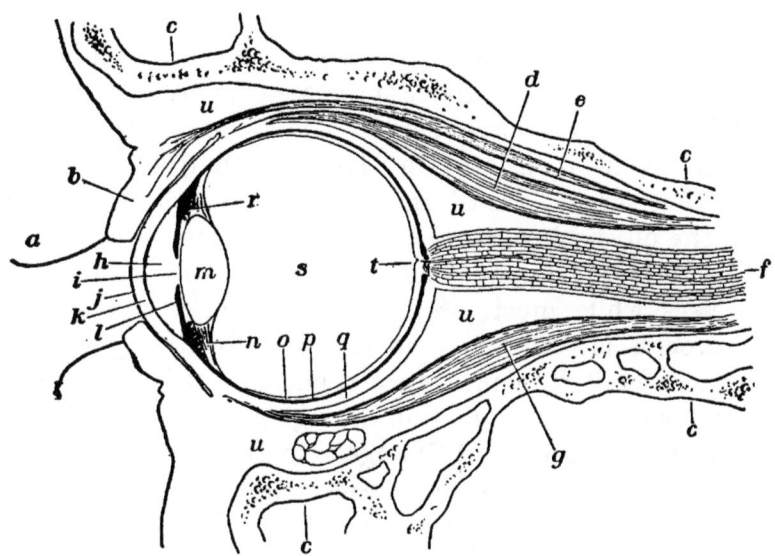

Fig. 65.—The vertebrate eye (human), vertical section *in situ*. *a*, eyelash; *b*, lid; *c*, bony orbit; *d*, *e* and *g*, muscles; *f*, optic nerve; *h*, anterior chamber filled with aqueous humor; *i*, pupil; *j*, conjunctiva, a transparent membrane continuous with the lining of the eyelid; *k*, cornea; *l*, iris; *m*, lens; *n*, suspensory ligament of lens; *o*, retina; *p*, choroid coat; *q*, sclerotic coat; r, muscles to ligament suspending lens; *s*, posterior chamber containing vitreous humor; *t*, point of entrance of optic nerve; *u*, fatty tissue. (From Woodruff.)

squid it is a projection from the margin of this layer. They are remarkable analogous organs, and an excellent example of convergence.

Blood. In the blood of the same three classes a similar diversity prevails. Insect blood consists of a fluid plasma in which float leucocytes similar to those of the vertebrates. The blood does not carry oxygen to the tissues, however, and there are consequently no red corpuscles. Oxygen reaches the tissues through the independent tracheal respiratory system. The blood of the mollusca, unlike that of the insects, must carry the animal's supply of oxygen, hence it contains a blue pigment, haemocyanin, which contains copper and iron, the former in relatively large quantities. Haemoglobin, the iron compound which gives the red color to vertebrate blood and serves as a vehicle for transporting oxygen in the body is also present in mollusca to a limited degree. Thus the blood of the three classes is similar as a conveyor of absorbed food and blood cells, but in only two cases is it a conveyor of oxygen, and in these two it accomplishes its respiratory function by means of different substances.

Metabolism. For a more generally applicable case we have only to refer to the several metabolic processes mentioned in Chapter VI. All animals must have energy. The response of the organism to its environment invariably involves the controlled release of energy in some form. Transformation or release of energy is not limited in the physical world to any one process, and the fact that the iron bacteria and sulphur bacteria secure their energy through two processes while most organisms carry on a third entirely different process is sufficient evidence that living organisms are not necessarily limited to one source of energy. It is very strongly suggestive of similarity of function through similarity of origin.

The Cause and Process of Change. In these examples organisms are seen to differ in degrees corresponding to the remoteness of their relationship. If all species were wholly independent in origin, we might expect a variety of structures and functions no less than the number of species involved. That similar results would be accomplished in many cases because of similar needs we cannot doubt, for examples are before us, but it is equally impossible to believe that these independent species would follow in their individual development unnecessarily tortuous paths, or that they would produce even temporarily structures which were of no use.

Factors in Existence. The questions naturally occur, why should these differences have come about, and how? Evolutionists have attempted to answer these questions, although in these attempts we still find the purely theoretical aspects of the subject. The interpretation of relationships leaves no doubt of the reality of evolution, but the exact forces through which it is expressed are not yet proved. These problems must be considered in detail, but to establish completely the evidences of evolution it is necessary to consider in brief the possibility of such changes.

Returning to the analogy of individual relationship, the three determiners of individual existence are seen to apply to any phase of individual activity. Even the production of a new individual demands first of all inherent fitness, second, the proper environment, and third, the proper response. Some individuals are congenitally unable to produce germ cells. In most species reproduction occurs only under definite conditions. Finally some individuals reproduce in response to certain conditions while others do not. When once produced the individual has a certain range of possibilities. Within the same species these possibilities may vary, but in the main they fall between certain extremes in most individuals. The different responses of various individuals within this range are known as variation. It is so prevalent that no two individuals of any species are exactly alike, and so obvious that it has been called the most invariable thing in nature.

Most species come into contact with environment common to various other species. Not all species respond to the same conditions in the environment, or like individuals, they may respond to them in different ways, depending on their inherent powers. In some cases the possibilities of the species are such that any individuals may accomplish the same end in either of two different ways. So it is with *Euglena*, which can, if light is lacking but organic food is available, live as an animal, although under normal conditions it is equally able to live as a plant. The inherent possibilities of an organism may be such as to enable it under different environmental conditions to live in very different ways.

Variability of Organisms. That difference of response results in qualitative differences in the organisms is evident in the same organism, *Euglena*, which loses its chlorophyll when it lives in the dark as an animal. Within our own lives we see similar evidences in the development of calluses through constant friction or pressure, and the increase in size of muscles through use. The converse is equally true, that disuse of a part results in diminution of its powers. These effects of use and disuse were emphasized by Lamarck as a part of his theory of evolution.

The degrees in which organisms are able to respond to varying conditions are in themselves variable. It has been recorded, for example, that in streams where dipnoid fishes are found, the cessation of flow and stagnation of the pools remaining in the river bed are sometimes fatal to ordinary gill breathing fishes, while the Dipnoids, because of their ability to secure abundant oxygen at the surface, are able to live. Some species live under extremely limited favorable conditions, like the myrmecophilous insects; others seemingly thrive almost without restriction, like the ubiquitous English sparrow. If the characters of these species are such as actually to enable them to live only if the environment is favorable within narrow limits, they are said to be specialized; if such that they can live equally well under any of a variety of conditions, they are called generalized species. Fitness for life under definite conditions of all degrees is known as adaptation.

Variability of Environment. The second factor in existence, environment, is no less variable. From day to day and from season to season conditions of temperature, light and moisture change, and with them come changes in the food supply. Change in environment demands a change in the organism, hence mammals acquire thicker fur in the winter, and certain aquatic snails seal their shells with a mucus plug and lie dormant when drought robs them of their normal habitat. However, many of the more highly developed animals are able to exist in spite of extreme fluctuations, and to remain in the same locality in spite of seasonal or other changes.

The Time Factor. Although environmental change, or exposure to different environments, may result in different responses among the individuals of a given species, it has been found impossible until recently to produce such changes experimentally as a permanent modification of the species. Müller has succeeded in producing permanent modifications of fruit flies by subjecting adults to the action of X-rays. After this treatment mutations appeared in the progeny of the flies much more abundantly than in control cultures which had not been subjected to the rays. The results depended upon the length of treatment with X-rays, and in the most heavily treated lots of flies reached a maximum of 150 times the frequency of mutation in the controls. While these results are interesting and significant, especially in the field of genetics, they do not refute the fact that the manipulation of conditions such as occur in a natural environment has not yet produced permanent modifications. We must remember in judging experiments of the latter type, however, that even a factor operating since the beginning of recorded history would have had relatively little time to effect a permanent change in comparison to the vast periods which have come and gone during the existence of some species, and that the duration of experiments has been an infinitesimal part of this period. The mere fact that even individuals are capable of some degree of adaptive adjustment leads one to believe that the species which they constitute, since their fundamental qualities are rooted in adaptations, are also plastic in that respect. But since the species is a more permanent unit than the individual, it is not difficult to understand its obviously slighter susceptibility to change.

Interspecific Relationships. Interspecific relationships among organisms offer another convincing proof that such change has occurred. According to the doctrine of special creation, a succession of forms appeared ranging from simple to complex, with man as the last. It is not surprising, therefore, according to this or any other view, that many lesser beings were in existence when man and related mammals appeared. Parasitic species, however, of many phyla are dependent upon the human race for food. If all of these were created in a brief span of time, granted that they could exist until the creation of man, it has been suggested that Adam and Eve must have been beset with all of the creatures that now live on or in the human body. Such a ludicrous picture is contrary to reason. Obviously change in the ancestors of these organisms has led to their association with man since his appearance on earth. If independently created, there is, as already pointed out, no reason to expect them to resemble so closely other less specialized forms.

Cause and Effect in Variation. One other phase of organic change is that of so-called fortuitous variation. It is doubtful that any variation occurs independently of response to definite conditions, but a complex organism consists of many parts, any one of which may vary. The entire body of the organism is coördinated, and each part through its functions affects other parts. This is a primary result of specialization, but it is conspicuous in such effects as the endocrine glands produce. Persistence of the thymus results in gigantism; thyroid insufficiency is responsible for cretinism; removal of the gonads results in the absence of the normal secondary sexual characters. The over-development of parts, or their failure to develop, is in all cases a definite response to a definite condition. When, as in these cases, a condition appears without evident relation to the environment, it is said to be fortuitous.

While a stable condition persists, there is no reason to expect a change of response. The persistence of a response removed from the normal is a modification of the internal environment, and as such is capable of initiating further modifications in the organism, so that, whatever may be their source, modifications established in an entire species or in a large majority of the individuals of a species are at least a potential stimulus to permanent change. The possible behaviour of such factors is another of the problems of evolutionary theory.

Summary. We have seen that relationship implies not merely resemblance, but similarity of origin, and that consequently the various degrees of relationship which are found among the existing species of animals point to common derivations of varying remoteness. The details of relationship indicate definite successions which represent the probable ancestry of different groups. Change of individuals is possible in response to change of environment, the environment is known to change, and single species exist even now which are capable of responding in ways fundamentally different. It is especially significant that while existing organisms show the possibility of merely analogous structures meeting the same need, in many cases the same fundamental structure is modified to meet many needs. Evidences of the past occurrence and the present possibility of evolution are abundant, and the factors mentioned of variation, adaptation and environment furnish the materials for further inquiry into the processes involved.

CHAPTER VIII

EVIDENCES OF EVOLUTION (*Continued*)

2. THE GEOLOGICAL RECORD

If the changes which have occurred in organisms during past ages were recorded completely, the record would show us exactly how one species has given rise to others, and how the gradual change from group to group through increasing degrees of complexity has been accomplished. The result would be a phylogenetic tree rooted in the beginning of life, every branch complete in its connection with the whole. In the chronology of the various steps would be evidence of the infinite slowness of phylogenetic change, and in the characteristics of the organisms would be reflected the conditions of environment under which they lived.

Fortunately some twigs from this tree are preserved, and through the study of these fragments of the record and the many evidences of physical change in the earth's surface, the sciences of geology and paleontology give us a satisfactory, if incomplete, account of past evolution. In the findings of geologists there are evidences of great transformations of the earth, of the elevation of land masses and mountain ranges, the inundation of great areas by the oceans, frigid climates resulting from extension of polar ice caps and tropical conditions following their recession. The evidences of these changes involve almost incomprehensible movements of material which could not have been brought about in less than millions of years. Paleontology correlates with these facts the records of organic remains, and in the whole we find fragmentary evidence of phylogenetic change which is a valuable corollary to that shown in existing organisms.

Rock Formation. From evidence which we need not consider here, geologists have concluded that the earth was once in a molten state. The first rocks which appeared in the crust as it cooled were therefore igneous. By weathering and erosion of these primordial rocks soils were formed and sediments of various kinds were carried down to the oceans. Here they settled in layers of mud and sand. Minute organisms living in the sea also settled to the bottom as they died, and the calcareous parts of their bodies formed other layers. Through long periods of time these layers of sediments were transformed into rocks,—shales, sandstones and limestones, respectively,—which are called sedimentary rocks in contrast to the basaltic and granitic igneous rocks formed by the cooling of molten masses.

Succeeding readjustment of the unstable crust resulted in the elevation of these stratified sedimentary rocks above the waters in which they were laid down. They were repeatedly folded and broken by great cataclysms which left jagged remnants projecting thousands of feet above the level of the surrounding land masses, and thus formed our mountain ranges. When once elevated, the sedimentary rocks themselves were exposed to the forces which weather and erode, and were in turn cut away and carried down to the sea to play their part in the formation of new sedimentary strata lying upon those previously deposited.

Geological Time. The lengths of time involved in this succession of constructive and destructive processes is inconceivable, but some slight comprehension is possible when we consider the enormous thickness of the aggregate deposits and the slowness with which materials are now being removed from the land masses. In all, geologists estimate sedimentary deposits to have reached a mean thickness of fifty-three miles. As Schuchert graphically expresses it, "this means the more or less rapid wearing away almost to sea-level, one after another, of more than twenty ranges of mountains like the present European Alps or the American Rockies. During the incredibly long intermediate time, when the lands were planed to a low relief, there was very little erosion." And yet the continent on which we live is said to be undergoing denudation to the extent of only one foot in 8,600 years. Since the earliest record of history, North America has lost, according to these figures, less than a foot of altitude on the average, yet the strata found in the Appalachian mountains indicate that they once reached twenty thousand feet above sea level. More than fifteen thousand feet of rocks worn away and washed into the sea at a rate which is imperceptible within human experience!

Even in view of the fact that erosion proceeds at a much more rapid rate where rainfall is abundant and slopes are steep, these enormous movements of material prior to the coming of modern man are impressive. The Appalachian Mountains are the same to us as to our ancestors of Revolutionary times, yet they were once as great a range as the Rocky Mountains.

The hill on which Denison University stands is one hundred feet above the nearest stream and a thousand above the sea, but at its top there may be found the fossil remains of Brachiopods and Crinoids, animals which once lived in the ocean. The greatest cataclysms of modern times have produced no such movements of material, yet in the terrific forces of a great earthquake we may find some slight conception of the vastness of geological processes in the past.

The evidence is before us that these things have occurred. We may see in a mountain range the twisted, folded remnants of rocks which could only have been produced by sedimentation. We may walk in deep grooves ground by the glaciers in solid rock and select from undisturbed masses above them the remains of corals which once lived in the ocean. And with these facts we may consider the visible modifications of the earth's surface today, infinitesimal but relentless.

Time Divisions. On the basis of all changes including climate, organic development, and modifications of the earth's crust, the earth's history has been divided by geologists into eras, these into periods, and these finally into epochs. The first are named from the prevailing types of life, the second largely from regions where their characteristic deposits are found, and the last with various descriptive terms. Each division is determined by characteristic sedimentary rocks and the fossils included in them, and by the vertical relationship of these strata the relative chronology of the various periods has been determined with reasonable accuracy. It is obviously impossible to estimate such periods in terms of years with anything approaching exactness. Estimates of the age of the earth run up to 1,000,000,000 and more years, within which a few thousands are a slight margin of error. All of these facts are briefly expressed in the following geological table:

GEOLOGICAL TABLE FOR NORTH AMERICA
(Modified, from Lull's *Organic Evolution*)

Eras	% of Total Time	Major Divisions	Periods	Epochs	Advances in Life	Dominant Life
Psychozoic	? very small			Recent (Post Glacial)	Era of Mental development.	Age of Man
Cenozoic	5	Quaternary	Glacial	Pleistocene	Periodic Glaciation. Extinction of great mammals.	Age of Mammals and Modern Floras
		Tertiary	Late Tertiary	Pliocene	Transformation of man-ape into man.	
				Miocene	Culmination of mammals.	
			Early Tertiary	Oligocene	Rise of higher mammals.	
				Eocene	Vanishing of archaic mammals.	
Mesozoic	12	Late Mesozoic	Epi-Mesozoic Interval		Rise of archaic mammals.	Age of Reptiles
			Cretaceous	Lance	Extinction of great reptiles.	
				Montanian Coloradian	Extreme specialization of reptiles.	
			Comanchean		Rise of flowering plants. Oaks and grasses.	
		Early Mesozoic	Jurassic		Rise of birds and flying reptiles.	
			Triassic		Rise of dinosaurs.	

Eras	% of Total Time	Major Divisions	Periods	Advances in Life	Dominant Life
Paleozoic	28	Late Paleozoic or Carboniferous	Epi-Paleozoic interval	Extinction of ancient life.	Age of Amphibians and Lycopods
			Permian	Rise of land vertebrates. Rise of modern insects. Periodic glaciation. Cycads, ginkgoes, pines.	
			Pennsylvanian	Rise of primitive reptiles and insects.	
			Mississippian	Rise of ancient sharks. Rise of echinoderms.	
		Middle Paleozoic	Devonian	Rise of amphibians. First known land floras, club mosses, horsetails and ferns. Earliest ammonoids.	Age of Fishes
			Silurian	Rise of lung-fishes and scorpions.	
		Early Paleozoic	Ordovician	Rise of land plants and corals. Rise of armored fishes. Rise of nautilids. Earliest annelida. Earliest sea-urchins.	Age of Higher Invertebrates
			Cambrian	Rise of shelled animals. Dominance of trilobites. First known marine faunas, including sponges, coelenterates and echinoderms. Thallophyta.	

Eras	% of Total Time	Major Divisions	Periods	Dominant Life (Inferred)
Late Proterozoic	10	Algonkian	Great Epi-Proterozoic interval	Age of Primitive Marine Invertebrates (Fossils almost unknown, earliest recorded ones occurring in the Huronian.)
			Keweenawan	
			Major unconformity	
			Animikian	
			Major unconformity	
			Huronian	
Early Proterozoic	10	Neolaurentian	Epi-Neolaurentian interval Peneplanation of mountains and continents	
			Neolaurentian Revolution	
			Sudburian	
Archeozoic	35	Paleolaurentian	Epi-Paleolaurentian interval Profound erosion of mountains and continents	Age of Unicellular Life Protozoa and Protophyta (Fossils unknown.)
			Paleolaurentian Revolution	
			Keewatin Coutchiching Grenville	
		The Unrecoverable Beginning of Earth History		

Fossils. The fossils which enable us to determine the type of organisms in existence during any of these periods are buried remains of organisms or deposits which record in some other substance the original form of the organism.

In order that fossilization may occur, it is necessary that the organism first be buried in the medium which is to preserve it. An animal may be mired in swampy ground or quicksand, or trapped in asphalt deposits such as those near Los Angeles. In any of these cases, it would be quickly engulfed and protected from disintegrative forces. Failing such accidents, any body which is preserved must necessarily have been covered quickly enough by

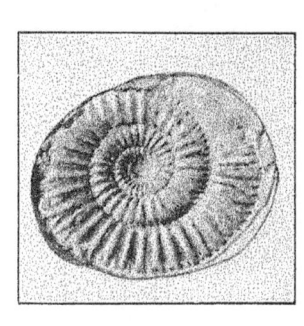

Fig. 66.—Types of fossils. *A*, a petrified tree trunk, fossilized by the replacement of its materials with other substances; *B*, the mold of a fossil shell; *C*, a natural cast formed in the mold shown in *B*; *D*, a fossil worm burrow; the rock was once sand on an ancient seashore and a worm burrowing through the sand left a tunnel, the sand was consolidated long afterward into sandstone and the burrow filled in with a more resistant material. (From *The Earth and Its History*, Bradley, with the permission of Ginn and Company.)

125

sediments or wind-borne material to afford it an effective degree of preservation. Once buried the original tissues of the organism may be preserved, their form may be reproduced by the addition or substitution of minerals (petrifaction), or the mold of the body in the material which surrounded it may persist after the tissues have disintegrated. Molds are sometimes filled in with

FIG. 67.—Beresovka mammoth, *Elephas primigenius*, discovered frozen in the soil. Specimen as it now appears in Petrograd. (From Lull.)

other mineral matter, forming casts of the original (Fig. 66). In view of these processes, it is only natural that the sedimentary rocks contain the majority of the known fossils, since they are formed from materials which are the most likely to preserve organic remains.

The perfection of fossils of the several kinds varies greatly. Records are available of the preservation of animals in Siberia by a natural cold storage in ice or soil. The remains of a mammoth (Fig. 67) preserved in this way were found in 1901 at Beresovka, Siberia, 60 miles north of the Arctic Circle. "This creature evidently slipped into a natural pitfall of some sort, possibly an ice crevasse covered with soil and vegetation. A fractured hip and fore limb, a great mass of clotted blood in the chest, and unswallowed grass between the clenched teeth all point to the violence and suddenness of its passing. Almost all of the animal was preserved, though the hair of the back has disappeared and the trunk had been eaten off by dogs before the specimen was discovered" (Lull). Fossils of this type, and any others buried in substances with preservative qualities, are, of course, not unlike preserved specimens. The most numerous are the amber fossils of the Oligocene (Fig. 68). The ambers are formed by the transformation of resinous exudates from coniferous trees, and in their initial state were of such consistency that even the most delicate insects

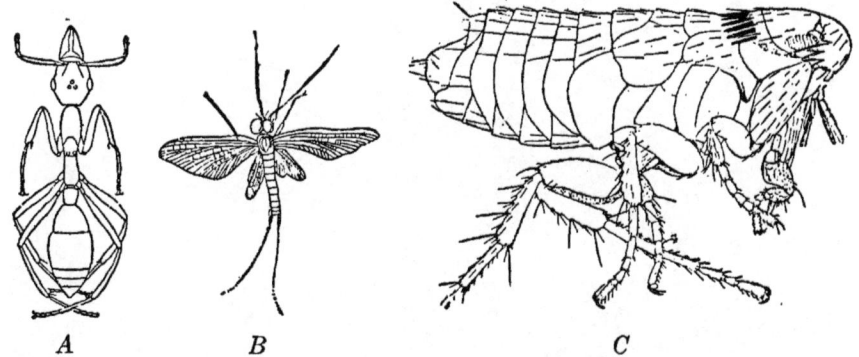

FIG. 68.—Amber fossils. *A*, an ant; *B*, a mayfly; *C*, a flea. (From Zittel.)

could be caught and embedded by them with little damage. As the resins hardened and turned to amber the contained organisms were preserved as in a modern microscopic mount. Thousands of specimens of Arthropoda, chiefly insects, have been discovered in the Baltic amber deposits of Germany, along the coast of the Baltic Sea. Although even such delicate bodies as those of insects have also been fossilized in shales, the amber fossils and frozen and preserved remains of animals are naturally more complete and perfect than any other kind.

Fossils preserved in rocks are often broken and distorted by the subsequent movements of the strata in which they lie. Consequently they are more often incomplete or imperfect, and demand more careful study and interpretation.

Interpretation of fossils. *Environment.* The value of fossils is not alone the record of past life which they give us, but also evidence of climatic conditions. In living organisms we can see that certain structures are correlated with certain qualities of

Fig. 69.—An ancient climate. (From Knowlton's *Plants of the Past*, with the permission of the Princeton University Press.)

environment. When fossils possess similar structures we can be certain that they met similar needs, and that the environment of the extinct animal was therefore similar to that of the one now living. Plants are very closely linked with the physical environment, hence fossil plants are valuable evidence of the temperature and rainfall of the period to which they belong (Fig. 69). The teeth of an animal are equally indicative of the kind of food which it eats, and if a fossil has shearing teeth, we know at once that it was carnivorous, while broad, grinding teeth indicate grazing forms. Thus aridity, forestation, temperature, and various habits are disclosed by these remains, and a careful study of all available evidences has pieced out the record of prehistoric life to a remarkable degree.

Succession of Forms. The general illustration of evolution derived from such sources is. adequately expressed in the geological table. If evolution has occurred, the most primitive creatures would necessarily have existed in the earliest periods of the process, and successive degrees of complexity would become evident as time passed. This proves true to such a high degree that we are justified in looking upon it as a further proof of evolution, and in adding to the chronological succession the earliest stages, which can only be inferred, since the organisms must have been too small and too delicate to be preserved readily as fossils. Fossils have been reported even from the Archeozoic, but they are open to doubt. With this initial step alone based only upon estimate, we find that the Proterozoic rocks contain only primitive invertebrates, and that a gradual ascension of forms proceeds through the remaining eras. Just as we see in the species now living the order proceeds through the higher invertebrates to the fishes, the most primitive of vertebrates, thence through the Amphibia and the reptiles to the birds and mammals, and finally man. Since animals are entirely dependent on the green plants for their food supply, it is significant that in this record the first known land floras precede the true terrestrial animals in development, and that in many other known details the various parts of the record coincide.

By reference to the table, it will be seen that the late Proterozoic, while it is evidently the period during which the marine invertebrates were the dominant form of life, is characterized by a scarcity of fossils. For this reason it is impossible to judge the phylogenetic association of species which must have existed.

The next geological division, the early Paleozoic, records so many forms of invertebrates that we are able to see in the fossils many evidences of phylogenetic succession. However, this sudden appearance of many forms deprives us of the opportunity to see how and from what they arose. We are forced to begin in the middle of the record, and by the steps which are clearly disclosed in its more complete portion to judge as best we can what processes took place in the periods which are forever closed to us.

Through the remaining periods we are confronted by similar difficulties. It seems that while hard parts are necessary for effective fossilization, the development of such parts is characteristic of specialized organisms, and not of those generalized species which might be expected to give rise to specialized forms. The record of detailed phylogeny is therefore confined to minor groups of relatively high development. We are able to trace descent from genus to genus in the ammonoids, for example, in the Devonian and later periods, and in the related nautiloids, which probably arose in the Cambrian. In the Ordovician, however, there are fossils of armored fishes of whose ancestors we have no exact record, and in the Devonian, we find evidences of amphibia, although no intermediate form connecting them with the fishes is known.

However, the major succession is established even by these abrupt transitions in the geological record. The lung fishes, able to breathe air, are known to have occurred in the Silurian, prior to the Amphibia, the Amphibia precede the wholly terrestrial vertebrates, reptiles precede mammals and man is the last species of all to appear. While a chapter is gone here and there, the entire record, together with such details as are well preserved, is convincing evidence of gradual succession.

Succession Within Animal Phyla. Among the invertebrate phyla, the status of individual groups also shows definite succession. The Protozoa, now recognized as the most primitive existing animals, are too small and delicate to be readily fossilized. Some are specialized, however, in the production of calcareous, siliceous, or chitinous tests, and consequently are readily preserved (Fig. 70). The Porifera, also very low in the systematic scale, are preserved because of their hard spicules, and are found among the earliest Paleozoic fossils in the Cambrian. The same is true of such Coelenterates as are preserved. Worms, because of their soft bodies, are not abundant among fossils, but the protective tubes and hard teeth of marine Annelida occur even as early as the Ordovician. The Bryozoa, now relatively unimportant, were very numerous during the upper Cretaceous. The genera then living are now extinct, however, while those which followed during the Pliocene are still represented, in some cases by the same species which then existed. The Brachiopods show a very similar course in their development, but reached the climax of diversity in the Ordovician and Silurian, and have since become less numerous. Most of the Tertiary species are congeneric with those now living. The Echinodermata reached their climax in the Paleozoic. They were represented by three classes which are now entirely extinct. One, the Crinoidea (Fig. 71), is represented by a few living species but was once well diversified, and four, including the sea cucumbers, sea urchins, starfishes and brittle stars have come down from the early Paleozoic. All of these are among the lower phyla of animals, and it is to be noted that they extend well back toward the beginning of the fossil record.

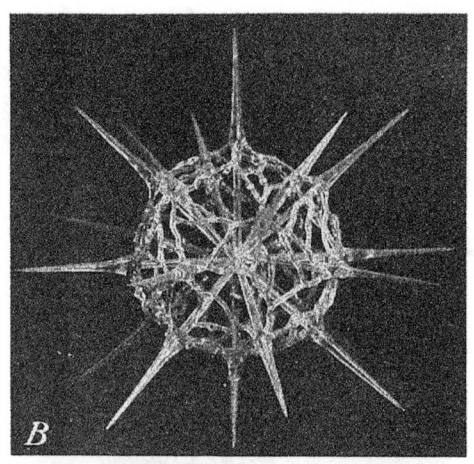
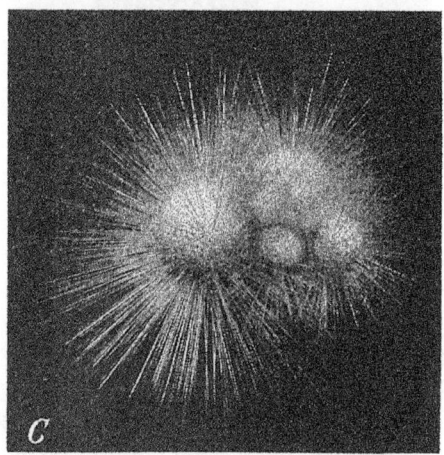

FIG. 70.—Skeletons of typical Protozoa. B, siliceous skeleton of a radiolarian, *Stauraspis stauracantha*, x 170; C, calcareous skeleton of a typical foraminifer, *Globigerina bulloides*, x 30. (From *The Origin and Evolution of Life* by Henry Fairfield Osborn, courtesy of Charles Scribner's Sons.)

The Mollusca. This is the highest group of unsegmented invertebrates that has yet appeared. They are characterized by (1) the absence of appendages, (2) a ventral "foot" which is usually associated with locomotion, (3) the mantle, a dorsal fold of the body wall which often secretes a shell and encloses a space in which lie the gills, (4) a heart consisting of two auricles and a ventricle, (5) a tubular vascular system associated with spaces called lacunae in which the blood also circulates, and (6) developmental stages of which one resembles some larval annelids. The Cephalopoda are apparently the most highly developed class. They include the four existing species of *Nautilus*, the cuttlefishes, squids, *Octopus*, *Argonauta*, etc., all of which have the highly developed eye described in a previous chapter.

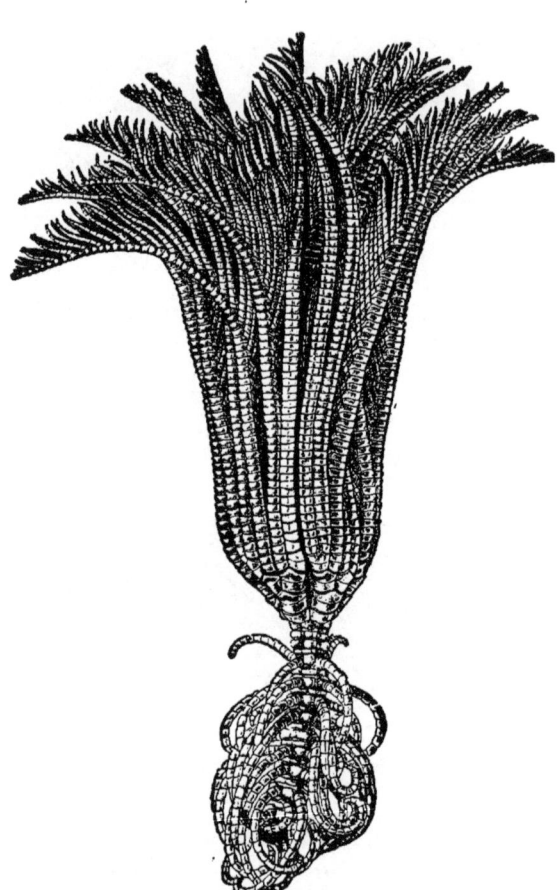

Fig. 71.—A crinoid, *Pentacrinus maclearanus*. (From the Cambridge Natural History.)

The shell is, of course, the part of a molluscan body most easily preserved as a fossil. Consequently it is on the change of structure of the shell that our knowledge of phylogeny is based. The phylogeny of the group is indicated in the following diagram, which has been modified to indicate the change of shell structure.

EVIDENCES—GEOLOGY 133

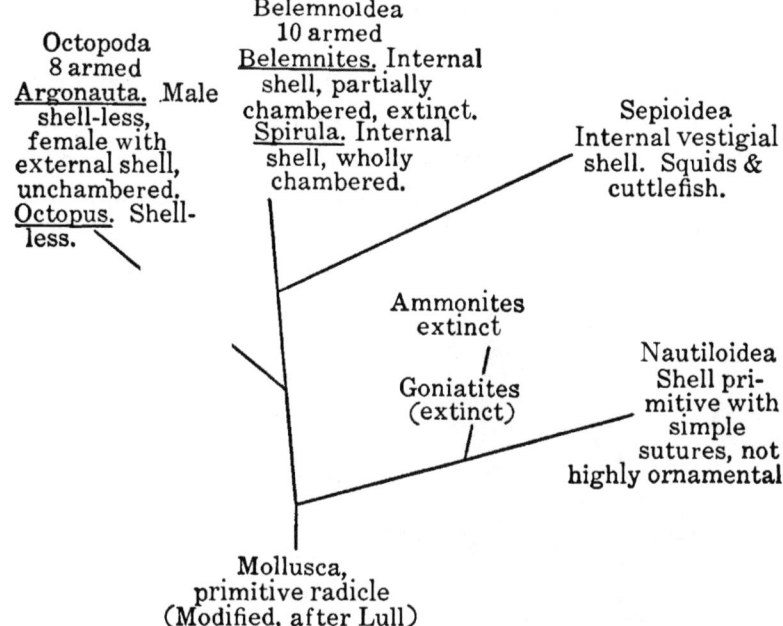

(Modified, after Lull)

By comparing this diagram with the table it will be noted that Mollusca of other groups occurred in the Cambrian. The earliest Nautiloidea, derived at some unknown point from the molluscan stem, are the oldest group of cephalopods, and appear first in the Ordovician. From this stem arise the more highly specialized Ammonites, which first appear in the Devonian (Fig. 72). The Ammonites then become extinct, while the nautiloids continue to the present, although with decreasing abundance and diversity. A divergence of the ancestral stem gives rise to three other forms, the Belemnoidea, Octopoda, and Sepioidea. Of the main branch, the belemnoids, many fossils are found in the Jurassic and Lower Cretaceous rocks, apparently derived from another group during the early Triassic. The squids, however, appear first in the Jurassic, while the origin of the Octopoda is not indi-

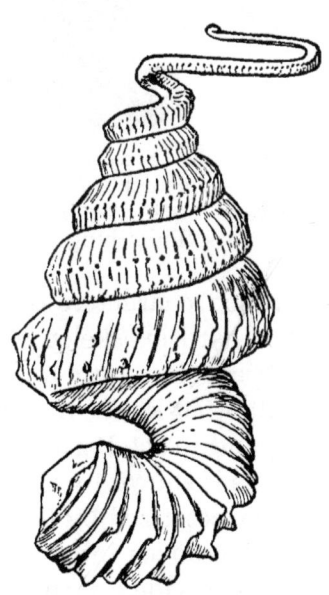

FIG. 72.—A shell of *Heteroceras*, a cephalopod mollusc. (From Lull, after Schuchert.)

cated by fossils. The belemnoids persist in the one genus, *Spirula*.

The Arthropoda. Another significant group of invertebrates is the phylum Arthropoda (Fig. 73). In this phylum a chitinous exoskeleton is well developed, favoring fossilization. Another distinctive characteristic, the presence of jointed appendages, gives the phylum its name. The body is segmented. Because of the high development of the group, it is looked upon as the culmination of evolution of segmented invertebrates.

The arthropods are at present more successful than any other group of animals for three reasons; the same may be said with equal truth of one of the included classes, the insects. Of these there are three times as many species as of all other animals together. Many of the species produce immense numbers of individuals, and finally, there are almost no habitats on earth of which the arthropods have failed to avail themselves.

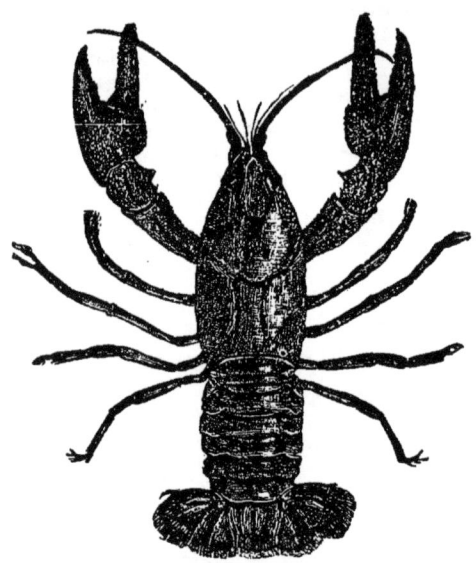

FIG. 73.—The crayfish, an arthropod. (From Comstock's *Introduction to Entomology*, with the permission of the Comstock Publishing Company.)

Arthropod Ancestors. The earliest organisms which show arthropod characters of segmented body and jointed appendages, with a hard exoskeleton, are the trilobites (Fig. 74). During the Cambrian they were numerous, exceeding in abundance and diversity all other kinds of animals. In the Devonian they began to decline, and in the Permian became extinct. They were aquatic animals, as is shown by their association with marine deposits. They are characterized by the division of the dorsal shield by two longitudinal grooves into three parts, hence the name *trilobite*. The body is divided transversely into an anterior head called the cephalon, a series of segments forming the thorax, and a tail piece or abdomen. The jointed appendages are biramous, as in the Crustacea, with the exception of the anterior pair, the antennae.

The outer branch of the appendage, or exopodite, has a long fringe which apparently converted it into a swimming and respiratory organ. In all of these characters and in others, the trilobite corresponds closely with fundamental crustacean

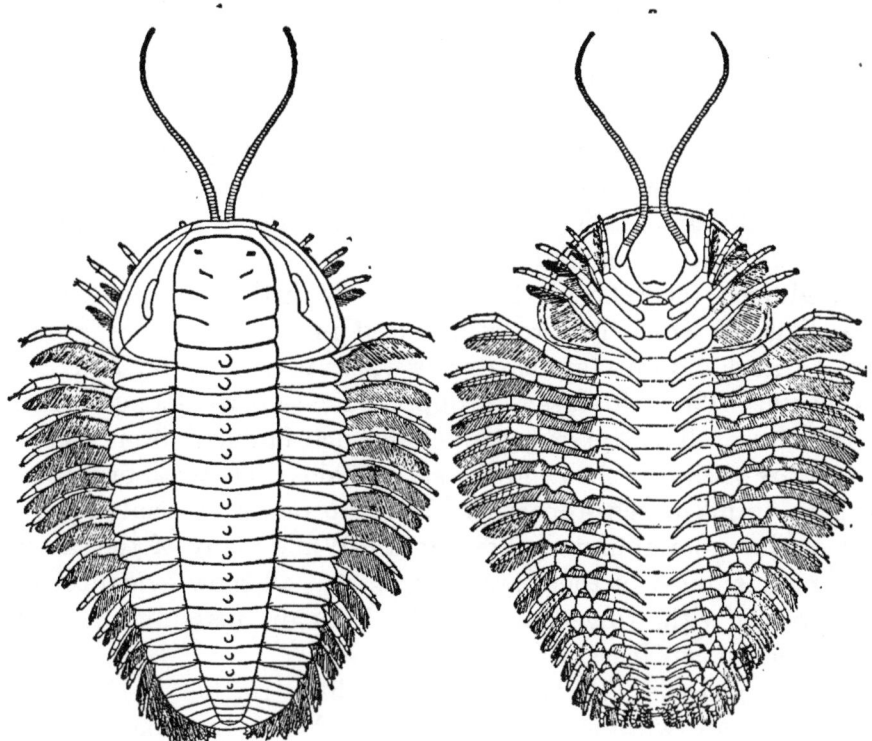

FIG. 74.—A trilobite, *Triarthrus becki*, restored. 1, dorsal aspect; 2, ventral aspect. x 2. (From Lull, after Beecher.)

characters which may be expected in ancestors of the latter organisms.

Aquatic Arthropoda and Their Descendants. On this basis the Crustacea are judged to have arisen from the trilobites very early in the Paleozoic. Some species long extinct show unmistakable similarity to the trilobites. One closely related group of primitive Arthropoda, the Palaeostraca, is still represented by a single species, the horseshoe crab. This group included forms, now extinct, from which developed the primitive Arachnida in the Silurian. The class Arachnida persists in the numerous modern spiders, mites, ticks, scorpions, etc., all terrestrial species.

Terrestrial Arthropoda. The source of the Chilopoda and Diplopoda (centipedes and millipedes) is obscure, but they seem to have come from trilobites.

Handlirsch interprets the insects as descendants of a trilobite whose second and third thoracic segments bore extensions of the lateral lobes which later developed into wings.

Several more obscure classes have also been worked out by Handlirsch from his studies of fossils. Of these the Onychophora are significant as the most primitive air-breathing or tracheate Arthropoda. They are worm-like animals which show definite evidences of segmentation in the paired appendages and nephridia. They show evident annelid affinities.

Handlirsch incorporates his analysis of arthropod phylogeny in a diagram which is here reproduced in part, with the group names translated into familiar terms where possible (Fig. 75).

Insects. The two insect groups indicated are looked upon as subclasses. The Apterygota include such familiar forms as the fish moths or silver fishes and the spring-tails. They are primitive wingless forms which apparently have an origin associated with that of the more abundant Pterygota. While many of the latter are wingless, most of them are winged, and all show evidences of relationship with winged ancestors. It is in this subclass that most well-known insects belong.

Metamorphosis. The two subclasses differ further in that the Apterygota are without metamorphosis. When they hatch from the egg they resemble the adult, and the only conspicuous change is growth. The Pterygota when hatched may resemble the adult to some degree but important changes always take place before maturity is reached. Three types of metamorphosis are recognized in the Pterygota. The Paurometabola are not unlike the adult when hatched although they lack wings. They live like the adults and develop gradually into mature insects, the chief transformation being in the development of wings. The Hemimetabola are wingless naiads when hatched and are adapted for a mode of life very different from that of the adults. After a period of growth their last transformation brings them suddenly into the adult stage which is normally winged (Fig. 76). The Holometabola emerge from the egg as larvae, very different in appearance from the adults (Fig. 6). They are wingless and often have differently formed mouths. Some are caterpillars, some grubs, some maggots;

EVIDENCES—GEOLOGY

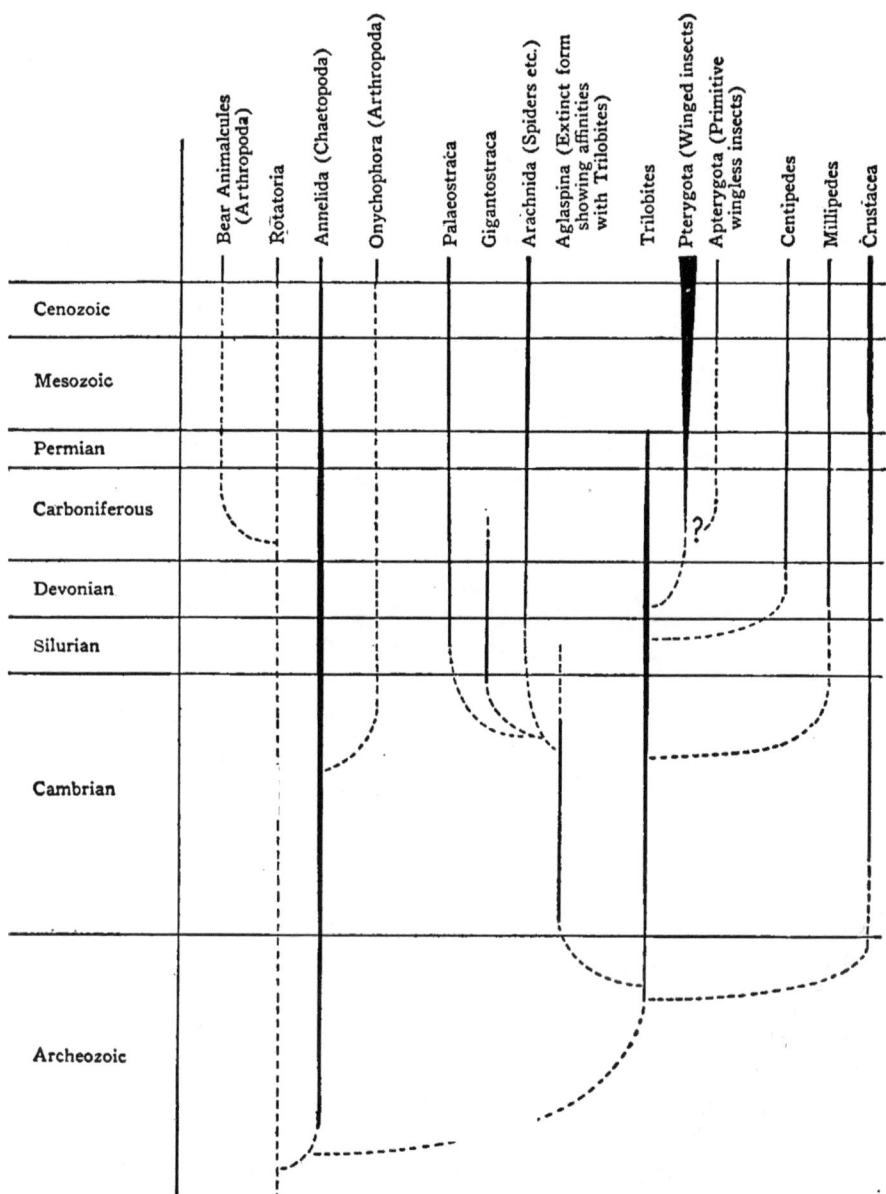

Fig. 75.—Diagram of the evolution of the major groups of Arthropoda. (Modified after Handlirsch.)

138   EVOLUTION AND GENETICS

all represent the growing stage of the insect and all transform ultimately into a more or less different resting or pupal stage. During the pupal stage the transition from larval structure to the very different adult structure is accomplished. The pupa transforms in due course into the adult, which undergoes no change of form and does not grow. The development of metamorphosis is evidently a factor which enabled the insects to meet the increasing rigors of climate during the period of their evolution,

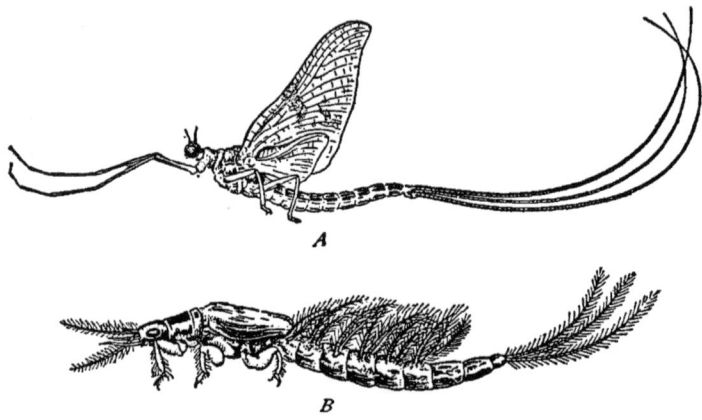

Fig. 76.—A may-fly, *Ephemera varia*. *A*, adult; *B*, naiad. (After Needham, from Comstock's *Introduction to Entomology*, with the permission of the Comstock Publishing Company.)

since it provides for periodical growth and dormant stages in a very effective manner.

Diversity of Insects. Structural modifications of insects are, as might be expected in a class containing more than 500,000 species, extremely diverse. Every part of the body displays some modification, but the main changes are in the mouth parts and wings.

The Mouth Parts. As in all Arthropoda, the mouth parts of the insects are apparently modified jointed appendages. In the primitive state they are formed for biting and chewing, but various modifications give rise to suctorial and lapping mouths of the kinds found in true bugs, flies, butterflies and moths, and bees. The biting mouth consists of a pair of mandibles behind which lies a pair of maxillae, and behind the maxillae there is a labium, originally paired. The mouth parts of a cockroach illustrate this type (Fig. 77). Suctorial mouths consist of a trough-like structure formed of the labium (in bugs and flies) or a more or less complete tube formed of the maxillae (butterflies and moths; the honey-bee) in which other modified parts may slide back and forth. The long, slender, bristle-like mandibles and maxillae operate in this way in the bugs (Fig. 78). In the flies a similar modification occurs. In both orders the slender parts may

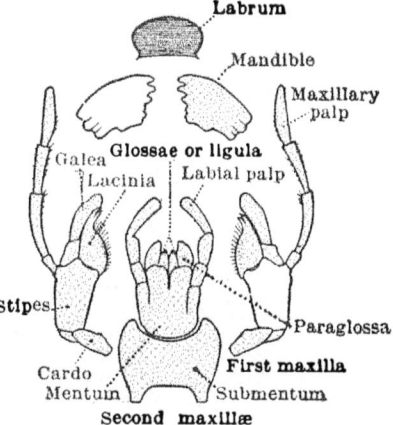

Fig. 77.—The mouthparts of a cockroach. (From Hegner, after Kerr.)

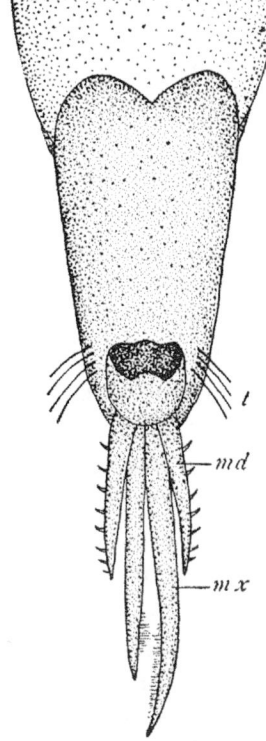

Fig. 78.—Last segment of the beak of *Lethocerus*, a bug. *md*, mandibular setae; *mx*, maxillary setae; *t*, tactile hairs. (From Comstock's *Introduction to Entomology*, with permission of the Comstock Publishing Company.)

be developed into sharp piercing organs. The maxillae of butterflies form a closed tube through which liquids are drawn by muscular suction. In the honey-bee they form a partial sheath for the hairy ligula, a slender part of the labium, which slides back and forth between them and raises liquids to the mouth.

The Wings. Some of the primitive ancestors of the insects had two pairs of well developed wings on the second and third thoracic segments and a pair of broad lobes on the prothorax which appear to have been rudimentary or vestigial wings (Fig. 79A). In the existing species the prothorax has no trace of wings but the persisting wings of the remaining divisions supply some of the finest evidences of evolution. They vary in texture from thin membranous structures with heavier supporting veins to heavy chitinous shields, in some cases so hard that they must be drilled before a pin can be passed through the insect. In some insects they are reduced to two, usually the anterior pair (Fig. 80). Other insects have lost them entirely.

The venation of the wings is based on a definite plan which shows modification in several ways toward greater complexity or simplicity (Fig. 136).

In all of these structures and in their metamorphosis insects show the possibility of almost infinite modification to fit varied environmental conditions. Their diversity is exceeded by no other animals and consequently furnishes an excellent illustration of an actual phylogenetic series.

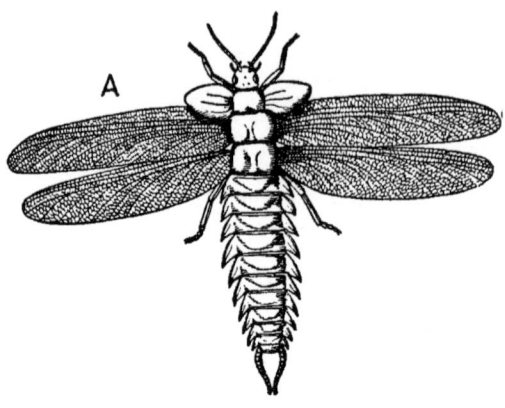

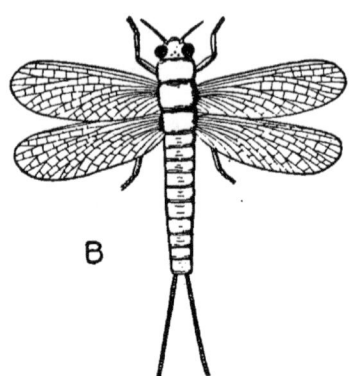

FIG. 79.—Palaeozoic insects. A, *Stenodictya lobata*; B, *Eubleptus danielsi*. Both belong to the primitive Palaeodictyoptera. (From Lull, after Handlirsch.)

Since Handlirsch's studies of insect phylogeny are more nearly complete than those of any other entomologist, his table is here reproduced with the popular equivalents for the ordinal names indicated wherever such terms are available (Fig. 81). It will be noted that the extensive fossil remains of the Carboniferous and Permian are in only four cases referable to modern orders. All of these creatures are insects, beyond a doubt, but they have almost completely disappeared, leaving a more or less different fauna to take their places. As the names of the primitive orders indicate, some resemble modern orders. Others are wholly unlike existing insects except in fundamental structure.

Such diagrams as this are necessarily based on all available knowledge. Comparative anatomy, embryology, and palaeontology alike contribute to the formulation of a reasonably complete result. Since the actual record of evolution is broken we must be content with such information as we can piece together from the available fragments, and it is gratifying that it is so often adequate. The mere fact that fossils prove the past existence of forms no longer living although often related to existing forms is in itself an evidence of the actual course of evolution, and the transition from form to form through the ages is even more significant.

In the phylum Chordata the record covers a briefer span and is more complete. Both for these reasons and because we our-

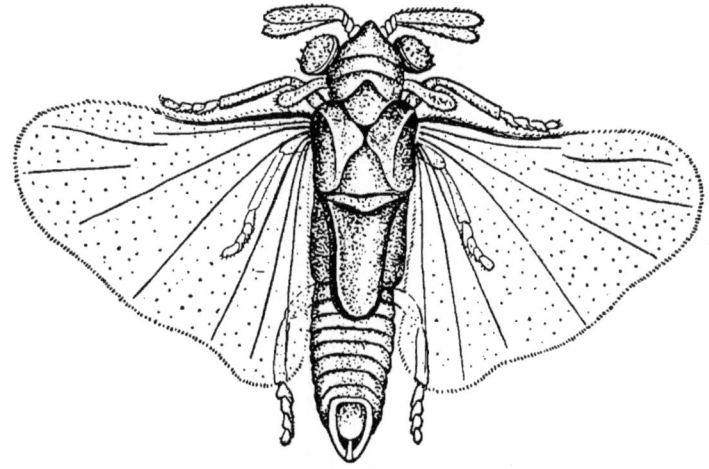

FIG. 80.—A male stylopid, *Opthalmochlus duryi*. The front wings are short club-shaped appendages while the hind wings are used for flight. Most two-winged insects retain the front wings. (After Pierce, from Comstock's *Introduction to Entomology*, with the permission of the Comstock Publishing Company.)

selves are vertebrates it is of great interest and deserves special treatment in succeeding chapters.

Summary. The present state of the earth is the result of a succession of physical phenomena. The ancient igneous rocks have been supplemented by layers of sedimentary rocks formed from the hard parts of minute organisms under water and from materials washed down from land masses. These changes have occupied an enormous span of time. Organisms of the various periods have been buried and preserved as fossils. From the combined study of the rocks and fossils it is possible to learn of the conditions prevailing at different times, as shown by the comparison of organic remains with existing organisms and the conditions under which they live. The succession of organic forms is disclosed in the same study. The result is a relative and incomplete record of evolution, but it is sufficiently complete to show that a gradual

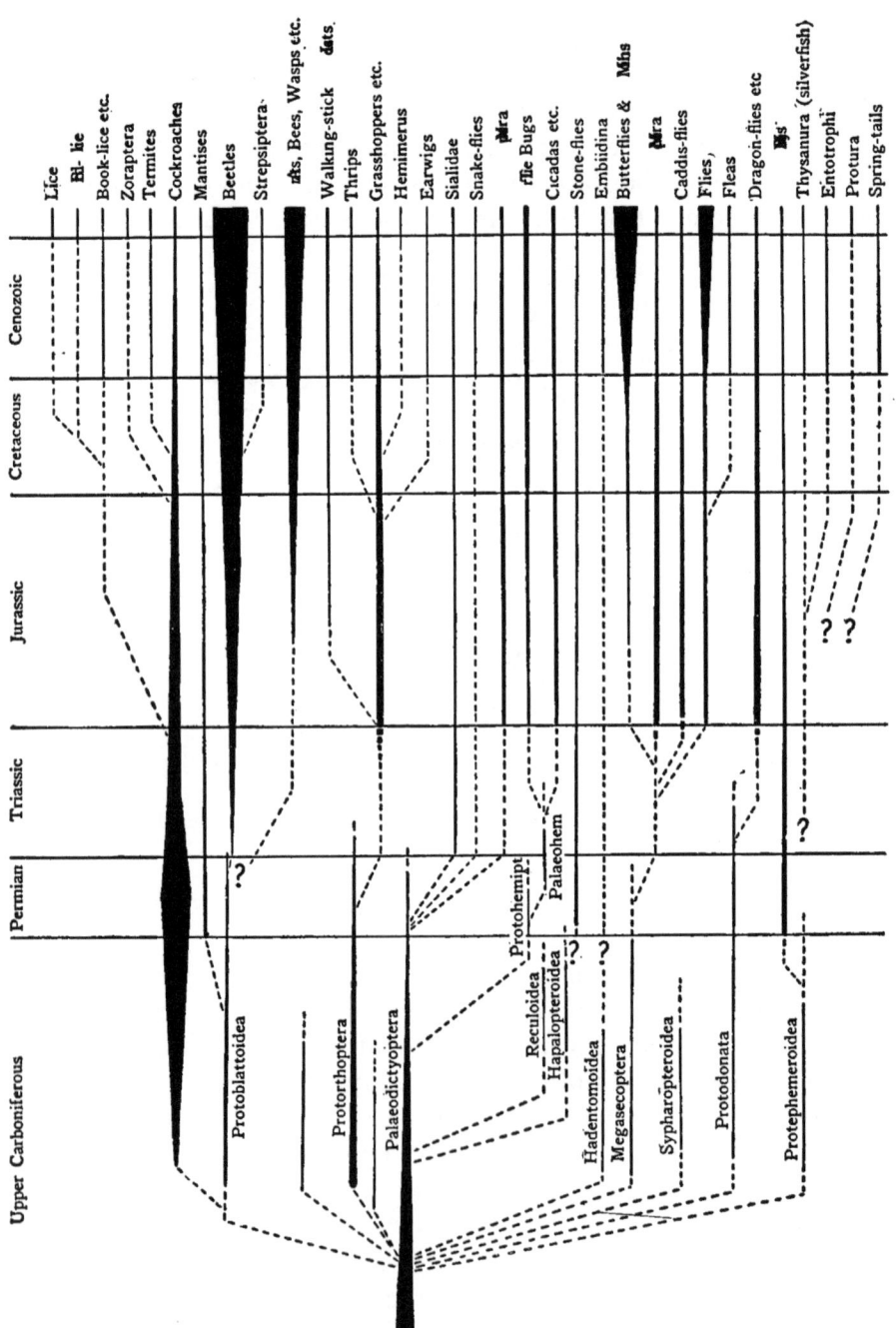

Fig. 81.—Diagram of the evolution of the insects. (Modified after Handlirsch.)

addition of more and more complex organisms to the population of the earth has occurred as time has passed. The succession corresponds to the major sequence of existing phyla and even within minor groups is often a valuable indication of phylogeny.

REFERENCES

VON ZITTEL, K. A., *Textbook of Palaeontology*, translated by Eastman, C. R., 1900.
SCOTT, D. H., *The Evolution of Plants*, 1911.
THOMSON, J. A., *Outline of Zoölogy*, 6th edition, 1914.
LULL, R. S., *Organic Evolution*, 1917.
LULL, R. S. (editor), *Evolution of the Earth and Its Inhabitants*, 1918.
HANDLIRSCH, A., Schröder's *Handbuch der Entomologie*, 1925.
KNOWLTON, F. H., *Plants of the Past*, 1927.

CHAPTER IX

EVOLUTION OF THE VERTEBRATES

In no phylum is the course of evolution more clearly or completely indicated than in the phylum Chordata, particularly the subphylum Craniata, commonly called vertebrates. As has already been shown, the details of comparative anatomy and embryology in this group are remarkably clear evidences of relationship of the various classes in a succession of stages which point clearly to a chronological as well as merely comparative series. In addition the geological record is excellent. The bony endoskeleton and many hard integumentary structures, such as teeth, scales and protective armor are well fitted for preservation as fossils, and conditions have often been favorable for the fossilization of many individuals.

The Origin of the Vertebrates. In spite of these conditions within the phylum, we are faced at the outset by a break in the paleontological record second to no other. The vertebrates appear in the Ordovician as well developed fishes without any indication of their invertebrate ancestors. In the contemporary invertebrate fossils and those of the Cambrian are found many forms that may have been ancestral or derived from the same source, but the gulf is great. We still have only theories to explain the origin of the vertebrates. While each theory finds facts to support it both in the study of living forms and in the geological record, as might be expected of organisms which have some degree of common relationship, none establishes that completeness and certainty of phylogeny which is to be desired.

The three leading theories of vertebrate descent are the *Amphioxus* theory, the annelid theory, and the arthropod theory. Several minor theories have been formulated which are of interest only to the special student.

Since *Amphioxus* is a chordate, the probability that it is similar to the ancestor of the vertebrates is merely one step in establishing their origin and does not indicate a connection with the invertebrate phyla. The annelid theory brings out facts which, with the

Amphioxus theory, afford a very satisfactory hypothetical connection of vertebrates and invertebrates.

The arthropod theory is at fault chiefly in tracing the vertebrates to invertebrates which were themselves highly specialized and has not been given general support. The fishes through which the connection is traced are looked upon as specialized types and such resemblance as they show with fossil arthropods may therefore be due to convergence.

Hypothetical Ancestors. It is generally agreed that the ancestors of the vertebrates must have been free-swimming, active,

FIG. 82.—*Amphioxus lanceolatus.* (From Newman, after Willey.)

aquatic animals. They are equally certain to have been bilaterally symmetrical with well developed heads, and definitely metameric in structure. They must also have possessed the three fundamental characters of the chordates, viz., tubular dorsal nervous system, notochord, and pharyngeal clefts.

The fishes are recognized as the most primitive of the true vertebrates, and such an ancestor could not have been very different from the typical fishes. The existing lancelets (*Branchiostoma* or *Amphioxus*) of all animals below the vertebrates best fit these specifications. They are small fusiform creatures with a dorsal tubular nerve cord, well developed segmental musculature, a notochord, and numerous pharyngeal clefts. They are able either to lead a sedentary life or to swim about.

Amphioxus has some characters which are not primitive, but it corresponds so nearly in its general structure with the hypothetical ancestor of the vertebrates that we are justified in considering it an approximation of their true progenitors. In the course of time the lancelets of the present have apparently taken on only a few new characters while the vertebrates have been attaining their remarkable diversity (Fig. 82).

The Amphioxus Theory of Vertebrate Origin. The possibilities of development of lancelets or similar ancestral forms is suggested by the habits of the existing species. They live along sandy shores and pass some time buried in the sand, but in response to tidal action they are able to swim rapidly against the currents. Their food concentrating mechanism is of an unusual type, well adapted to sedentary life. The animal is one of those peculiar forms mentioned in the last chapter, able to respond with equal facility to widely different conditions of environment. As pointed out before, the constant action of one environmental condition would be expected to call forth only one response. Hence a shift into the still waters of the ocean might develop its sedentary characters and produce something akin to the tunicates, while the opposite influence, constant motion of the waters in swift flowing streams, might bring out the characters which fit *Amphioxus* for active resistance to such motion. Here the geological record supplies the interesting information that the earliest fossil fishes show association with fresh waters, and not with marine deposits.

The Annelid Theory of Vertebrate Origin. The annelid theory is based on the resemblance of embryonic characters of vertebrates to certain annelid structures, and on the resemblance of the adult vertebrates in general plan to an *inverted* annelid, with a different mouth and anus (Fig. 83). Embryology shows that the mouth and anus of vertebrates actually develop secondarily, and are not at the original ends of the primitive gut, as in the annelid. Moreover, several highly developed invertebrates of the present live normally in an inverted position. The theory is thus less peculiar than it seems at first thought.

Reversal of the annelid results in the nerve cord lying above the alimentary tract as in the vertebrate. Formation of a ventral mouth would then cut off the anterior end of the alimentary tract, which in the annelid passes through the central nervous system. The entire central nervous system would then lie on the same side of the alimentary tract. In the annelid the blood flows forward in a dorsal vessel, down through paired connectives, and back through a ventral vessel; by reversal this results in a flow similar to that of the vertebrates, forward from the heart through the ventral aorta, dorsad through the paired aortic arches, and back through

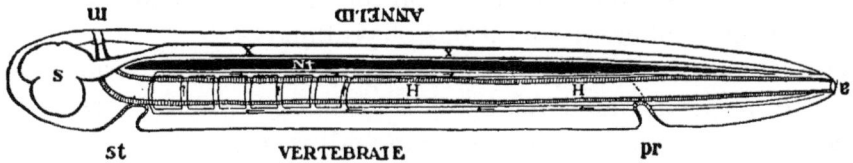

Fig. 83.—Reversible diagram illustrating the annelid theory of vertebrate origin. Index letters applying to both forms: S, brain; X, nerve cord; H, alimentary canal, applying to annelid only; m, mouth; a, anus; applying to vertebrate only; st, stomodaeum; pr, proctodaeum; nt, notochord. (From Wilder's *History of the Human Body*, with the permission of Mrs. H. H. Wilder and Henry Holt and Company.)

the dorsal aorta. The typically vertebrate notochord finds its counterpart in the *Faserstrang*, a fibrous structure with similar anatomical relations which appears in the annelid.

While the idea is no more than a theory, we cannot fail to see in it the possibility that vertebrates may have arisen through a form similar to Amphioxus from some invertebrate with annelid characters.

Origin of Vertebrate Classes. Given the existing classes of vertebrates, there is much evidence to explain their common sources. The cyclostomes are probably derived from remote and unknown ancestors. The remaining vertebrates are known as the *Gnathostomata*, or hinge-mouthed animals, and come undoubtedly from a common source. The sequence from fishes to amphibia, thence to reptiles, and from this class by different lines of descent to the birds and mammals is well marked. Whether we trace entire groups with all available evidence, or single organs and systems alone, the succession is in most cases clear and well substantiated.

Emergence of Terrestrial Vertebrates. From the most primitive vertebrates, the fishes, the initial step is one of the greatest to be taken anywhere in the development of the higher classes. The transition from aquatic to terrestrial life between fishes and amphibia, or better still between fishes and reptiles since the amphibia are still in a transitional state, involves fundamental changes of no slight degree.

Water as an environmental medium contrasts with the air in several important particulars. (1) Because of its higher specific gravity it buoys up the organism with greater force. (2) It prevents the loss of water from the body by evaporation. (3) It offers greater resistance to motion on the part of immersed bodies. (4) It presents different conditions of visibility.

The Demands of Terrestrial Life. While removal from the water to the air demands relatively slight modifications in response to some conditions, to others extreme adjustments are necessary. The body is no longer buoyed up completely by the surrounding medium, but is of so much higher specific gravity that it must rest on the ground. This demands a different type of locomotion involving even in primitive forms limited points of support for the body, and consequently a more rigid skeletal structure. The integument must be modified to conserve moisture within the body in proportion to the dryness of the air. While aquatic animals need streamline bodies if they are to move rapidly, in the air the resistance is so much less that body form is of relatively little importance. The difference of visibility in the air makes possible some modifications of the eye. In addition, the two media demand entirely different respiratory organs, and correspondingly different circulation.

For complete separation from the water as a habitat, the development of fetal membranes is an apparent essential, due no doubt to the delicacy of embryonic tissue and the resulting necessity for protection against dessication and mechanical injury.

Transitional Forms. A first step in tracing the theoretical portion of vertebrate evolution is to decide what characters would be present in a species capable of developing into terrestrial forms. To substantiate the hypothetical ancestor, it is necessary to determine whether or not such organisms have ever existed. Finally we must determine whether environmental conditions during the geological period of the transition were favorable for the change.

Since the earliest vertebrates were fishes, the ancestors of the terrestrial forms must have belonged to this class. In order to accomplish the shift of habitat, these ancestral forms must have been able to breathe air, and because of the extreme changes in locomotion, they must have had structures adequate to move their bodies over a solid substratum. The other conditions of change need not have been met at once. If the primitive terrestrial vertebrates remained in a moist environment, as would be expected, protection against dessication would be unnecessary. Changes in the special senses and in the skeleton would be valuable, but in the absence of competition with other terrestrial species, would not necessarily be of vital importance. This transitional state is nicely illustrated by the existing Amphibia, particularly the tailed species, the newts and salamanders.

When we seek fishes with some capacity for terrestrial life, we find that two groups are capable of breathing air by a diverticulum of the alimentary tract fundamentally similar to a primitive vertebrate lung. While other fishes are able to exist out of water, the blennies even to the extent of leaving it to escape their enemies, they are adapted in ways different from the usual terrestrial forms. One of the two significant groups is the subclass *Dipnoi*, the lung-fishes. The other is the order *Crossopterygii* of the subclass *Teleostomi*. The Dipnoi are an interesting group but for the purposes of this study they may be set aside. Their paired fins, which alone are in a position to aid in locomotion on a solid substratum, are very different in structure from the pentadactyl appendage and cannot be looked upon as precursors of such appendages.

The Crossopterygii. In the order Crossopterygii, on the other hand, the structure of the pectoral fins shows a surprising degree of resemblance to the terrestrial limb. The fishes of this order use the pectoral fins not merely as balancing organs, but also as paddles in swimming, and when resting on the bottom as support for the anterior part of the body. Like the pentadactyl appendage they have a single proximal bone, two bones distal to it, and several series of smaller bones distal to the two. To the small bones is attached the fringe of dermal rays which support the aquatic part of the appendage. However the fins are not the only point of resemblance. The fishes breathe air at the surface of the water by means of a double air sac connected by a single tube to the ventral side of the pharynx, a condition resembling the lungs of Amphibia and the early embryonic lungs of other classes. The larvae are said to resemble those of Amphibia. In short, the Crossopterygii have a number of the characters of the hypothetical ancestor of terrestrial vertebrates.

The Period of Transition. In the geological record a very significant fact is the abundance of these fishes in the Devonian, the period which also saw the rise of the Amphibia.

During this time, the paleontologists tell us, climatic conditions were such as to force upon animals the change from aquatic to terrestrial life. As Lull expresses it, "Diastrophic movement during the Silurian period initiated a widespread aridity which culminated in the latter part of the period, continued with varying intensity into and through Devonian time, and rose again to greater severity in the latter part of that period. This meant, as in Australia today, the reduction of rivers and other bodies of fresh water and the entailed concentration of their fauna, which is borne out by the mode of occurrence of the Lower Devonian (Old Red Sandstone) fishes—innumerable specimens in very restricted areas. Add to this the diminution of aëration of these waters and it will be seen that a high premium would be placed upon powers of air breathing or of aestivation. Still further dessication would necessitate some sort of activity during the increasingly long droughts, for the periods of torpor would otherwise bear too great a ratio to the creature's life span. Thus a premium would be placed upon ability to crawl ashore and maintain an active life, while the less fit would sleep the sleep that knows no waking, to their racial extinction."

The Amphibia. Once able to exist on land there would be abundant reason for animals to continue their development toward greater fitness for the new mode of life. In the waters would be concentrated all of their ancient enemies; on land would be freedom up to the limits imposed by their own structure. The story of vertebrate succession, already touched upon many times, shows by what means this has led up to the maximum development of the class. The geological record shows how many species, once successful, have fallen before unfavorable conditions because they were unable to meet the requirements of a changing environment.

The development of the Amphibia as terrestrial organisms probably began in the Middle Devonian period and extended through the Carboniferous when the drying of the earth's surface produced vast swampy areas. Restriction of bodies of water and

EVOLUTION OF THE VERTEBRATES 151

the persistence of abundant wet regions were ideal conditions for amphibian evolution. It is recorded of the earliest transitional types that they retained both lungs and gills, and had both limbs and a tail fin. During the Carboniferous they attained a considerable variety of forms, but the Permian brought an extension of continental areas, with relatively dry surfaces and seasonal

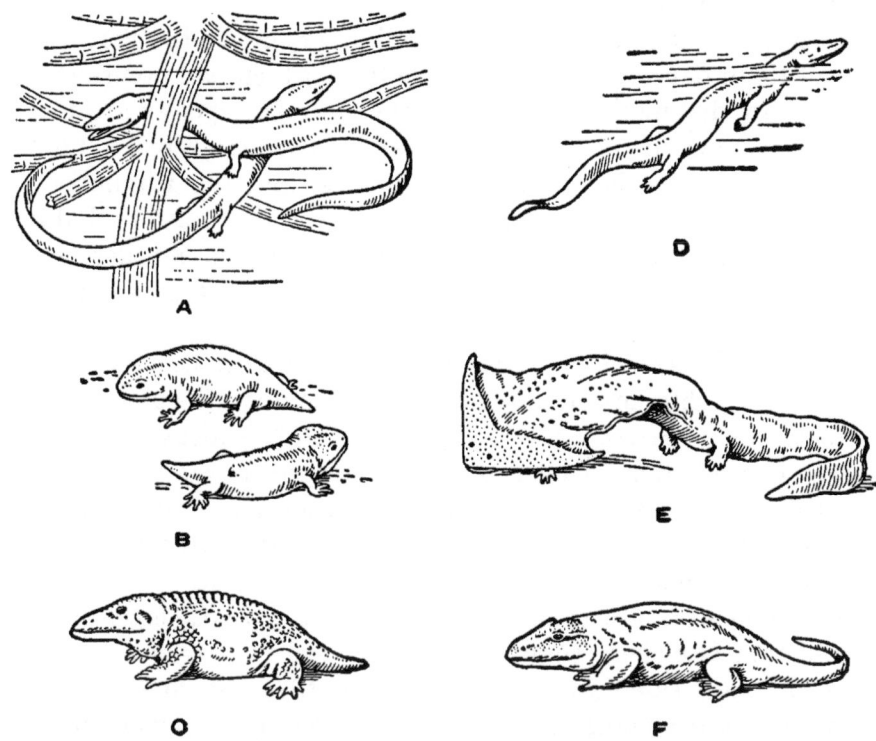

Fig. 84.—Group of extinct Amphibia. A, B, D and E from the Carboniferous; C and F from the Permo-Carboniferous. A, *Pylonius;* B, *Amphibamus;* C, *Cacops;* D, *Cricotus;* E, *Diplocaulus;* F, *Eryops.* (From Newman, after Osborn, based on restorations by Gregory and Deckert.)

changes in the bodies of fresh water, which were unfavorable to amphibian life. They have been able to persist, but beset by the limitations of two environments and the narrowness of a transitional life zone they have been unable to rise above a corresponding limitation of forms.

During their ascendency some time in the early Carboniferous the Amphibia developed a number of terrestrial forms which were made possible by the alternation of arid seasons with periods when the streams were full to overflowing and herbaceous flood-plain vegetation was Abundant. Among such Amphibia, the Stegocephalia or "solid-headed" forms, were the ancestors of the reptiles (Fig. 84). Already well adapted in many ways to the conditions of terrestrial life, they were forced to make use of these adaptations or perish when the increasing aridity of the later Carboniferous deprived them even of the seasonal opportunity to return to the water.

The Reptiles were abundant and well diversified in the Permian. Some of them developed peculiar structures which had no use that we can now distinguish, and all of the Permian species are now extinct. Among the orders represented by Permian reptiles, however, are probably the Chelonia or turtles, and the Rhynchocephalia, both still in existence. Primitive crocodiles also probably appeared, but the other great orders of extinct reptiles and the snakes, lizards and crocodiles of the present arose much later.

Among the adaptations by which the reptiles are differentiated from their amphibian ancestors are the resistant integument and the foetal membranes, both associated with life away from the water. The integument even of the early species was tough, and provided with scales or bony exoskeletal structures. Through this adaptation the animals were able to remain constantly in dry air without losing more water from the body than could be replaced. The foetal membranes, as already noted, are apparently essential to reproduction elsewhere than in water.

Beyond these essential structures the reptiles also show a great variety of adaptive possibilities which resulted in the development of fish-like aquatic forms, flying forms, the dinosaurs, cynodonts, and a few of less importance (Fig. 85). They attained their greatest diversity in the Mesozoic, where they were represented by some of the greatest animals that the world has ever seen. The aquatic ichthyosaurs and plesiosaurs included highly specialized animals, some enormous in size. Newman graphically describes one, *Trinacromerion*, as "a creature with all the earmarks of an aquatic speed demon, and doubtless as much of a terror to the fishes as were the dinosaurs to the smaller denizens of the dry land." These latter were of two fundamental types, the carnivorous and herbivorous dinosaurs. They ranged from small species to creatures as large as the modern whales, and from heavy, sluggish forms, protected by massive bony armor, to active preda-

EVOLUTION OF THE VERTEBRATES   153

cious bipedal giants which must have been a terror to the other inhabitants of the earth. The pterosaurs flew by means of folds of skin extending from the hind limb to one enormously elongated digit of the fore limb. In the cynodonts, or dog-toothed reptiles, we find the probable ancestors of the mammals, and in some an-

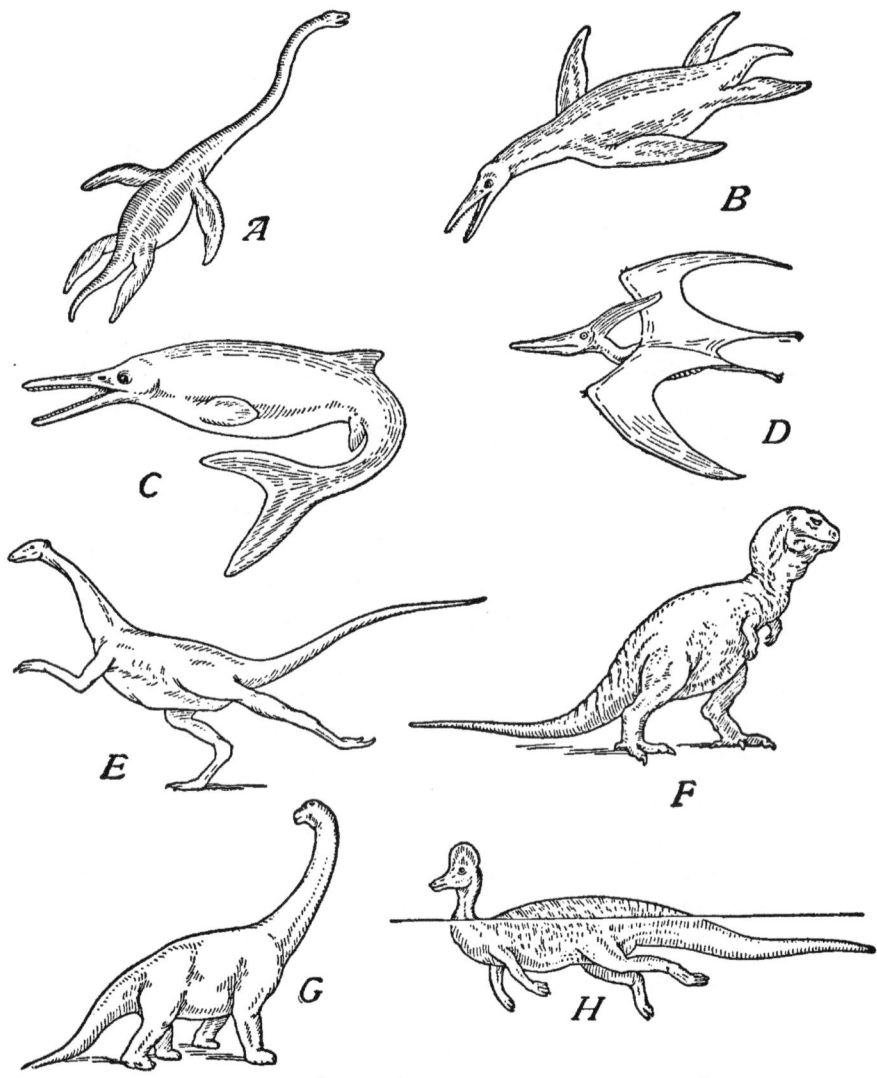

FIG. 85.—Group of Mesozoic reptiles. *A*, long-necked plesiosaur, *Elasmosaurus;* B, short-necked plesiosaur, *Trinacromerion;* *C*, ichthyosaur, *Baptanodon;* *D*, pterodactyl; *E*, ostrich dinosaur, *Struthiomimus;* *F*, carnivorous dinosaur, *Tyrannosaurus;* *G*, giant herbivorous dinosaur, Br*achiosaurus;* *H*, hooded duck-bill dinosaur, *Corythosaurus.* (From Newman, after Osborn.)

cestral dinosaur or an even more primitive form, the forerunner of the birds.

The Origin of Birds. The origin of the birds is another lost step in the evolution of the vertebrates. That they came from some extinct reptilian form is certain, but paleontology has no definite information to offer concerning these ancestors. Some of the dinosaurs show characteristics similar to those of birds in skeletal structure, bipedal locomotion, and the development of a beak. While these facts do not point to such dinosaurs as ancestors of the birds, they do harmonize with the paleontologists' theory that both birds and dinosaurs came from the same stock. According to this theory the ancestor of the birds must have been either a primitive dinosaur or an unknown common ancestor.

Flight Adaptations. The chief characteristics of the birds are associated with flight. By comparing them with the reptiles we find that no other distinctive characters are present. The avian skeleton is highly specialized for lightness and rigidity by the development of hollow bones and fusion of separate parts. The scales are restricted to the legs and feet, where they are the same as reptilian scales; elsewhere the body is covered with feathers which serve the double purpose of light planes for flight and a warm covering for protection against the low temperature of the upper air. Although much more complex, feathers are seen by their development to be closely related to scales. The digestive system is specialized to provide the abundance of energy required by flight. The lungs are so constructed that air enters the alveoli through one passage and leaves through another, so that every breath brings an entirely fresh supply of air into contact with the respiratory epithelium. The vascular system is completely divided into pulmonary and systemic circulations. High and constant body temperature is maintained by a vaso-motor system. It serves to promote the rapid metabolism demanded by flight and to protect the bird against low temperature.

Theories of the Origin of Flight. In view of the high degree of specialization indicated by these characters, and the lack of definite fossil ancestors, it is not surprising that all explanation of the origin of birds should rest upon theories of the origin of flight. These theories include cursorial origin, arboreal origin, and diving origin. All assume the development of broadened limbs

EVOLUTION OF THE VERTEBRATES 155

through extension of the scales, and the use of these broadened limbs as aids to locomotion. Many birds now flap their wings for this purpose while running. Such primitive wings would also have been of aid in jumping from branch to branch or from tree

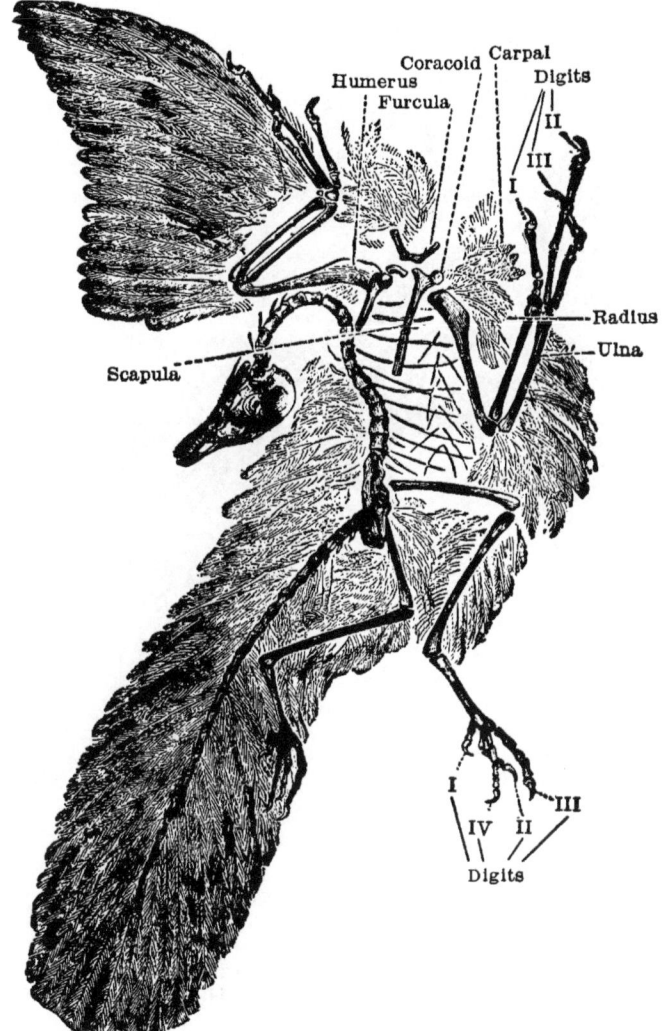

Fig. 86.—*Archaeopteryx lithographica*, as it appeared in the fossil specimen. I-IV, digits. (From Hegner after Steinmann and Doderlein.)

to tree, or in sailing out over the water to dive for fish. None of the theories has any particular advantage over the others, although the one fossil pro-avian, *Archaeopteryx*, was probably a climber. However, this may indicate either that it climbed trees or that it climbed rocks or cliffs in order to glide or flap out over the water as it dived for fish.

Archaeopteryx. This pro-avian is represented by well preserved fossils which give an excellent idea of its structure (Fig. 86). It

Fig. 87.—Nestling hoatzin, climbing with thumb and forefinger. (From *Jungle Peace* by William Beebe, published by Henry Holt and Company, with the permission of author and publisher.)

lacked a number of the characteristic flight adaptations of birds. The bones, for example, were not hollow, and fusion had not occurred to the extent noted in modern birds. The sternum lacked the deep keel which provides attachment for the powerful flight muscles of birds. The tail was not reduced to the rudder-like condition found in modern birds, but consisted of a series of vertebrae and bore lateral feathers. The wings were well developed, but the first three digits remained free as long clawed grasping appendages. Finally the jaws bore teeth, though this condition was found in other extinct birds as well. *Archaeopteryx* is therefore not truly a bird, and yet is so definitely bird-like as to be beyond the reptilian ancestor whose discovery is still to be made.

The Hoatzin. One existing species of bird is reminiscent of this ancient pro-avian condition. This is the hoatzin of tropical South America. In young hoatzins the first and second digits, unlike the vestigial remnants found in other birds, are functional and are used effectively as grasping organs in climbing (Fig. 87). The young hoatzin is also an able swimmer.

Mammalian Evolution. The mammals differ less conspicuously from the reptiles, since their development is not associated with any extreme specialization. The skeleton shows certain peculiarities, including triple centers of ossification in long bones and paired occipital condyles. Hair, a characteristic mammalian structure, has been definitely associated with scales in proof of transition between the two (Fig. 58). The teeth undergo great differentiation and are of several kinds, i.e., the dentition is heterodont. Both the process of reproduction and the nourishment of the young by milk, a glandular secretion, are generally distinctive, but the lowest mammals, the Prototheria or Monotremata, lay eggs. In many ways the mammals are similar to the birds in degree of development, and they correspond closely in the four-chambered heart and vaso-motor apparatus, but in most respects the two classes are divergent.

Mammalian Ancestors: The Cynodonts. Because of structural peculiarities the mammals were once supposed to be more closely related to the amphibia than to the reptiles, but the evidence of paleontology now points definitely to the Cynodontia, the dog-toothed reptiles, as their ancestors.

The cynodonts are distinctly reptilian animals in most particulars but they agree with the mammals in two distinctive ways. The skull is articulated with the first vertebra by two occipital condyles instead of one as in other reptiles, and the dentition is heterodont, while in other reptiles all of the teeth are primitive conical structures.

Among the cynodont characteristics which are favorable for

development may be included the evolution of the four limbs for rapid locomotion and the differentiation of the teeth. The first is supposed to have been associated with the development of intelligence and change of location through migration. Differentiation of the teeth would fit the animal for eating different kinds of food, which would tend to stimulate the development of powers of observation and choice, and consequently intelligence. We are unable to associate the warm-blooded condition of the mammals so directly with external factors, but Osborn supposes that it may have appeared in some of the cynodonts; if so, it would no doubt also have favored the evolution of mental powers by maintaining a high and constant rate of metabolism and thus freeing the animals partly from their dependence upon the physical environment.

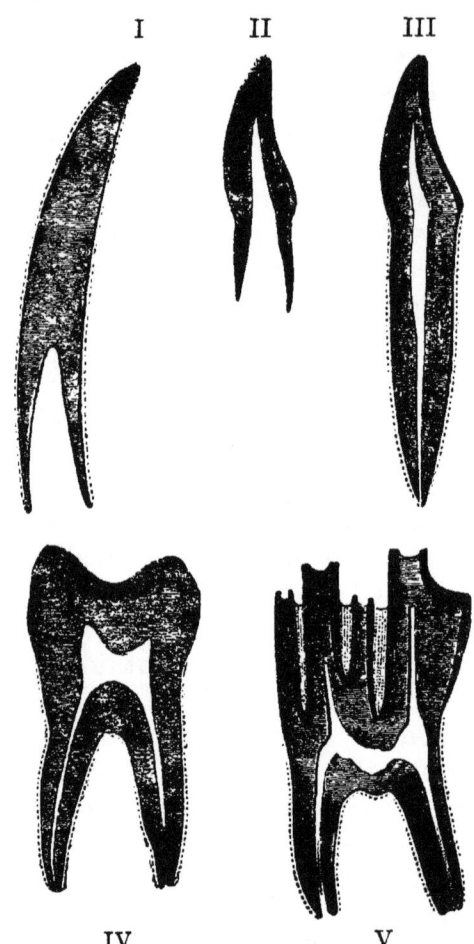

FIG. 88.—Diagrammatic sections of various forms of teeth. I, tusk or incisor of elephant; II, human incisor during development; III, completely formed human incisor; IV, human molar; V, molar of ox. In all figures the enamel is black, the dentine shaded with horizontal lines, the pulp white, and the cement stippled. (From Parker and Haswell, after Flower and Lydekker.)

The Course of Mammalian Evolution. With independence of the water for reproduction and the maintenance of body moisture, came greater dependence for existence upon the varied factors of the terrestrial environment. Since food habits establish the most important contact of the individual with his environment, and the teeth of mammals are specialized for various food habits (Fig. 88), the evolution of the teeth is important in mammalian evolution. On the food habits of himself or his neighbors also depends an animal's need for keen senses, defensive or aggressive structures, and the development of powers of locomotion for escape from enemies or for the capture of prey. Mammalian reproduction, while it is distinctive, and to a limited degree varied within the class, is by no means as important an indication of the course of evolution as these other characters.

Teeth. The fairly primitive teeth of man are an excellent illustration of the possibilities of heterodont dentition. The incisors are sharp edged for cutting through tough tissues or biting off pieces of food; they are effective in cutting flesh. The canines, on the other hand, are of little or no use for anything but holding and tearing. The broad molars are effective grinding structures by which tissues of all kinds may be crushed and reduced to smaller parts before swallowing. Carnivorous animals obviously have greater need for cutting and tearing than for grinding teeth, hence we find that even their molars are sharp edged shearing teeth. Herbivorous species, on the other hand, have need for grinding teeth in proportion to the harshness of their food; the grazing animals therefore have broad teeth with hard ridges for chewing grains and grasses. Still other animals, like the anteaters, have no need for teeth, which are correspondingly reduced or lacking.

Structures and Habits. Correlated with food habits are other structural characters. The carnivores have keen senses and speed in order to detect and reach their prey, and powerful jaws and sharp claws to aid in its capture. Herbivores find their teeth poor defensive structures so their limbs are highly specialized for speed without limitations imposed by other needs. If an animal finds food or protection in a different environment, its limbs show the need of that environment, like the powerful fossorial front legs of the mole, flippers of aquatic mammals, wings of the bats, and arboreal adaptations such as the sharp claws of the squirrels, hooked claws of the sloth, and prehensile appendages of the primates (Fig. 56).

Classification. The classification of the Mammals is based largely on such structural differentiation, although they are first divided into the Prototheria, Metatheria and Eutheria partly on a basis of reproductive functions. Members of the first group lay eggs. The second are marsupials, including such animals as the kangaroo and the opossum, and are viviparous. The young are, however, usually without a placental connection with the mother, and are in an imperfect state of development when born. They are then carried in an abdominal pouch and are temporarily attached to the long tubular teats by a special oral sucker. In the last group occur those animals whose young are connected to the uterine wall of the mother by a placenta. Through this the developing animal receives nourishment from the blood of the parent until birth.

The Eutheria include most mammals. They are supposed to have originated from the cynodont stem later than the Metatheria, while the Prototheria are supposed to have sprung from an even more primitive reptile. During the Tertiary they attained the differentiation which they have since maintained. Some species of immense size appeared, although in this direction the mammals were surpassed by the reptiles; the whales of the present are among the largest animals ever developed. While this period in mammalian development produced some species belonging to existing genera, all are now extinct. Most of the Tertiary mammals were less closely related to existing species.

Dispersal. In writing of the rise of the mammals during the Tertiary, Lull calls attention to the similarity of the North American and European faunas, and to the existence of a land bridge across what is now Bering Strait as indicative of their circumpolar origin. He adds that the climate of this region was at first warm and favorable and speaks as follows of the influences of the period:

"There is always a tendency on the part of every group of animals, as their numbers increase, to spread from their ancient home along lines of least resistance, provided no climatic or other insuperable barriers are to be overcome, and that may well have been one very potent cause for the southward migration of the modernized hordes. But there was an additional incentive, for throughout the early Tertiary there is evidence of climatic variation and of a very gradual cooling of the northern climate and a consequent southward retreat of the higher plants and mammals which occurred as a succession of migratory waves. In this way there came, first, the least hardy like the insectivores and primates, the latter especially depending so largely upon the tropical forests for their sustenance that any change either in extent or character

EVOLUTION OF THE VERTEBRATES 161

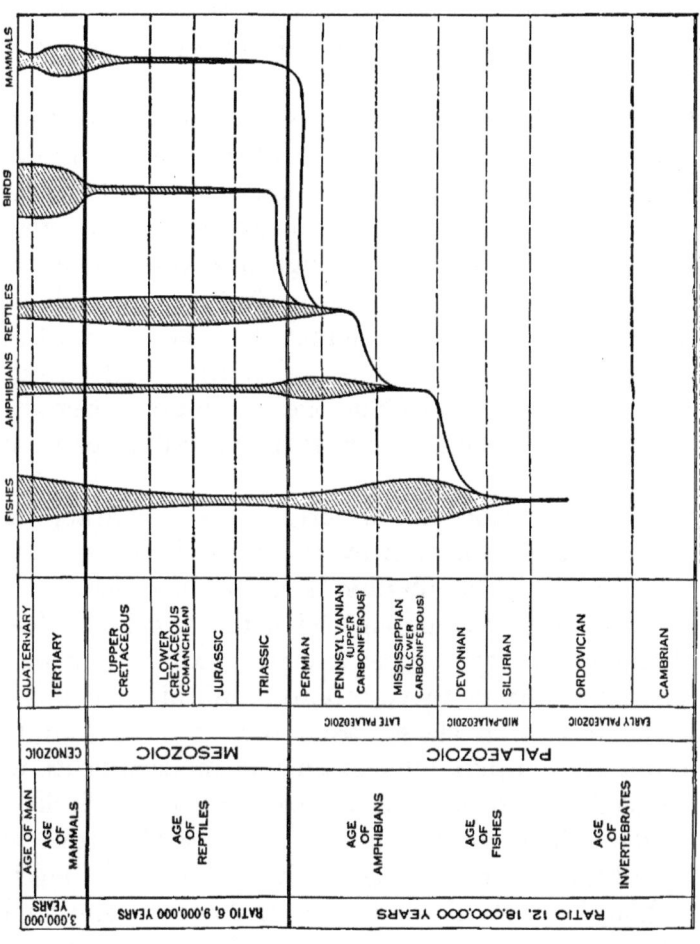

FIG. 89.—The evolution of the vertebrate classes. Chart showing the origin and degree of development of the five classes of vertebrates, correlated with the geological time scale. (From *The Origin and Evolution of Life*, by Henry Fairfield Osborn, courtesy of Charles Scribner's Sons.)

of their habitat would be reflected in their distribution at once . . ."

So definite is the geological record of vertebrate evolution that these changes of fauna, correlated with the varying climates of the world, tell us of the influences which were active in shaping the group to which we belong. To these influences we owe not only the present state of the fauna of the world, but even our own existence.

Step by step, through the successive conditions imposed upon the phylum by swift waters, increasing aridity, the drying up of streams and bodies of water, and finally changes of climate we have come from these remote invertebrate ancestors to our present state (Fig. 89), not suddenly, but through a wonderful inherent power to meet varying conditions successfully. By such easy steps have the vertebrate classes originated. Each in itself has made some adjustment to the various possibilities of the modern world, but only in the whole do we find the maximum power of the phylum as it is thus expressed. Upon future possibilities we can only speculate.

Summary. The evolution of the vertebrates is characterized by the completeness of all kinds of evidence and is therefore well established. The origin of the phylum is, however, obscure. It is explained by several theories, among which the annelid and *Amphioxus* theories combine to suggest a logical explanation of the transition from invertebrate to vertebrate. Within the phylum the origin of the terrestrial classes from their primitive aquatic ancestors is a major step, but the existence of modern fishes with some adaptations for both habitats is proof of the possibility of such a transition. The most primitive terrestrial forms were Amphibia. Conditions of increasing aridity favored the development of the wholly terrestrial reptiles which later became highly diversified. The birds and mammals probably originated from different groups of reptiles and independently acquired the important vaso-motor apparatus and circulatory system in which they are so similar. Development of the birds, however, has been concentrated upon the perfecting of flight while development of the mammals has resulted in structural diversity involving a considerable degree of evolution.

REFERENCES

Von Zittel, K. A., *Textbook of Palaeontology*, translated by Eastman, C. R., 1902.
Patten, W., *The Evolution of the Vertebrates and Their Kin*, 1912.
Scott, W. B., *A History of Land Mammals in the Western Hemisphere*, 1913.
Lull, R. S., *Organic Evolution*, 1917.
Lull, R. S., (editor) *Evolution of the Earth and Its Inhabitants*, 1918.
Osborn, H. F., *The Origin and Evolution of Life*, 1918.
Newman, H. H., *Vertebrate Zoölogy*, 1920.

CHAPTER X

ELEPHANTS, HORSES, AND CAMELS

The fossil remains of these three groups of vertebrates and their ancestors are so abundant that they afford us an almost complete record of the evolution of modern species since the Eocene. The record is so remarkable that it has deservedly been described many times, and has had unsurpassed influence in the establishment of the theory of evolution. The first extensive series of fossils of any of these animals was that assembled at Yale University, largely through the efforts of Professor Marsh, to illustrate the evolution of the Horse. This collection has been characterized as the "first documentary record of the evolution of a race" (Lull). It was regarded by Huxley as conclusive evidence of evolution, and would have been visited by Darwin had his health permitted. The fossil records now available afford many such examples of an actual phylogenetic series, but none surpass that of the horses. The records of elephants, horses, and camels are among the most striking and complete that we have.

Systematic Position. The three forms of animals are included in three orders of Eutheria, the Proboscidea, Perissodactyla and Artiodactyla respectively. All belong to the section Ungulata, the hoofed animals. The first order is characterized by the elongation of the nose and upper lip, the occurrence of five functional digits on all feet, the development of the upper incisors into tusks, and the highly developed grinding molars. The second order represents a very different type of adaptation, although the animals of both orders are herbivorous. The Perissodactyla have not more than four digits on the front feet and not more than three on the hind feet. In all species the third digit is the most important and in the family Equidae, including the horses, asses and zebras, the third alone is functional. Their teeth, like those of the elephants, are ridged grinding structures, but are less extremely specialized. The camels are specialized for life in arid regions. They are even-toed ungulates, i.e., the feet retain hoofs on two or

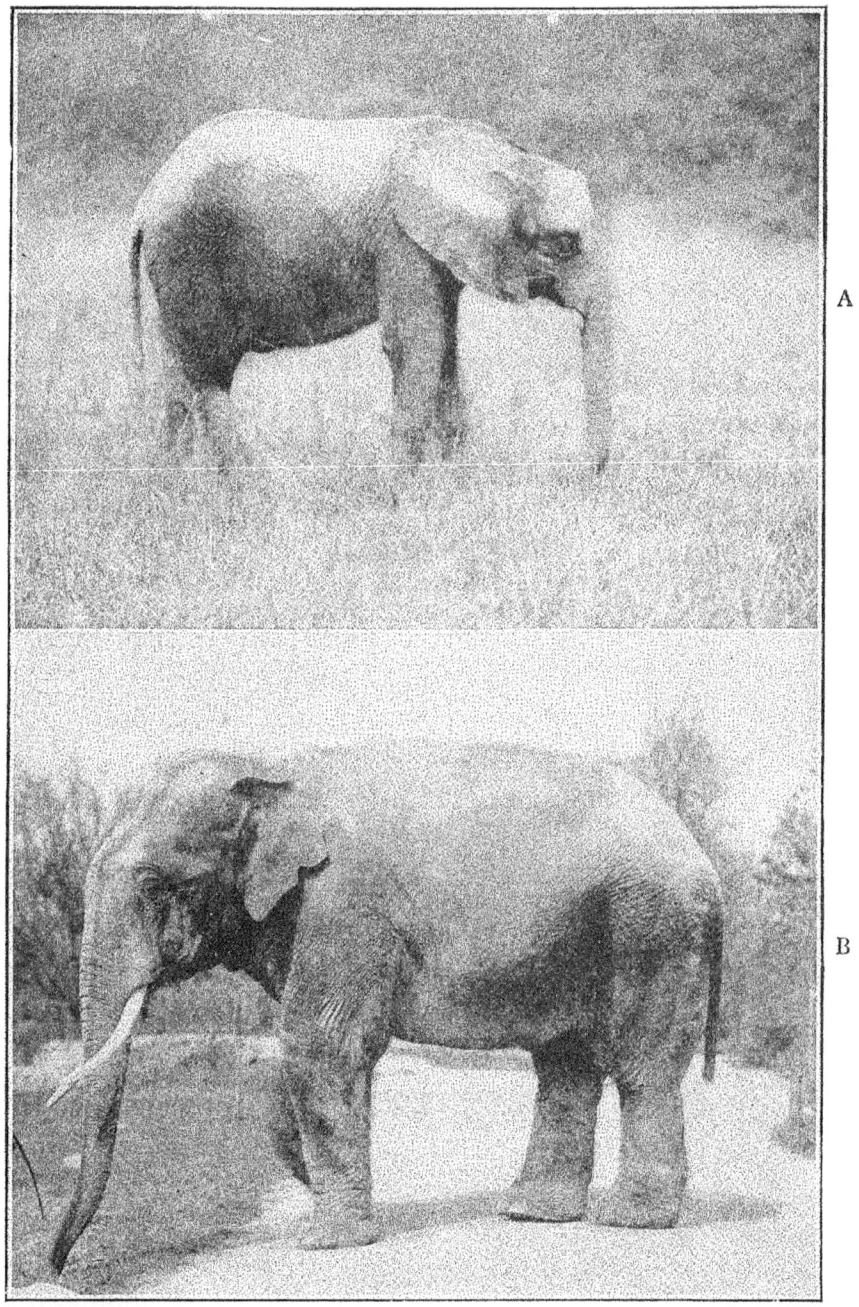

Fig. 90.—Modern elephants. A, African; B, Indian. (From Lull, after photographs from the New York Zoological Society.)

four toes, and they are herbivorous animals like those of the two preceding orders.

Elephants. *Size and Structure.* The elephants are in general more primitive animals. While highly specialized in the development of the ridged grinding molars and tusks, and in a number of ways associated with their great bulk, they retain several distinctly primitive characters, including the five functional digits. Their specializations are mostly associated with their size.

The existing elephants belong to two species, the African and the Indian (Fig. 90). A height of thirteen feet has been reported for some African individuals, and eleven feet actually recorded, while a weight of six and a half tons for the famous "Jumbo" is the maximum on record. In these points they rival all but the largest of the extinct species. Lull states that a skeleton of the extinct *Elephas meridionalis* in the Paris Museum measures about fourteen feet in height at the shoulder.

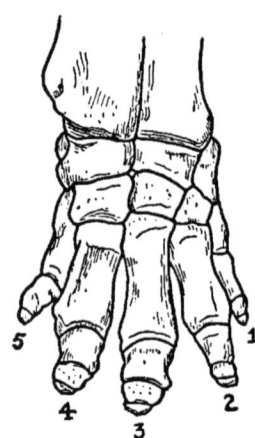

FIG. 91.—Forefoot of the Indian elephant, anterior aspect, showing the compact skeletal structure. (From Lull, after Flower.)

Limbs. To support this enormous bulk, the limbs of the elephants are developed in such a way that stresses are placed on the longitudinal axes of the bones. Throughout the limb the bones are aligned with each other in such a way that at no point is the stress applied obliquely or transversely. The result is an appendage aptly characterized as pillar-like. Anyone who has watched a circus parade is familiar with the peculiar shuffling straight-legged gait which is the result of this structure, and with the strange feet which seem little more than the blunt termination of the limbs themselves, with five nail-like hoofs set along their anterior margins. The skeletal strength is supplemented by this compactness of the feet, whose digits are scarcely divergent (Fig. 91). The greater part of the sole of the foot is made up of a thick pad which lies behind and below the digits and receives most of the weight. The hoofs, unlike those of the horse, are unimportant in this respect.

The Trunk. While in most animals with long limbs the neck is correspondingly long, a compensation evident in the horse, the elephants' proboscis, or trunk, serves the same purpose and the short neck permits an efficient arrangement of structures for the support of the massive head. The proboscis, composed of the elongated nose and upper lip, is a powerful yet delicate prehensile organ. Because of its relative lightness and flexibility it is superior to the jaws for handling objects, and with the head and neck formed as they are becomes an exceedingly strong and efficient

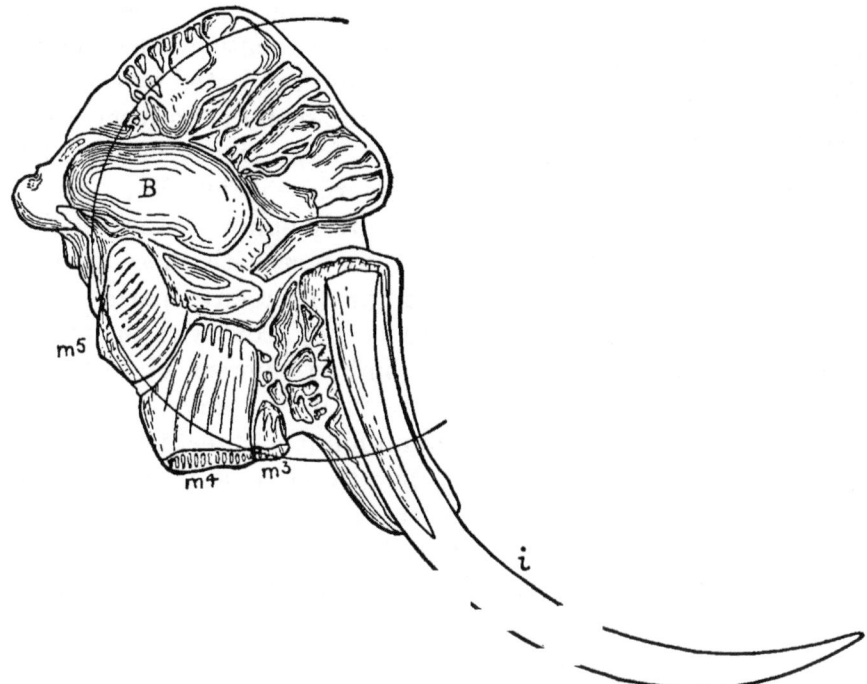

Fig. 92.—Skull of Indian elephant, in longitudinal section. B, brain cavity; i, incisor or tusk; m 3–5, the third to fifth molars. (From Lull, after Owen.)

organ. In contrast the elongation of the neck in other animals to compensate length of limb is of limited use.

The Skull. In the elephant the skull is much shorter and higher than in most mammals (Fig. 92). The change in form is accomplished by the thickening of the bones, whose lightness is preserved by the formation of large enclosed spaces. The skull of an animal acts as a lever of which the occipital condyles are the fulcrum, the longitudinal axis the work arm, and the vertical height above the condyles the power arm. In most animals the power arm is relatively short, but in the high skull of the elephant it is much longer and the leverage available for the support of the heavy head and trunk is correspondingly increased. By this adaptation the animal is also freed from the necessity of supporting the weight of its head at the end of a long neck, although it has an ample reach and great power in the combination of short, powerful neck, short skull, and long, flexible proboscis.

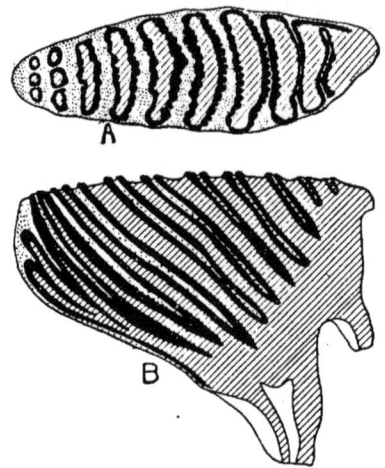

FIG. 93.—Molar tooth of Indian elephant. A, crown; B, longitudinal section. Enamel is black, dentine shaded with oblique lines and cement stippled. (From Lull.)

Teeth. The molar teeth of elephants are made up of relatively thin plates of dentine surrounded by enamel and connected by cement. They lie in such a position that their ends form ridges across the grinding surface (Fig. 93). Since the teeth grow obliquely toward the plane of contact with those of the opposed jaw, successive parts come into use as they grow, with the result that a limited number of teeth or parts are in use at any time. The maximum number of teeth is twenty-eight, but these come into use and are worn out and shed in such a way that not more than two molars in each half-jaw are functional at the same time, making eight molars in all. In addition to these the tusks or upper incisors are present from their first appearance; they are preceded by a pair of small milk tusks which are shed early in life. Lull records a pair of tusks of an African elephant which were 10 feet ¾ inch and 10 feet 3½ inches long respectively, and weighed 224 and 239 pounds, an almost incredible weight for an animal to carry in two teeth alone. The tusks are composed of dentine, excepting a small enamel tip, and grow throughout life.

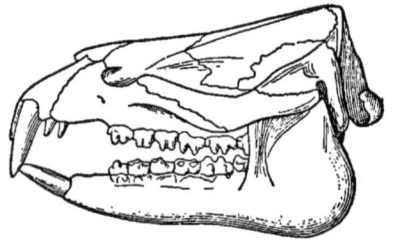

FIG. 94.—Skull of *Moeritherium lyonsi*, one tenth natural size. From the Eocene, Africa. (From Lull, after Andrews.)

Specialization in three directions is shown by this dentition: (1) departure from the primitive number of sixty teeth by reduction to a total of twenty-eight, (2) modification of the primitive mammalian molar to form a finely ridged grinding tooth, and (3) development of two incisors into enormous tusks, used in fighting and digging.

Elephant Phylogeny. *Remote Ancestors and Divergent Forms.* The earliest fossils of proboscidean ancestors are found in the upper Eocene deposits of Africa. They belong to the genus *Moeritherium*, which is characterized by the elongation of incisors in both jaws, enlargement and recession of the nasal openings in the skull, formation of air cells at the back of the skull, and transversely ridged molars, although it shows little further resemblance to the Proboscidea (Fig. 94). It is supposed to have existed until some time in the Oligocene, when it became extinct.

The Oligocene also produced the genus *Palaeomastodon* (Fig. 95), of which fossil remains are found in Africa and Asia. The members of *Palaeomastodon* show a

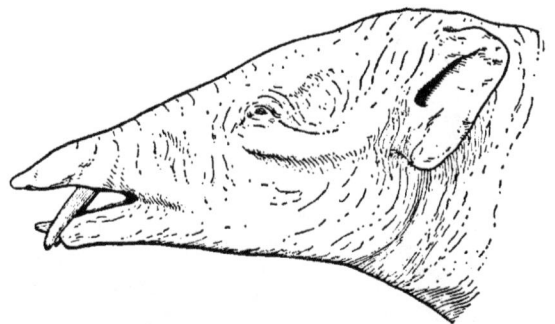

Fig. 95.—Head of *Palaeomastodon*, restored by Lull. (From Lull.)

marked advance in the development of the characters mentioned, and were probably descendants of *Moeritherium*. The molars are little more advanced, but the incisors of the upper jaw are well developed tusks, and the skull is higher, with a greater development of cancellated bone.

In the Miocene another probable descendant of *Moeritherium* appeared in Europe, existing until the Pliocene, when it became extinct. This genus, *Dinotherium*, displays a peculiar development of deflected lower jaw and tusks, while the tusks of the upper jaw appear to have been lacking. The size of these animals was about that of the mastodons. While they show evidences of the development of a proboscis, and are in general well advanced over *Moeritherium*, they gave rise to no existing forms.

The Miocene, however, saw the rise of three other genera which appear in Europe, Asia, Africa and North America. The first, *Trilophodon*, is supposed to have descended directly from *Palaeomastodon*. Its distribution includes the four continents mentioned and it persisted "until the extinction of the mastodon in post-Glacial time." Aside from the elongation of the lower jaw, and the presence of lower tusks, it was distinctly like the elephant in appearance (Fig. 96). Its teeth and skull show a greater development of the typical proboscidean character than any of its predecessors.

Mastodon and Its Contemporaries. During the same period *Mastodon* arose (Fig. 97), to persist in America until its extinction

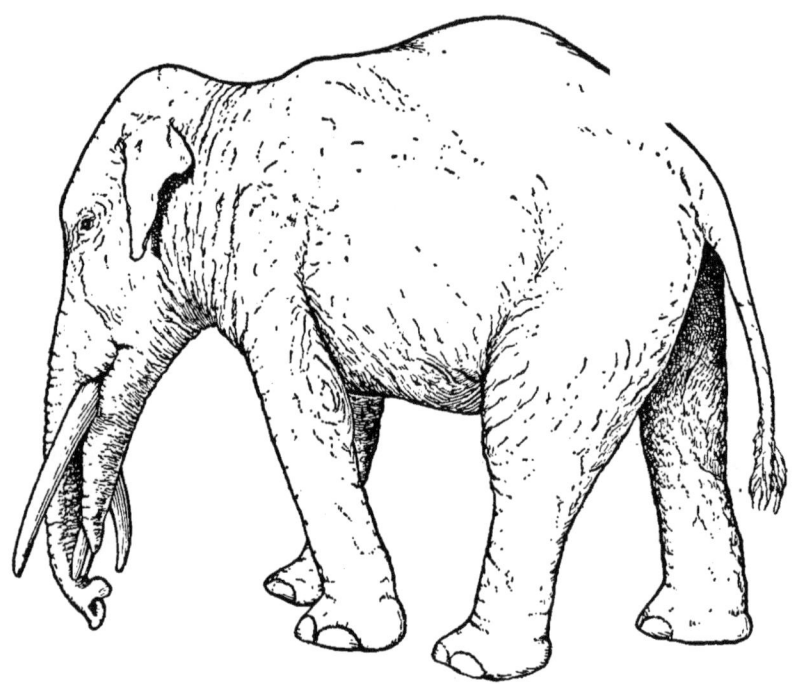

Fig. 96.—Restoration of *Trilophodon*. (From Lull, after the British Museum Guide to Elephants.)

in the Pleistocene, and *Tetralophodon* marked the beginning of another line which later produced *Dibelodon*. *Dibelodon* migrated to South America and became extinct at about the same time as the mastodons.

Tetralophodon (Fig. 98) represents the extreme development of the lower jaw and the four-tusked condition in the *Proboscidea*, although its descendant, *Dibelodon*, had the lower jaw shortened and without tusks. The two are different from the main line of descent in the more complex structure of the molars.

Mastodon included a number of different species, and has been subdivided into other genera by some paleontologists. In general these animals had the form of elephants, but they were more

primitive. The molars were transversely ridged, but the ridges were made up of associated cusps (Fig. 100C). They were so small that several were probably in use at the same time. Tusks were sometimes present, although poorly developed, in the lower jaw. Writing of the American Mastodon, Scott says that there is evidence to show that "it had a covering of long, coarse hair, and that it fed upon the leaves, shoots and small branches of trees, especially of conifers." The genus became extinct in the Old World before the end of the Pliocene, but migrants which entered America by way of the land bridge between Asia and Alaska persisted until the middle of the Pleistocene, and were probably contemporaneous with the early human inhabitants of the continent. Remains indicate that the heavy animals were often mired in sloughs and marshes.

FIG. 98.—Head of *Tetralophodon lulli*. The lower jaw, the longest recorded in any proboscidean, measured at least six feet in length. (From Lull, after Barbour and Kunz.)

An unusually fine skeleton was taken from such a situation in the summer of 1926 at Johnstown, Ohio. The bones were very near the surface and in a well settled region, conditions which emphasize one reason for incompleteness of our knowledge of extinct species, viz., the element of chance in the discovery of fossils.

From Mastodon to the Elephants. *Mastodon* produced another genus, *Stegodon*, which appears in Asiatic deposits of the early Pliocene. *Stegodon* was much like the modern elephants. Its molars bore more and finer ridges than those of the true mastodons, and the derivation of these ridges from rows of conical eminences is less evident (Fig. 100B). The worn surface shows dentine plates surrounded by enamel, but the cement is not abundant as in the true elephants. The lower jaw is short and without tusks, and other characters are so close to those of the elephants that *Stegodon* has been looked upon as congeneric with them. Fossil remains of *Stegodon* have been found only in southern and southeastern Asia, which may therefore have been the region in which the true elephants developed.

From *Stegodon* it is only a step to the genus *Elephas*, including many extinct species and the existing Indian elephant. The African elephant is included by some authorities in the same genus and by others in the genus *Loxodonta*. All may be considered as true elephants. From Asia they probably reached North America during the Pliocene, and Africa during the Pleistocene. In all regions except Africa and Asia they became extinct during the latter period.

Through the frozen mammoths of Siberia we have detailed knowledge not only of the skeleton but also of the soft parts of

Fig. 99.—The Woolly Mammoth (*Elephas primigenius*). (From Lull.)

these great mammals. The species preserved in this way is *Elephas primigenius*, the hairy mammoth (Fig. 99). It was provided with a coat of coarse hair covering a close woolly vestiture which enabled it to resist the cold of high latitudes. It ate grasses and the tender parts of trees. Other extinct species include *Elephas antiquus* and *Elephas meridionalis* of Europe, and *Elephas imperator* and *Elephas columbi* (Fig. 208) of North America, all of which inhabited warmer regions. *Primigenius* inhabited the northern parts of both continents as well as Asia. The chief differences in fossil remains of these animals are the development of the molars and the degree of curvature of the tusks. In *Elephas antiquus* the latter are nearly straight, while in the Columbian elephant they spiralled to such an extent that in old age the tips crossed. Drawings on the walls of caverns in Europe show that early man was familiar with the mammoths.

174 EVOLUTION AND GENETICS

The chief features of the direct line of evolution of the elephants are graphically illustrated in Figure 100. Here the steps leading gradually from teeth with distinct cusps to those with fine trans-

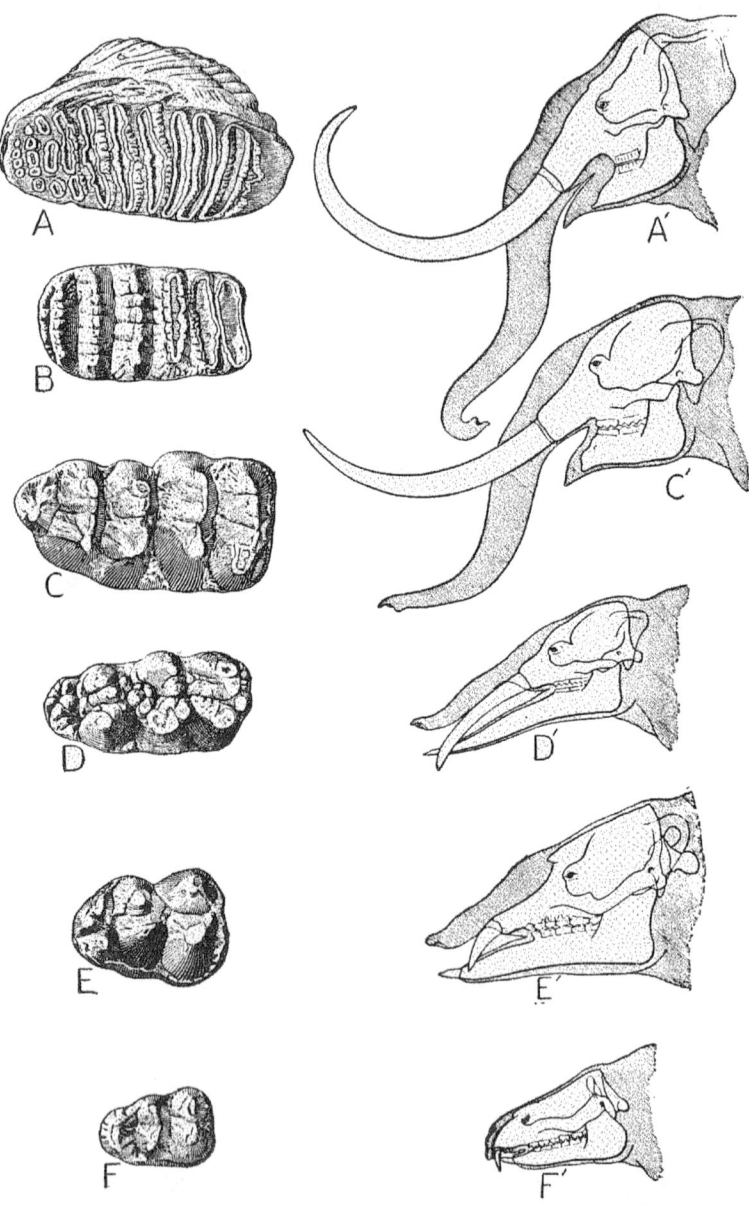

Fig. 100.—Evolution of the head and molar teeth of proboscideans. A, A'; *Elephas*, Pleistocene; B, *Stegodon*, Pliocene; C, C', *Mastodon*, Pleistocene; D, D', *Trilophodon*, Miocene; E, E', *Palaeomastodon*, Oligocene; F, F,' *Moeritherium*, Eocene. (From Lull.)

ELEPHANTS, HORSES, AND CAMELS 175

verse ridges are shown as they occur in the fossil remains of extinct species. Here also the transition from the elongate head of *Moeritherium* to the high short head of the elephant is illustrated from actual remains of a chronological succession of species. Such a series is not merely interpreted as evidence of evolution; it is the actual record of evolutionary change.

Adaptive Structure of the Horse. *Locomotion.* The evolution of the horse is in a number of ways more extreme than that of the elephants. With respect to speed, the legs are elongated and slender, retaining only one functional digit. As in many other animals, the feet no longer rest flatly on the ground, but in no other animal is the elevation to the toes more extreme than here, for not merely the tip of the toe, but the hoof alone, the homologue of claws and nails, comes in contact with the ground. This lifting of the body and elongation of the lower parts of the appendages results in a relative shortness and concentration of the leg muscles, and consequent rapid movement and lengthening of stride. The function of propulsion is largely relegated to the hind limbs, in which elongation of the foot and concentration of leg muscles is extreme.

Elongation of the legs is compensated in the horse by lengthening of the neck, so that the head of the animal can reach the ground.

Teeth. The food habits of the horse and other grazing types demand special development of the teeth for chewing harsh vegetation. Grazing habits demand no canines, hence they are reduced. The incisors are important for cropping low vegetation and are elongated. The molars and three premolars are similar. All are

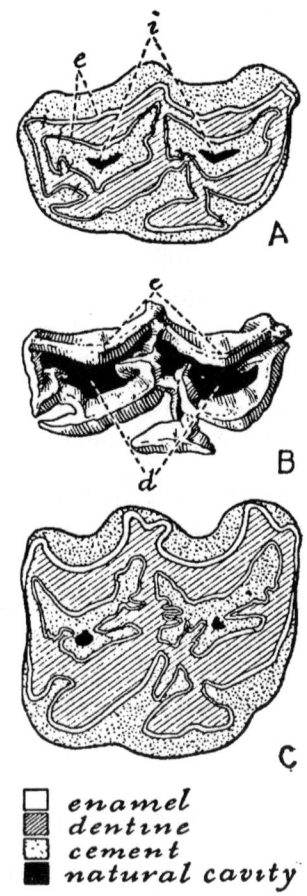

FIG. 101.—The grinding surface of horses' teeth. A, worn surface of milk molar of colt about six months old; B, unworn surface of milk molar before birth; C, premolar of horse eight or nine years old; e, enamel; i, natural cavity in cement; d, cavity, later filled with cement. (From Lull, after Chubb.)

high crowned. The worn surfaces show a complex pattern of dentine enclosed by enamel, which is in turn surrounded by cement (Fig. 101). Such a structure presents the same fundamental characteristics as the teeth of the elephants, but in the horse wear is compensated by the height of the crowns, while in the elephants the teeth are used successively. Growth of horses' teeth continues for a little more than five years. At the end of this time all of the permanent teeth are in use. Subsequently the teeth are pushed farther out of the jaws as the crowns are worn down, and when of no further use, are shed (Fig. 102). The total number of teeth which develop in the horse is forty, but canines are not found in mares, and the first molar is vestigial or lacking. Two incisors, three molariform premolars, and three molars are normally produced in each half-jaw.

Phylogeny of the Horses. *The Direction of Change.* The evolutionary changes which are made evident by the known fossil horses are summarized by Lull under the following heads:

1. Increase in size.
2. Lengthening of the limbs.
3. Reduction of ulna and fibula, with a consequent limitation of the range of movement.
4. Change of the foot posture from plantigrade to unguligrade.
5. Reduction and loss of digits from five to one.
6. Perfection of the hoof.
7. Perfection of the dental battery in elongation and complexity of teeth.
8. Premolars becoming molariform.

Eohippus, the American genus, and *Hyracotherium,* the European genus, belong to the Eocene epoch and are the oldest known ancestors of the horse. The connection of the two is uncertain. *Hyracotherium* is more primitive, but no ancestral forms are known to indicate the origin of the closely related forms. While the European genus has left no known descendants, there is a succession of genera from *Eohippus* to *Equus* in North America, although the line at last became extinct in the Pleistocene, and modern horses were introduced from Europe by the early explorers. From time to time ancient forms migrated from America into other continents, where they either became extinct or gave rise to the limited number of modern species of *Equidae.*

ELEPHANTS, HORSES, AND CAMELS 177

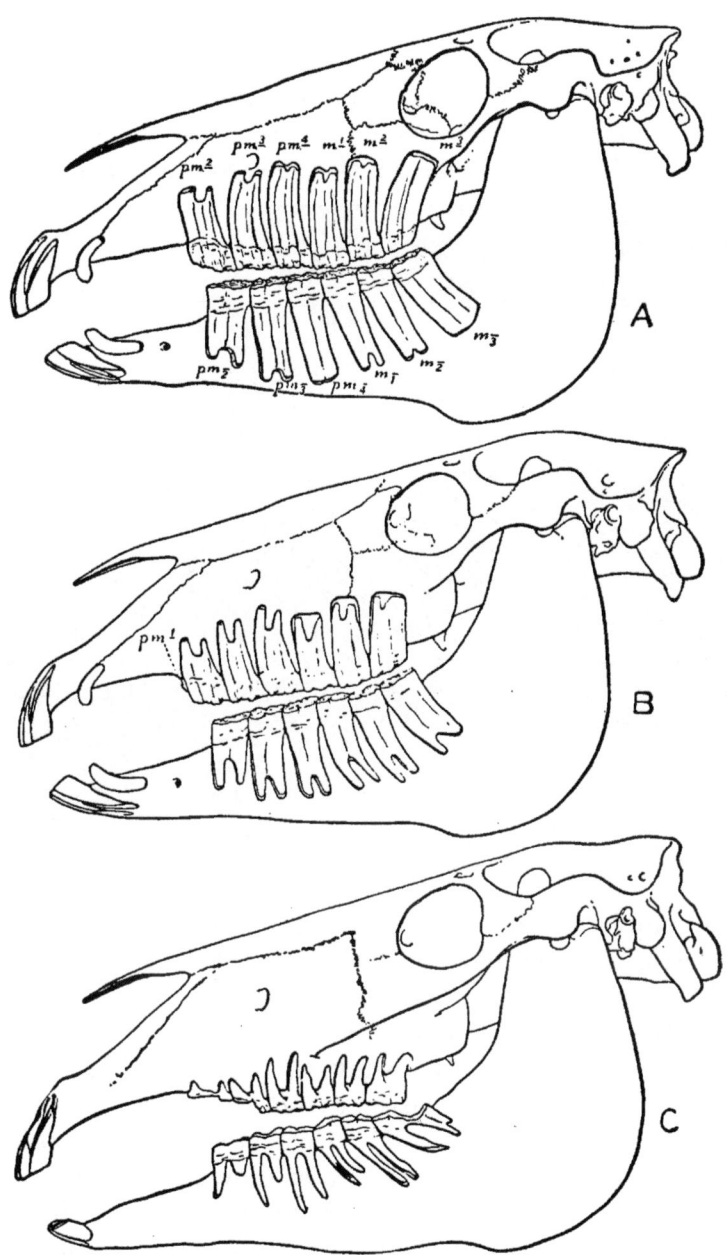

FIG. 102.—Dental battery of horse, to show growth and wear of the teeth. A, five years old, permanent teeth all in use. B, eight years old, crowns reduced by wear and roots longer. Vestigial upper premolar, pm^1, present. C, thirty-nine years old, lower molars incline forward. Canines absent (female). (From Lull, after Chubb.)

Eohippus, the "Dawn Horse," was a small animal, about the size of a fox terrier (Fig. 103). Its head was elongate and rather horse-like, although the eye was much farther forward than in modern horses. The feet were digitigrade, and the legs only moderately elongated, hence the neck was also moderate. The fore limbs had four functional digits, the second to fifth, while the

Fig. 103.—Restoration of *Eohippus* sp. of the lower Eocene. (From Scott.)

first was completely lost. The hind limbs were three-toed, with minute vestiges of the first and fifth (Fig. 104). Each lower limb retained the two primitive bones, the radius and ulna of the fore limbs and the tibia and fibula of the hind limbs. The teeth were relatively advanced, foreshadowing the modifications to come, but the premolars were still distinct from the molars and the first premolars were present (Fig. 105). Correlated with this primitive condition of the teeth is the relative shallowness of the jaws and the position of the eye. The conditions under which the species lived were such as to encourage the development of grazing habits because of the abundance of meadows and grassy plains. In spite of such facts, however, these little creatures are conspicuously unlike the modern horses. It is only through the intermediate forms that the two extremes can be associated.

Orohippus, the next genus to appear, followed *Eohippus* in the Eocene and shows only slight changes. The vestigial bones of the first and fifth digits of the hind limb had disappeared, the fifth digit of the fore limb was shorter, and the third premolar was molariform. It included somewhat larger species than its predecessors.

Epihippus, a third genus, occurred later in the Eocene. Complete skeletons have not been found, but the last two premolars were molariform, the first still persisted, and the teeth were somewhat higher crowned than in the older genera.

Mesohippus, an Oligocene genus, included several species which varied in size but scarcely attained the size of a sheep. The upper incisors differed from those of earlier species in the "mark," an enamel ridge behind the cutting edge, which is characteristic of true horses. The premolars were molariform with the exception of the first, which shows a tendency to

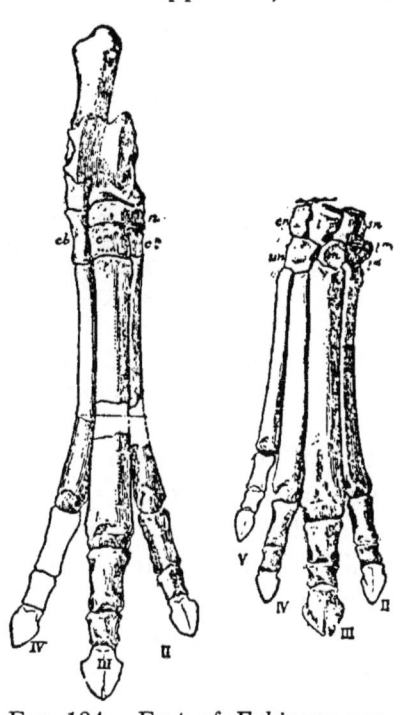

FIG. 104.—Feet of *Eohippus venticolus*. (From Matthew.)

disappearance. The teeth had not yet acquired the high-crowned form of the true horses (Fig. 106). The eye was set farther back than in *Orohippus*, and its orbit was closed behind, unlike the preceding forms. The bones of the feet of *Mesohippus* are still further reduced (Fig. 107). All four feet had three functional digits, and only the anterior pair retained a vestige of the fifth. The ulna of the fore limb and fibula of the hind limb were very slender, but complete. The same epoch produced *Miohippus*, a genus of similar forms.

FIG. 105.—Upper teeth of *Eohippus*. Premolars visibly smaller and simpler than molars. (From Lull, after Matthew.)

In his recent review of the evolution of the horse Matthew records changes in the relationships of the carpal and tarsal bones which are of importance in the development of the one-toed foot.

In the earlier species the cuboid bone received only the metatarsal of the fourth digit. As the third digit enlarges the cuboid articulates in part with it, and so gives the functional toe greater lateral support and tends to prevent rocking. This change first appeared in *Miohippus*.

Environmental Conditions. This epoch was a time of increasing aridity due partly to continental uplift. Such conditions re-

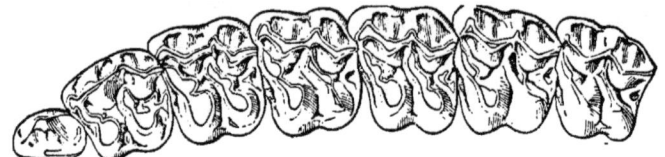

Fig. 106.—Upper teeth of *Mesohippus*. Three premolars like the molars. (From Lull, after Matthew.)

sulted in the decrease of bodies of water and the extension of prairie areas, although the persistence of forests and meadows, as well as dryer areas, favored the development of several types of primitive horses which have since disappeared. The continuation of climatic change in the Miocene gave rise to the great prairies of North America and was accompanied by the development of large numbers of grazing species. These animals must necessarily have had the characters which we see in the modern horse, viz., teeth adapted to harsh grasses and legs adapted for rapid locomotion over hard open ground.

Miocene Horses. The divergence which began in the Oligocene culminated in two lines during the Miocene in North America. These were *Parahippus* and *Hypohippus*. Both had low crowned teeth, obviously developed for browsing on soft herbage rather than for grazing, and spreading feet which must have been fitted for walking on soft ground. They were undoubtedly forest animals. Most of the species were fairly large, a little more than three feet in height. Migrants from the divergent types also gave rise to an Asiatic genus, *Anchitherium*, which likewise became extinct.

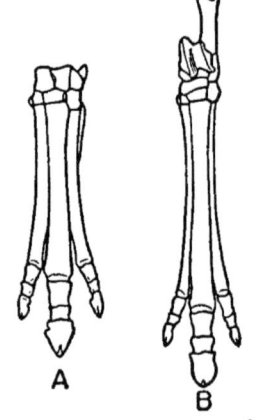

Fig. 107.—Feet of *Mesohippus*. A, anterior; B, posterior. (From Lull, after Marsh.)

In the direct ancestry of the modern horses, the period produced two significant genera, *Merychippus* (Fig. 108) and *Protohippus*. The molars of *Merychippus* mark the transition from the browsing to the grazing type, for paleontology records that the milk denti-

FIG. 108.—Restoration of the prairie horse, *Merychippus*, from the Miocene. (From Lull.)

tion is of the low crowned primitive type while the permanent molars are high crowned grazing teeth (Fig. 109). The feet were three-toed, but the middle toe was so much more highly developed than the others that it alone supported the body, while the others did not reach the ground. *Protohippus* differed from *Merychippus* in the fact that both milk and permanent molars were high crowned. These genera include the first highly specialized grazing horses.

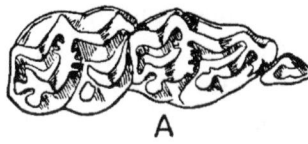

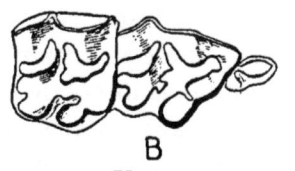

FIG. 109.—Upper premolars of *Merychippus*. A, milk teeth, without cement; B, permanent teeth, with cement. (From Lull.)

From the same stem another genus, *Hipparion*, originated during the late Miocene in North America, whence it migrated into Asia and Europe. During the Pliocene it became extinct. Matthew states that *Hipparion whitneyi*, from South Dakota, was a very slender and graceful horse, except for its large head, and was adapted by the complexity of the enamel ridges of its teeth to eat the harshest herbage. It is therefore looked upon as a very fleet species, fitted to live in semi-desert country. It was of about the same size as the other Miocene species.

Pliohippus, a North American genus of the early Pliocene, was the last progenitor of the modern genus *Equus*. The included species were no larger than the few immediate ancestors, i.e., about forty inches high. They differed conspicuously, however, in the reduction of the second and fourth digits on all feet to splints; thus *Pliohippus* is the earliest known one-toed horse. The third digit was highly developed and bore a well formed hoof. This genus gave rise to *Plesippus*, from which it was only a step to *Equus*.

South American Horses. Either *Pliohippus* or *Protohippus* gave rise to the genus *Hippidion* of South American horses which existed during the Pliocene. This genus, its derivate *Onohippidion* of the Pleistocene, and some migrants from North America belonging to the genus *Equus* are all of the horses known to have occurred in South America. All were extinct by the end of the latter epoch. (See diagram, page 183.)

The Genus Equus in North America. *Equus* of North America is not well known. Scott's statement quoted below, gives the paleontologists' opinion of these species.

"In the latest Pliocene, and no doubt earlier, species of the modern genus *Equus* had already come into existence; and in association with these, at least in Florida, were the last survivors of the three-toed horses which were so characteristic of the early Pliocene and the Miocene. However, little is known about these earliest recorded American species of *Equus*, for the material so far obtained is very fragmentary. In the absence of any richly fossiliferous beds of the upper Pliocene generally, there is a painfully felt hiatus in the genealogy of the horses; and it is impossible to say from present knowledge, whether all of the many species of horses which inhabited North America in the Pleistocene were autochthonous, derived from a purely American ancestry, or how large a proportion of them were migrants from the Old World, coming in when so many of the Pleistocene immigrants of other groups arrived. It is even possible, though not in the least likely, that all of the native American stocks became extinct in the upper Pliocene and that the Pleistocene species were all immigrants from the eastern hemisphere; or the slightly modified descendants of such immigrants; but, on the other hand, it is

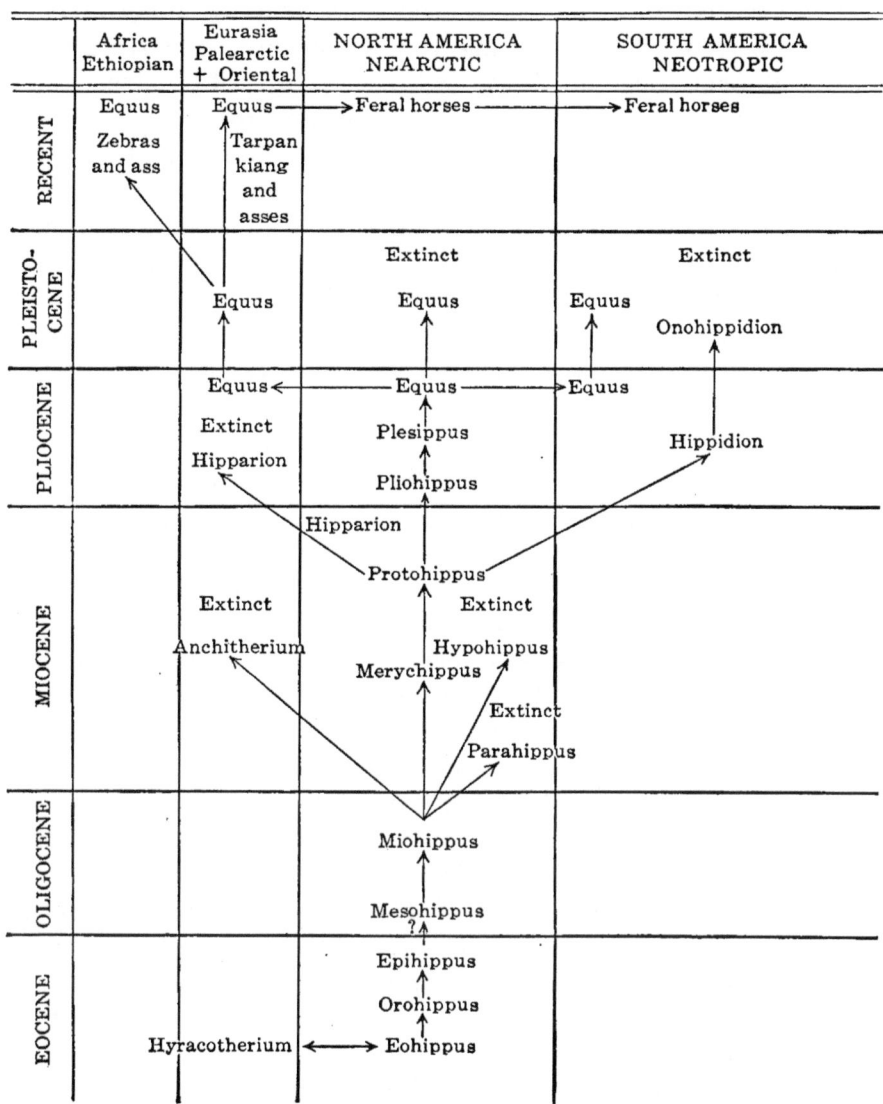

Phylogeny of the horses. (Modified, after Lull.)

altogether probable that some of these numerous species were intruders."

Whatever may have been their source, all species living in the Americas became extinct during the Pleistocene, and from the Palearctic stock developed the zebras, asses, wild horses and domesticated races of the present. Conditions are now so favorable for their existence that feral horses are found in the new world. On the western ranges of our own country they have even multiplied to such an extent that they are becoming a problem to stock raisers. What may have been the cause of their ancient extinction becomes a difficult problem. Lull offers as a theory the introduction of some virulent insect-borne disease, but of course no such theory can be proved with evidence now available. All that we can know is that they died, and in some places immense numbers are preserved as fossil skeletons.

Camels. In still another group, the camels, an extraordinarily complete series of fossil forms are available. These animals are now represented only by domestic and feral individuals, and do not occur in North America. They are Artiodactyla, or even-toed ungulates and represent the course of adaptation for speed and for subsistence on harsh and scanty herbage. The modifications of the legs, however, show clearly the animal's fitness for progress over a soft substratum such as the loose sand of many deserts, and in existing forms the nostrils and eyes are equally eloquent of ability to resist desert conditions. The ability of camels to store water beyond their immediate needs, and the storage of fat in the hump, are probably the best known adaptations.

The evolution of the camels and their relatives, the llama and alpaca of South America, is so similar to that of the horses that Figure 110 will be an ample presentation for the purposes of this work. It is significant that their development occurred at the same time as that of the horses and in response, apparently, to the same conditions of gradually increasing aridity. Although they no longer occur in North America the remains of ancestral species show that they were once common over the greater part of the western United States and it was here that the true camels were evolved. The camels of the Old World were apparently derived from ancestors which migrated from North America over the Bering Isthmus, leaving the main stock to perish.

Such cases as these are the last word in evidence for evolution.

That we can see relationship in living forms, particularly relationship which can be interpreted only on the basis of common deriva-

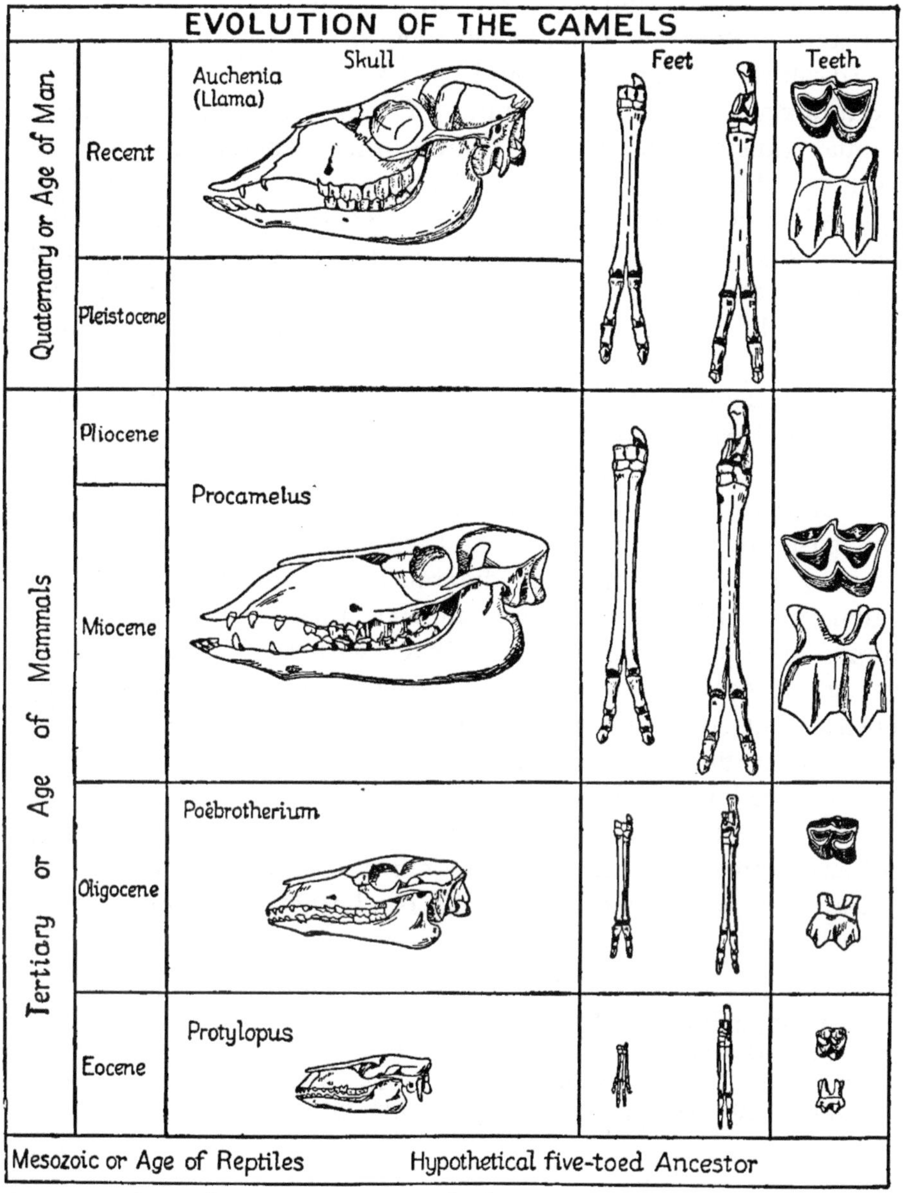

		Skull	Feet	Teeth
Quaternary or Age of Man	Recent	Auchenia (Llama)		
	Pleistocene			
Tertiary or Age of Mammals	Pliocene	Procamelus		
	Miocene			
	Oligocene	Poëbrotherium		
	Eocene	Protylopus		
Mesozoic or Age of Reptiles		Hypothetical five-toed Ancestor		

EVOLUTION OF THE CAMELS

Fig. 110.—Evolution of the camels, as indicated by the skull, feet, and teeth. (From Lull, modified after Scott.)

tion, is conclusive in itself, but the discovery of actual remains of creatures now extinct in such series as these admits of no other logical explanation. A transition by gradual steps from a relatively primitive state, such as that of *Eohippus*, to the highly specialized modern horse, correlated with geological time succession and evidence of climatic change, is a part of that natural record which, if complete, would be the whole story of evolution.

Summary. The fossil remains of many vertebrates show a gradual transition in structure from primitive ancestors to existing species. The evolution of the elephants, horses, and camels is especially well demonstrated by these records. Evolution in the elephants is chiefly linked with the development of great bulk and browsing habits, and is shown by the development of pillar-like limbs, finely ridged grinding teeth, and characteristically high short heads. Their length of limb is accompanied by the development of a trunk. Horses are more lightly built. Their feet are elongated, their leg muscles bunched, and their toes reduced to one on each foot. Every characteristic of the limbs shows adaptation for rapid locomotion over hard ground. Length of limb is compensated by elongation of the neck and the teeth are adapted for grazing. The camels are differently adapted for life in arid regions. In all three groups primitive ancestral species of the Eocene are known which can be linked with the existing animals through a chain of other species occurring in the intervening epochs. The structures of these species in all cases show a gradual transition leading up to the highest stage. Divergent species are also recorded which became extinct without leaving known descendants. The correlation of structural transition, chronology, and environmental conditions is significant evidence of evolution. These cases are, in fact, fragments of the actual record of past evolution.

REFERENCES

Scott, W. B., *A History of Land Mammals in the Western Hemisphere*, 1913.
Lull, R. S., *Organic Evolution*, 1917.
Matthew, W. D., "The Evolution of the Horse. A Record and Its Interpretation." *Quarterly Review of Biology*, I, 139–185, 1926.

CHAPTER XI

THE EVOLUTION OF MAN

The differences between man and the lower animals are such that we can hardly avoid being prejudiced judges of our place in the world. We know relatively little either of the world or of ourselves and there are many factors in our lives as human beings which tend to influence our evaluation of such things as we do know. Science tries to set aside these prejudices and to judge the characteristics of man as impartially as those of other organisms. It has succeeded to some degree and therefore gives us a more logical account of ourselves than any other field of knowledge, but in this as in other fields we must be constantly aware that our information is incomplete.

Man's Systematic Position. Of man's position among other organisms, fortunately, we need not be in doubt. Although he stands well above the other animals in some ways, he is animal in structure and in functions and shows in his anatomy as definite relationships as are displayed by the forms already studied. Among his animal structures he has the dorsal tubular nervous system, the vertebral column which replaces the embryonic notochord, and at one stage evidence of the pharyngeal clefts which stamp him a chordate and a vertebrate. Within this phylum he has the hair, the circulatory system, and the mode of reproduction of the mammals, and in the highly specialized connection of parent and embryo he shows the most conspicuous character of the Eutheria.

The Eutheria include four groups, the Unguiculata or clawed animals, the Primates or animals with nails, the Ungulata or hoofed animals, and finally the Cetacea, made up of such highly specialized marine creatures as the whales, dolphins and porpoises. Obviously man is most nearly like those species which have nails on the digits, and so we find him a member of the order Primates, in the group of the same name.

The Primates. *Characteristics.* In general the primates are arboreal animals. They have prehensile appendages with the thumb and great toe more or less opposable to the remaining digits. Such structures are effective in the plantigrade position for walking but are especially fitted for locomotion in trees. Instead of claws or hoofs they have nails. They are further characterized by having the body covered with hair except the palms, soles and parts of the face; the mammae are reduced to a single pectoral pair, the eyes directed forward and the orbit surrounded by bone, a clavicle always present, and the brain relatively large and well convoluted. The body of every individual verifies man's possession of most of these characters.

Classification. Classifications of the Primates differ. The following is that of W. K. Gregory, which has been widely used. Recent publications dealing with the phylogeny of the group do not agree with it in detail, but as an indication of the general subdivisions it is wholly adequate:

Suborder 1. Lemuroidea. Lemurs or "half-apes."
Suborder 2. Anthropoidea.
 Series 1. Platyrrhini. New World apes.
  Family 1. Hapalidae. Marmosets.
  Family 2. Cebidae. Capuchins, howler monkeys, spider monkeys, etc.
 Series 2. Catarrhini. Old World apes and monkeys.
  Family 3. Cercopithecidae. Monkeys, baboons, macaques, etc.
  Family 4. Simiidae. Man-like or anthropoid apes.
  Family 5. Hominidae. Man.

The series Platyrrhini is distinguished by a broad nasal septum, a reduced and non-opposable thumb, and other characters. In contrast, the Catarrhini have a narrow nasal septum, and as the name suggests, the nostrils point downward. They have thirty-two teeth, opposable thumb, and non-prehensile tail, which is often rudimentary and not developed as an external appendage. The latter group corresponds in these things with the structures of man but the two lower families differ from man in the opposable great toe. In general, then, man harmonizes closely in structure with the other members of the Catarrhini. The differences which exist are traceable almost entirely to differences in habits, since man is an erect terrestrial species while the others are arboreal, and at the most only semi-erect.

The Man-Like Apes. The family Simiidae includes four existing genera: *Hylobates*, the gibbons; *Simia* or *Pongo*, the orang; *Pan*, the chimpanzees (Fig. 111); and *Gorilla*, the gorilla (Fig. 112). All of these animals are tailless. Though they can walk in a semi-

Fig. 111.—Chimpanzee, *Pan pygmaeus*. (From Lull, photograph from the New York Zoological Society.)

erect position, touching the knuckles to the ground to aid their progress, they are predominantly arboreal with the exception of the gorilla. The opposable thumbs and great toes provide them with four grasping appendages which are very effective for moving through the trees. So much has been said of the habits of these apes, particularly of the chimpanzee and orang, that their intelligence and imitativeness are almost common knowledge. The

Fig. 112.—Gorilla. (From a specimen in the American Museum of Natural History mounted by Carl E. Akeley. Through the courtesy of the Museum.)

half-human antics of certain simian performers in the movies has probably done more than anything else to acquaint man with these interesting relatives, and there are numerous excellent accounts of their structure and habits in print.

The anatomy of the man-like apes is much like that of man (Fig. 113). The apes have stronger jaws and teeth and a relatively low cranial capacity, their mouths are not formed in such a way as to permit articulate speech, their hands and feet alike are grasping appendages, and the skeleton is not sufficiently modified to allow a fully erect posture. Within the range of their own group, however, they show much greater anatomical differences than are evident between the highest apes and man. Man shows all of the modifications incidental to intellectual development and erectness. These structures and the probable reasons for their development are treated in the following paragraphs; detailed comparison with the existing apes is unnecessary for these animals are, at the most, merely similar to some of the remote ancestors of man.

The Arboreal Origin of Man. The fact that man is in so many ways like these great apes, and that they are highly developed arboreal animals, suggests that man himself is derived from arboreal ancestors. In analyzing this possibility it is first necessary to consider how the assumption of arboreal habits might affect the normal quadrupedal form characteristic of the lower vertebrates. With the condition of the arboreal primates established, it is then necessary to inquire into the possibility of return to the ground, and into the effects of such a return upon the arboreal organism.

Arboreal Quadrupeds. The assumption of arboreal life by quadrupeds is not at all uncommon. Squirrels are a familiar example. In all of the many mammalian orders represented by such species, however, sharp curved claws are the effective means of locomotion except among the lemurs, the lowest primates. Here the thumb and great toe are opposable, and the animal is able to grasp the limbs of trees instead of merely clinging to them or hanging from them as by a series of hooks.

Hands versus Claws. The result of opposability is a much more effective appendage for arboreal progress. Claws, if sufficiently long, as in the sloth, are most effective for suspension of the body from branches, but they are a hindrance to any other use of the appendages. If short and sharp, they provide sure footing, but as organs of prehension they are of little or no use, and for suspending the body from branches, of very limited use. A grasping structure, however, as we can determine from personal experience, is useful in many ways. It provides sure footing for locomotion above the substratum; it is an effective organ for suspending the body from branches within the limits of muscular strength; lastly, the

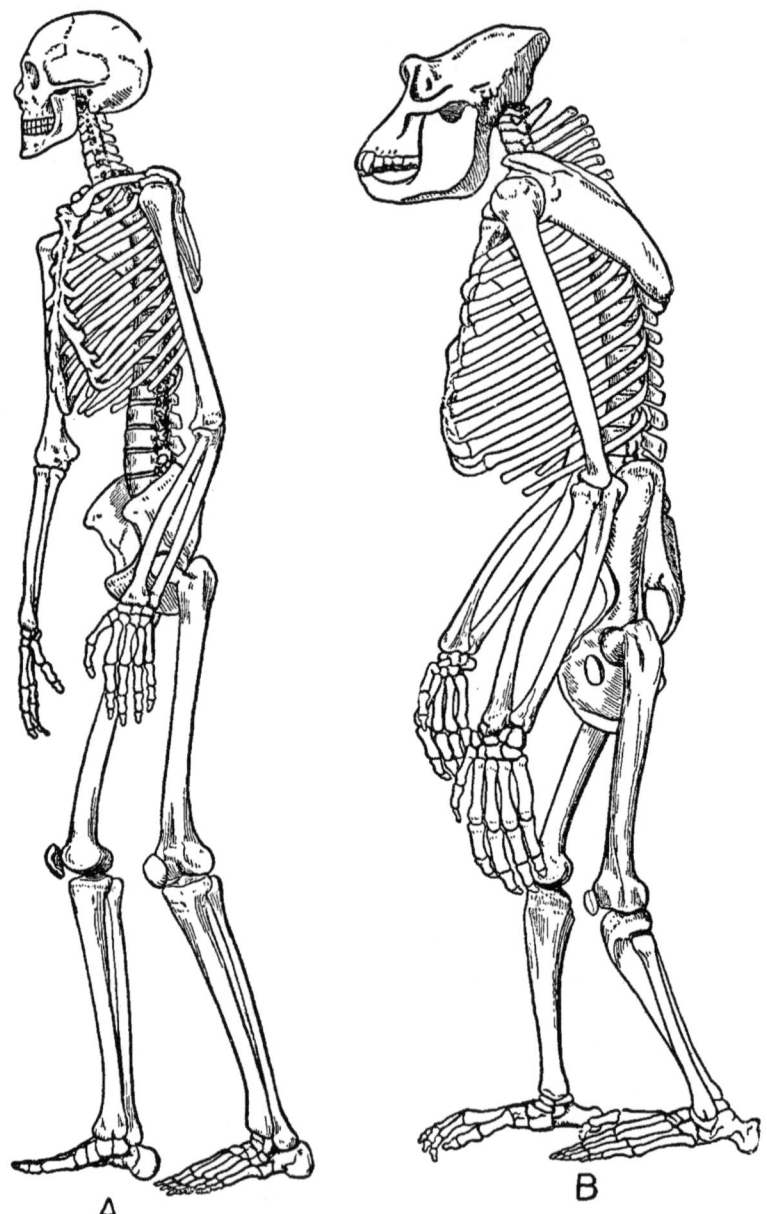

Fig. 113.—Skeletons of man (A) and gorilla (B). (From Lull.)

ability to grasp objects enables the animal to move them about and bring them before its eyes for examination.

The Effects of Brachiation. Two important changes may result from the use of such appendages. Physically, the ability to hang from a limb by one or more appendages, makes possible the type of locomotion, known as brachiation, i.e., swinging from branch to branch by the fore-limbs, as is done by the arboreal primates. The effect of gravity would then be felt by the body as a straightening pull, while the hind-limbs would be resolved functionally into supports for the body while resting upon branches or moving about on the ground. Such division of locomotion would be functional specialization of great importance to arboreal animals. Jones writes, "As arboreal life becomes more complete, the search for a new foothold will become a far more exacting business than it is in the mere clambering we have pictured (of quadrupedal animals). The more exacting the search becomes, the more will there tend to be developed that most important factor—*the specialization of the functions of the fore- and hind-limbs.* While the animal reaches about with its fore-limb, the hind-limb becomes the supporting organ. With the evolution of this process there comes about a final liberation of the fore-limb from any such servile function as supporting the weight of the body; it becomes a free organ full of possibilities, and already capable of many things. This process I am terming *the emancipation of the fore-limb,* and its importance as an evolutionary factor appears to me to be enormous." Restriction of the supporting function to the hind-limb and the straightening of the extended body alike would tend to develop the erect posture from the quadrupedal. The transition is, of course, not a slight one, but in the peculiar locomotion of the apes we see that such transitional development may exist.

Brachiation and Mind. The "emancipation of the fore-limb" could not fail to be an active stimulus to nervous development. Swinging from branch to branch high above the ground would require keenness of the senses, especially of vision, nice coördination, and exactness of judgment of contributing physical factors, such as wind. We can best understand these things by considering the feats of acrobats. Since the penalty of inaccuracy would be death or serious injury in most cases, there would be no question of perpetuation of the unfit. In this way and through the freedom of the fore-limbs for handling and examining any object within reach the development would react upon the senses. Not only would inherent keenness of the mind and senses be emphasized; a wide range of stimuli would also be brought to bear upon the animal by its increased ability to investigate other things.

Mind and Environment. Conditions at present show us that only tropical forests are favorable to the existence of such arboreal types. Elsewhere they would find food too scarce, and would be forced to seek it on the ground as well as in the trees. In the tropical forests, however, there is a constant supply of fruits, which are well adapted to mastication by their rather primitive teeth. Insects are also available as a part of their diet. We have already seen in the heterodont dentition of more primitive mammals and of the cynodont reptiles an initial stimulus to development of intelligence through the variety of reactions in securing food. Such food habits as those of the primates might well be correlated with this type of development, but in other ways as well the tropical forests would provide diversity of contacts. Their richness in all forms of life is unsurpassed by any other terrestrial environment.

From the Trees to the Ground. Anything which might force these arboreal creatures to the ground, such as scarcity of food or the thinning of the forests, would find them well able to meet competition. They would be less specialized for attack or defense than the carnivores, less specialized for flight from their enemies than the herbivores, and consequently unable to meet either of these groups in direct competition within their limited fields of activity. However, they could easily escape from terrestrial enemies by climbing, and through their omnivorous habits would find an abundance of food without the limitations imposed upon the more specialized animals.

On the ground the animal would have a new set of tools at his disposal. His hind-limbs, developed for perching upon branches, would find the ground a more stable substratum, and his hands would then be freed from all but occasional use in locomotion. Their grasping power, turned to new uses, might readily accomplish many things. The use of stones and clubs as weapons suggests itself as a simple result of his inferiority in competition with carnivores, but whatever might be the beginning of his use of implements it could hardly fail to open up new possibilities.

All of the things which might logically follow upon descent from the trees point toward ever increasing diversity of activities and stimuli. Once the ability is acquired to make use of other things than those provided by the body itself, the possibilities are without limit. Inventiveness in the human race has gradually brought us to our present state. We cannot yet see the limits of our powers in this direction, but our attainments are only the gradual accumulation of the ages.

Evolution of Terrestrial Primates. Such diversity of activities and its resulting diverse reactions upon the organism suggest that subsequent physical modification would be slight. With adequate locomotion, a free pair of grasping appendages, and the ability to make use of inanimate objects to compensate inherent deficiencies, everything would favor continued development along the same lines, and increasing mental power would be the result.

One physical modification begun during arboreal life might be expected to continue. Brachiation does not emancipate the forelimbs to the extent that is possible in terrestrial life, and consequently does not favor the development of completely erect posture. Terrestrial life supplies a dependable support on which the hind-limbs are adequate for locomotion. Increase in stability of equilibrium and the consequent freedom of the arms from accessory locomotor functions would make possible the maximum maintenance of the erect position, and would favor any changes in structure dependent upon it.

The Results of Erectness. Man is the only available example of a wholly erect animal. In his body are found all of the modifications which depend upon the change from quadrupedal to erect posture. Such a change involves primarily a shifting of the horizontal axis of the body to a vertical position with concomitant changes in anatomy, but Jones logically cautions against acceptance of this shift as an explanation. It seems to him rather an outcome of "an arboreal apprenticeship." "Walking upright upon the surface of the earth," he points out, "has produced its changes in the human body, of this there is no doubt; but we must be careful to distinguish between these 'finishing touches' and those other changes which are so much older and so much more important—the adaptations to arboreal life."

These "finishing touches," since they include a fundamental change in the axis of the body, bring about compensating changes in axial structure (Fig. 113). The spinal column of a quadruped is arched between the supporting appendages, and has flexible anterior and posterior parts, the neck and tail. Rotation through 90°

from a fixed attachment to the pelvic girdle brings about some curvature immediately above the point of attachment; this lumbar curvature is incipient in monkeys, and well developed in man. The head, supported without change of position upon this new axis, would point upward rather than forward. Its articulation with the vertebral column in quadrupeds allows only partial inclination, since flexibility of the neck provides a greater latitude of movement. The vertical position of the axis is compensated by a shift in position of the occipital condyles, which are ventrocaudal in quadrupeds. In arboreal primates, and to a greater degree in man, they have shifted along the formerly ventral surface of the skull to such a degree that this surface has also swung through 90° and become caudal instead of ventral. The face is thus brought forward.

Changes in the skull are not, however, entirely referable to the changed axis of the body. Reduction of the prognathous form is more directly correlated with the ability of the organism to use its hands for grasping, breaking or tearing, and bringing food to the mouth. The mouth in an animal of even semi-erect form is not its only facility for handling objects, and consequently is not used in the same way as the prognathous mouths of other animals.

Of the remaining parts of the skeleton the pelvic girdle and limbs alone undergo great change. The pectoral girdle and limbs are more primitive than in many other mammals. In comparison with the horse, for example, the hands of man are still in the primitive pentadactyl state, while the fore-limbs of the horse have lost four digits. The forearm in man contains the primitive bones, the ulna and radius, and they retain their primitive flexibility of movement. In the horse the ulna is reduced to a vestige which is combined with the radius, and the articulations are so modified that the limb moves in one plane, forward and back. Man also retains the clavicle, which is no longer present in the horse.

The pelvic girdle in man is compact and firmly articulated with the spinal column, as in all bipedal animals. Shift of the body axis places the stress of body weight differently upon its articulation, however, and the result is an elongation of the articular surfaces cranio-caudally, in contrast to the dorso-ventral elongation found in quadrupeds. The shift of stresses also throws the weight of the viscera toward the pelvis instead of toward the ventral body wall, a change which is compensated in part by the broad, basin-like pelvis.

The leg bones of man are modified to a relatively slight extent, but the femur is much straighter than in arboreal primates or quadrupeds. The foot is completely without the grasping power found in the primates, and is therefore much different from the hand which retains an opposable thumb. Its most interesting character is the predominance of the big toe. Specialization of the appendages usually results in emphasis upon one digit, but the most prominent digit of the primitive pentadactyl appendage is the third. In the horse we have noted the development of this one to the exclusion of all others. Man therefore is unusual in the great development of his first toe. A logical explanation is found in the significance of this digit in an arboreal animal. Development as a digit opposable to all others would find it already the most specialized of all when its possessor became terrestrial, and would provide the basis for further emphasis upon it during terrestrial life. In connection with its dominance, all other toes are reduced, and the little toe is even rudimentary in some races. Jones cites the Malays and Nubians as extreme examples. In these peoples the fifth toe is said to be stumpy and often without a nail. The foot has an arched skeleton, apparently for the absorption of shocks which would otherwise be transmitted through the entire longitudinal axis of the body.

A final specialization of man, the loss of hair from most of the body, and his delicacy of skin are probably associated with the development of intelligence. Matthew points out that man's retention of hair on the ventral surface of the body, where it is thinnest in other animals, is exactly what might be expected of long use of protective clothing. A simple garment, such as a skin thrown over the shoulders and tied around the waist, would protect just those parts of the body where the hair is most completely lost. It is significant that monkeys in our zoos make use of covering in this way. A number were wintered in outdoor cages as an experiment, and were provided with gunnysacks as protection against the cold. When evening came, each monkey helped himself to a sack, climbed to his perch, threw the sack over his shoulders and settled down for the night.

In most of his structures man is a primitive animal. In those mentioned, however, he is definitely specialized, and his specializations coincide very well with the effects of arboreal life. Or on a basis of known facts entirely we may say that anatomical conditions which fit existing primates for arboreal life point strongly toward the higher development of similar structures in man as a terrestrial descendant of arboreal ancestors.

The Geological Record. Inquiry into the fossil record of man's development shows first of all that primates were abundant during the Eocene, even in North America. They became extinct on this continent, but continued their existence in Eurasia, where the remains of several interesting genera have been discovered. Wilder describes two European species, *Pliopithecus antiquus* and *Dryopithecus fontani*, as Miocene apes. In Asia *Palaeopithecus sivalensis* and *Pithecanthropus erectus* are significant.

Palaeopithecus. This primate has been called a chimpanzee, but against this identification we are told: "In comparison with the chimpanzee its canine and lateral incisor teeth are much reduced, and the two lines formed by the lower molars converge anteriorly, this character lying midway between the condition in the chimpanzee, in which the two rows are parallel, and that found in Man, where marked anterior convergence of the rows of lateral teeth results in the formation of a gentle curve" (Wilder).

Pithecanthropus. *Pithecanthropus* was found in central Java in 1891 by a Dutch army surgeon, Eugen Dubois. The remains first uncovered consisted of a single upper molar tooth and the top of a skull, separated by about a meter in the same deposits. Later a second tooth, also a molar, and a left femur were discovered about fifteen meters away but also in the same deposits. These parts have been literally bones of contention. There seems little reason to doubt that they belonged to the same individual, although that possibility must be admitted. However they are of great importance, whatever our opinion of their relationship with each other, for conclusions based upon the single parts are in themselves significant.

The age of the deposits in which the bones were found has been placed at the early Pleistocene, but Osborn interprets the remains of mammals found in the same strata as late Pliocene. All evidence points to the fact that Java was connected with Asia at that period, and contemporary researches are being centered upon the search for fossils in the Siwalik Hills of India and central Asian regions.

THE EVOLUTION OF MAN

Pithecanthropus has been called the Trinil race and the Java ape-man. He is looked upon generally as more than ape, and yet less than man, so the latter term is apt.

The skull cap is fortunately complete enough to give an expert anthropologist data for the reconstruction of the cranium, while the teeth are indicative of additional characters of the jaws (Fig. 114). The cranial capacity was about two-thirds that of man, the forehead low, and the brow ridges prominent. The centers of touch, taste and vision were well developed in the brain, according to Osborn, and the "central area of the brain, which is the storehouse of memories of actions and of the feelings associated with them . . . but the prefrontal area, which is the seat of the faculty of profiting by experience or of recalling the consequences of previous responses to experience, is developed to a very limited degree." The known teeth are larger than human teeth, and differ in some particulars, but are manlike.

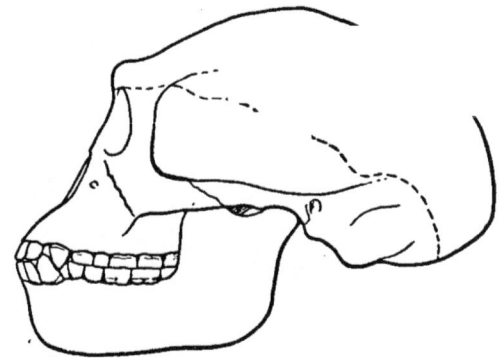

FIG. 114.—Skull of the Java ape-man, *Pithecanthropus erectus*, restored. (From Lull, modified after Dubois.)

The femur is not very different from that of man, indicating, if the bones belong together, that the owner of the skull cap was erect, and consequently that he had free use of his arms.

From the many discussions of the standing of *Pithecanthropus* in relation to man we may conclude that the species is undoubtedly a transitional form representing a very early stage in the evolution of the human beings of today. It is uncertain whether it is in the direct line of human descent or slightly removed from that line, but in either case it has many characters of our prehuman ancestors (Fig. 115A).

A more recent discovery of prehuman remains was made in 1925 in Bechuanaland, South Africa, by Professor Raymond Dart. This consisted of a brain cast bearing the bones of the face and part of the skull of a child of six years. The complete association of parts indicates that the creature was a transition form perhaps even more important than *Pithecanthropus*. Wilder gives a brief

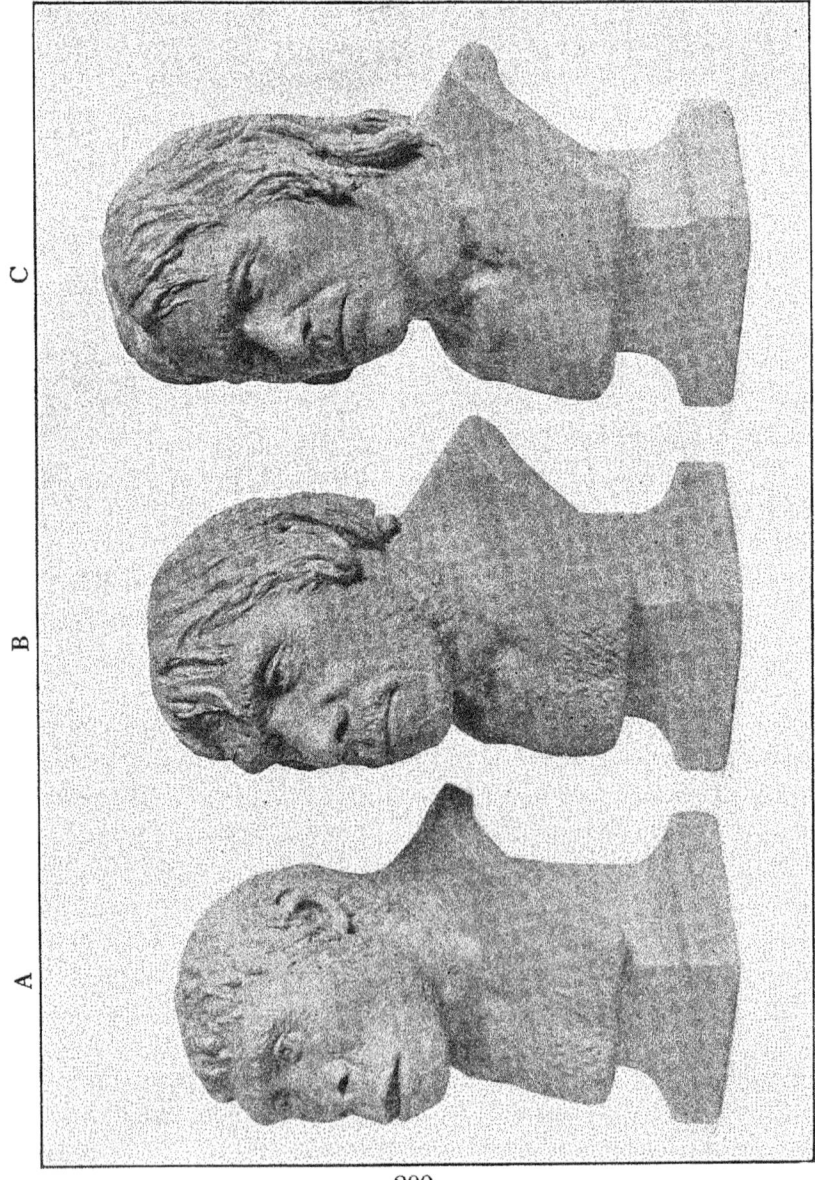

Fig. 115.—Restorations of prehistoric men, after models by J. H. McGregor. A, *Pithecanthropus erectus*; B, *Homo neanderthalensis*; C, *Homo sapiens*, Crô-Magnon man. (From Lull.)

account of the discovery in his book on *The Pedigree of the Human Race*. This species has been named *Australopithecus africanus*.

A single molar tooth discovered in Nebraska which bears the name *Hesperopithecus haroldcooki* has caused much comment. This tooth is said to indicate a stage intermediate between *Pithecanthropus* and modern man, but its standing is in dispute. Such a stage of human development is not to be expected in North America and so it demands very accurate analysis.

With these limited remains of prehuman species we cannot, of course, establish a continuity of descent comparable to the phylogeny of the horse. They are significant, however, for they verify many points in the theory of derivation from arboreal ancestors by the physical "finishing touches" of erect posture and the development of intelligence. All show greater brain capacity than the apes. All show the reduction of the jaws and teeth. Finally, within limits, all indicate increasing erectness. Such checks, however incomplete, can only strengthen our belief in evolution.

Climatic Factors. Not only are the structural features of these remains of prehuman species in accordance with the theory of descent, but also the climatic conditions under which they must have lived. Geologists tell us that the late Pliocene witnessed the first glaciation. Over the northern hemisphere the extension of ice sheets from the north in some regions accompanied a general lowering of temperature which marked the culmination of a process long in operation. The gradual cooling of the climate brought about a change of flora. Tropical forests could no longer flourish, and the trees which existed were more like those of the north temperate zone of the present, finally to be replaced by conifers in regions far south of their present range.

These changes tended first of all to destroy the arboreal habitat of the great anthropoids. They were faced with two alternatives as the tropical forests disappeared and the climate became more severe, namely, to migrate southward with the retreat of conditions favorable for their continued arboreal existence, or to remain where they were and meet the changed conditions with changed habits. As in the case of other animals which we have considered, there is every reason to believe that they could do either; very probably they did both. Those which migrated had no reason to change, but those which remained in the north must have been subjected to exactly those stimuli which we have considered.

Forced from their arboreal homes by the search for food and by inadequacy of the thinning forests for a wide range of movement, they would immediately encounter the conditions of a semi-terrestrial life, with its manifold advantages and demands. To these they must have responded if they were to exist, and by their remains we see that they existed.

Subsequent to the first glaciation a second occurred in the early Pleistocene, and a third and fourth later in the same epoch. These resulted in fluctuations of climate, but a relative abundance of fossil remains of man shows that his development had reached a point where changing conditions could be met by his improving intelligence. The remains with which we deal are not as highly developed as modern man; we are forced to look upon them as different species; but all authorities agree that they are definitely above the status of the apes. Moreover they are associated with implements of stone and other evidences of culture which have never been acquired by other animals.

Three species of fossil men are recognized in addition to remains referable to *Homo sapiens*. These are *Homo heidelbergensis*, *Homo neanderthalensis* and *Eoanthropus dawsoni*. Of these the second is represented by many specimens, while the other two are based on limited material.

Heidelberg Man. *Homo heidelbergensis* is based on a jaw found near Heidelberg in 1907. The bone was in excellent condition and contained a full set of lower

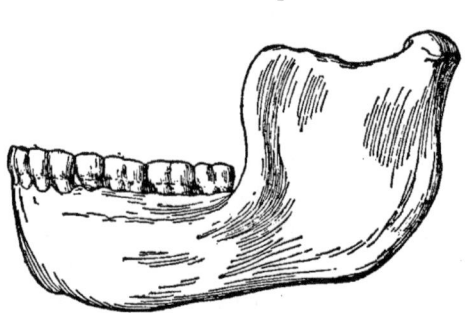

FIG. 116.—Jaw of Heidelberg man, *Homo heidelbergensis*. (From Lull, after Schuchert and Woodward.)

teeth (Fig. 116). Osborn, quoting from Schoetensack, the author of the species, gives the following account of the specimen: "The mandible shows a combination of features never before found in any fossil or recent man. The protrusion of the lower jaw just below the front teeth which gives shape to the human chin is entirely lacking. Had the teeth been absent it would have been impossible to diagnose it as human. From a fragment of the symphysis of the jaw it might well have been classed as some gorilla-like anthropoid, while the ascending ramus resembles that of some large variety of gibbon. The absolute certainty that these remains are human is based on the form of the teeth—molars, premolars, canines, and incisors in form, show no trace of being intermediate between man and the anthropoid apes, but rather of being derived from some older common ancestor. The teeth, however, are somewhat small for the jaw; the size of the border would allow for the development of much larger teeth; we can only conclude that no great strain was put on the teeth, and therefore the powerful development of the bones of the jaw was not designed for their benefit. The conclusion is that the jaw, regarded as unquestionably human from the nature of the teeth, ranks not far from the point of separation between man and the anthropoid apes. In comparison with the jaws of Neanderthal races, as found at Spy, in Belgium, and at Krapina, in Croatia, we may consider the Heidelberg jaw as pre-Neanderthaloid; it is, in fact, a generalized type." The race probably lived between the second and third glacial stages, i.e., during the second interglacial stage.

Piltdown Man. *Eoanthropus dawsoni*, the Piltdown man, may have been contemporaneous with the Heidelberg man but probably lived later. The remains were found near Piltdown, Sussex, in a gravel pit, and included enough fragments of a skull to make possible its reconstruction, half of a jaw, nasal bones, and a canine tooth (Fig. 117). The jaw has been interpreted as of a species of primate lower than primitive man, but the later discovery of additional bones shows conclusively that it belonged with the skull. Such ape-like characters as the jaw

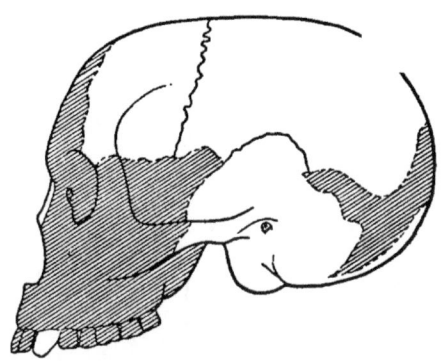

FIG. 117.—Skull of Piltdown man, *Homo* (*Eoanthropus*) *dawsoni*. (From Lull, modified after Woodward.)

displays may therefore be regarded as proof of the primitive nature of the species. The remains are generally assigned to the third interglacial stage, with an estimated age of 100,000 to 150,000 years.

The Piltdown race combined both primitive and fairly advanced characters. The jaw is distinctly ape-like, as shown by the single canine tooth and the mandible, yet the brows lack such prominent ridges as are found in the apes and most prehuman species. Even the Neanderthal race had much more prominent supraorbital ridges. The cranial capacity was probably 1100 c.c. or slightly less, in contrast with a range of approximately 1200 to 1500 c.c. in modern races.

Interpretations of this species vary greatly. It has been supposed that Piltdown man represents an ancestor of the Heidelberg and Neanderthal races, which must then be looked upon as degenerate. On the other hand, Osborn interprets the race as a side branch, while still other scientists look upon it as ancestral in varying relationships to other forms. There seems to be no adequate reason for interpreting the race as ancestral to Neanderthal man, but the development of the brows points rather definitely to such a relationship to *Homo sapiens*. It is therefore highly probable that *Eoanthropus* diverged from a remote ancestral stage of modern man, and that it is derived from the same line as *Homo neanderthalensis*, of which the Heidelberg race is supposed to be ancestral (Fig. 118).

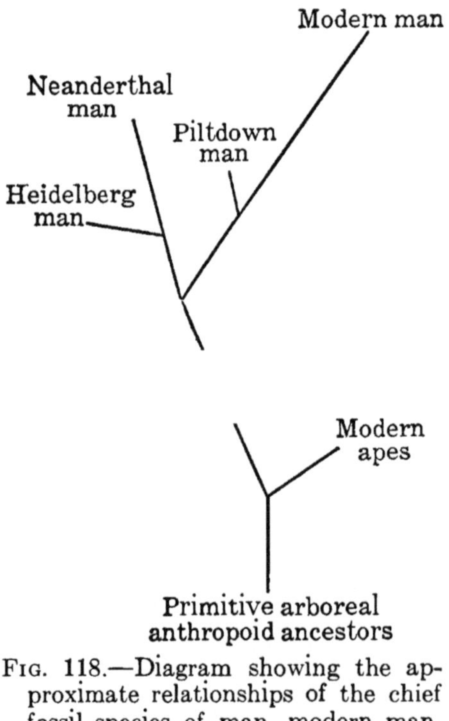

Fig. 118.—Diagram showing the approximate relationships of the chief fossil species of man, modern man, and the modern apes.

A most interesting circumstance regarding the Piltdown race is the association with it of crude chipped flints, which indicate a very primitive culture. The reader should consult Osborn's *Men of the Old Stone Age* for an account of the development of primitive industry and art.

Neanderthal Man. The Neanderthal race, as already indicated, is represented by numerous skeletons from various localities in Europe. These people lived after the third interglacial stage, and are looked upon by some scientists as degenerate. The skull is characterized by large orbits and heavy, prominent supraorbital ridges (Fig. 119). The cranial capacity varies, quite naturally, but is in most cases estimated as well above the minimum of *Homo sapiens* and in some well above the average for modern man. The lower jaw was powerful, but less so than that of Heidelberg man. There was no chin.

Other parts of the skeletons show that the race varied in height from a little less than five feet to over five and one-half. The chest

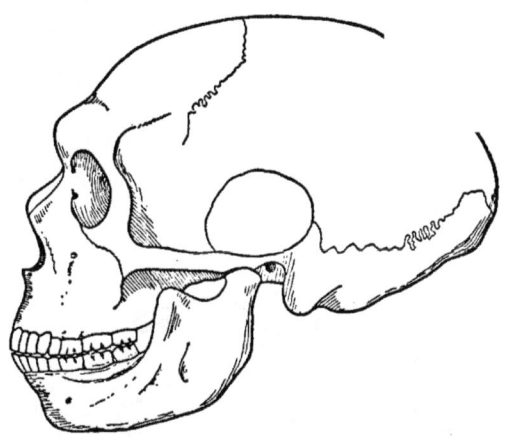

Fig. 119.—Skull of Neanderthal man, *Homo neanderthalensis*, from Chapelle-aux-Saints. (From Lull, after Boule.)

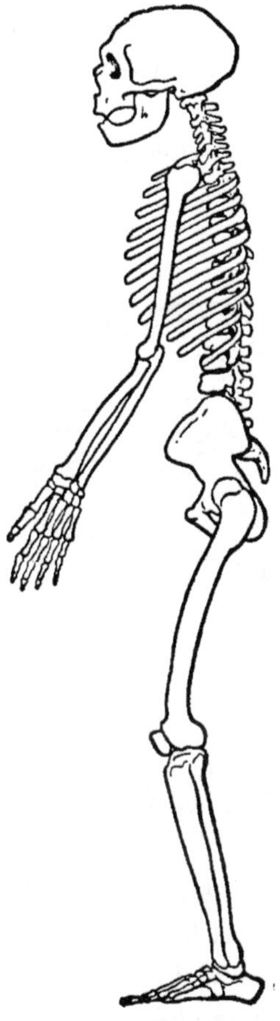

was large, the shoulders and arms were powerful, and the hands large. The thigh bones are curved in such a way that the race could not have been fully erect, a conclusion which is borne out by the absence of a cervical curvature of the spine and by the form of the knee joint (Figs. 120 and 115B).

The Neanderthals were cave dwellers. Their bones have been found associated with worked flints, bones of animals, and evidences of the use of fire. Skeletons have been found which indicated formal burial.

Fig. 120.—Skeleton of Neanderthal man. (From Lull.)

All known facts indicate that the Neanderthal race was human. While they were very primitive both in anatomy and in mental development, they had, no doubt, the power of articulate speech. Their burial customs indicate reverence for the dead, and therefore probably belief in some form of future existence. These things can hardly have failed to accompany departure from purely objective mental processes, a step in evolution limited to man. Their well-worked flints and use of fire indicate a degree of control over the environment not previously approached. No longer subject to the untempered vicissitudes of life in the open, they could make themselves reasonably comfortable in their caves during severe weather. Food and protection were assured by their powerful bodies, aided by the use of weapons, and their mental development was such as to guarantee gradual improvement of the means at their disposal.

The race lived for several thousands of years, but finally became extinct. Some authorities have believed that they developed into

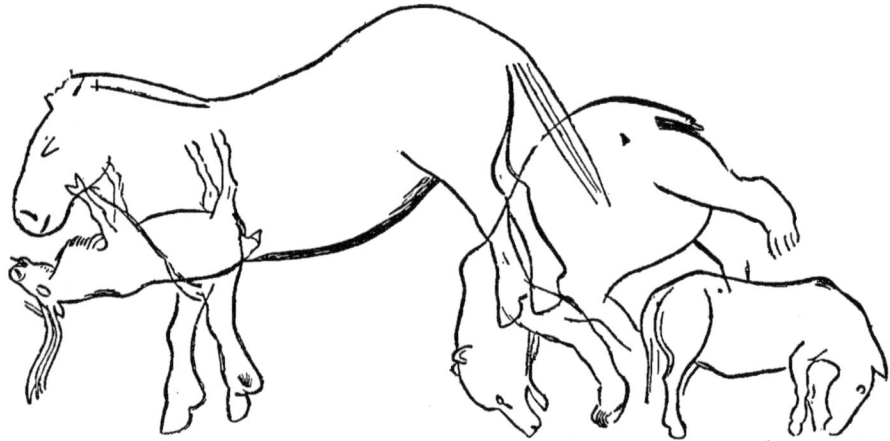

FIG. 121.—Reindeer, cave bear, and two horses, from rock engravings in the Grotte de la Mairie, Dordogne. (From *Men of the Old Stone Age* by Henry Fairfield Osborn, after Capitan and Breuil, courtesy of Charles Scribner's Sons.)

the lower races of *Homo sapiens*, but the opinion generally held is that they were exterminated through the arrival of a more highly endowed race, the Crô-Magnon.

The Crô-Magnon Race. Unlike the other fossil men, this was a race of *Homo sapiens*, physically and mentally equal to many existing peoples. It was first made known to science through the discovery of five skeletons at Crô-Magnon, France. They were tall people, males averaging over six feet and females almost five and one-half, and were fully erect. The forehead was high but narrow, the face broad, the chin well developed. The brain was large. Osborn says that the facial characters are most suggestive of Asiatic races of the present (Fig. 115C).

THE EVOLUTION OF MAN 207

The Crô-Magnon people are supposed to have produced the drawings and paintings on the walls of European caverns which are so beautifully reproduced in Osborn's book (Fig. 121). It is certain that they were mentally developed to a point which would have made this attainment in art possible. In addition there is evidence that their art included crude sculpture. Industrially they worked flint and bone, and probably developed weapons such as the spear and harpoon (Fig. 122). They were easily on a par in these respects with existing savage peoples.

What may have been the fate of the Crô-Magnon race we cannot know with certainty. Their head form is so nearly reproduced in

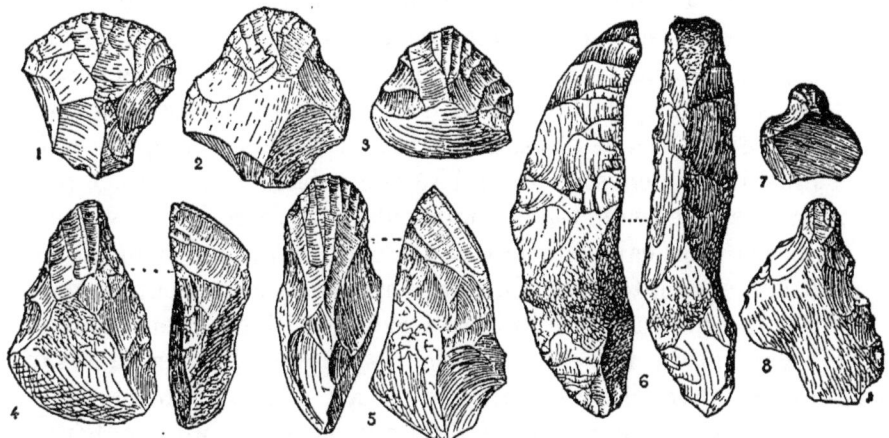

Fig. 122.—Chipped stone implements such as were made and used by the Crô-Magnon race. Numbers seven and eight are supposed to have been used for sculpture. (From *Men of the Old Stone Age* by Henry Fairfield Osborn, after Breuil, courtesy of Charles Scribner's Sons.)

the people of Dordogne that Osborn emphasizes the possibility that the ancient race gave rise to these modern inhabitants of their old land, and correlates with this the theory that the Basque language, different from all other European tongues and the most primitive of all, bears the stamp of early association with the Crô-Magnon tongue.

Recent Human Evolution. The fact that the ancient Crô-Magnon race was so highly developed anatomically places additional emphasis on the course of evolution in an intelligent species as already outlined. Although these magnificent representatives of our species lived at least twenty-five thousand years ago, they were structurally similar to ourselves. Mentally the gulf between them and modern man is equivalent to that between Neolithic cultures and modern civilization with its highly developed arts and complex industries. Crô-Magnon man had a brain apparently adequate for the gradual development of such things, but the difference is evident. Whether modern civilizations represent merely a gradual accumulation of individual experiences and discoveries, or in addition to this form of progress an actual increase in mental power it is difficult to say, but it is highly probable that the change in mental development which is so well demonstrated by primitive species of man has continued ever since. Although the ancient Greeks were intellectually our equals in so far as we can judge, we must remember that they lived only two thousand years ago, while primitive species of man occurred one hundred thousand years ago according to geological estimates. The first seventy-five thousand years of this period, leading up to the Crô-Magnon race, show evidences of physical change in our ancestors, but only those finishing touches incidental to the attainment of erect posture. In the last twenty-five thousand years there is no evidence of significant physical evolution. Such changes as have come about in man are associated with his intelligence; the development of writing and other facilities for the exchange of ideas, perfection of social organization, the control of food supply and other phases of environment have all contributed to the attainment of our modern state, and none has required any different physical equipment from that of the earliest members of our species.

In this record, fragmentary though it be, is the story of *Homo sapiens*. One hundred thousand years ago he did not exist. The mute remains of that period show us that other species did occur, half ape and half man, and that they were succeeded by still others which we can definitely call human, although much more primitive than any known race. By comparing the anatomical structures disclosed by these remains with existing arboreal primates—the great apes,—and man we find reason to believe that all came from a single source, a great arboreal primate in many ways like the apes of the present. This ancestral species probably existed in Asia, but change of climate and consequent change of flora resulted in his becoming in part terrestrial, and with his adaptation to terrestrial life he developed migratory powers which account for the appearance of fossil remains in Europe, and the later occurrence of man in all the world.

Like the more complete record of the development of the horse, man's progress has been due to changing environment, range of inherited possibilities and the interaction of the two. Unlike the other animals, his development has involved a shift from physical to mental modifications. Through his intelligence he is now able to control to some degree one of the fundamental factors of existence, environment, and on that biological foundation rests his future.

Summary. Man occupies a systematic position among the highest mammals, the Primates. He is ordinarily included in a separate family, the Hominidae. The anthropoid apes of the family Simiidae are the nearest relatives of the human species and so furnish the only indication to be found among existing species of his probable origin. Structurally man and the apes are similar but man differs in details associated with his intelligence, articulate speech, and erect posture. The arboreal habits of the apes and their structural resemblance to man suggest that he may be derived from arboreal ancestors, and a consideration of the effects of arboreal life show that this may well be true. Descent from the trees, which must have preceded the development of terrestrial man, was favored by the climatic conditions of the time when primitive man arose. Of these early creatures we have only scanty records but they are sufficient to show that species existed which were higher than the apes and lower than man, and that a gradual transition occurred leading up to the structural characters of modern man. Even the development of culture is indicated by artefacts associated with the remains of extinct species of man.

REFERENCES

HAECKEL, E., *The Evolution of Man*, 1905.
DRUMMOND, H., *The Ascent of Man*, 14th edition, 1911.
GEIKIE, J., *Antiquity of Man in Europe*, 1914.
JONES, F. W., *Arboreal Man*, 1916.
OSBORN, H. F., *Men of the Old Stone Age*, 1916.
LULL, R. S., *Organic Evolution*, 1917.
WILDER, H. H., *The Pedigree of the Human Race*, 1926.
OSBORN, H. F., "Recent Discoveries Relating to the Origin and Antiquity of Man," *Science* LXV, 481–488, 1927.
GREGORY, WM. K., "How Near is the Relationship of Man to the Chimpanzee-Gorilla Stock?" *Quarterly Review of Biology*, II, 549–560, 1927.

CHAPTER XII

ADAPTATION

In previous chapters we have considered the correlation of inheritance and environment in the determination of the organism, and various specific instances of the resulting adaptation. These things are of particular importance in the theories of evolutionary processes. If species change and give rise to other species, the results of their modification must be adaptations fitting the later generations to some definite type of environment. We can hope to understand the process of change only through extensive knowledge of changes which have already come about, since the duration of science has been too brief to afford us an actual view of evolution in progress.

Adaptations: Process and Result. The process of adaptation, for it is a process as well as a result, is visible in the lives of individuals, and is experienced by each of us. We spade the garden, and blister our hands, but if we continue such work day after day our palms form calluses which no ordinary amount of friction can blister. We train for sports, and our strength or endurance or skill increases day by day. After an athletic career in college we return gradually to the less active round of business or professional life, or suffer from a sudden change of habits. All of these things are adaptive processes.

In the individual adaptations can be observed as readily as the activities which give rise to them. They fit each being into the environment which he occupies, and whether they are mental or physical, they are no less real.

Adaptations are as conspicuous in all species, moreover, as they are in its component individuals, but the processes which brought them about are no longer evident. All animals of common experience show peculiar fitness for the lives which they lead. Squirrels have chisel-like teeth which are effective for opening nuts. Similar teeth serve the beaver for cutting down trees, but the beaver is otherwise fitted for swimming and the squirrel for climbing trees. The fitness of the horse's teeth for grazing has been mentioned, and the effectiveness of teeth and claws of carnivorous animals for seizing and tearing their prey. Such niceness of adaptation is universal. Several attempts have been made to explain its attainment; thëse we shall consider later.

Caenogenesis, Neoteny, and Paedogenesis. In some organisms metamorphosis enables the individual to occupy different habitats at different periods of its life. The adaptations of earlier stages may be wholly different from those present at maturity and are sometimes even more wonderful examples of fitness for a given environment. The development of such temporary adaptations has been called *caenogenesis*. In some cases caenogenetic modifications have apparently been of greater benefit to the species than adult adaptations and have been carried over into the adult stage. This condition is known as *neoteny*. A more extreme emphasis upon the value of caenogenetic adaptations is found in species which attain sexual maturity while still in an immature stage morphologically. This phenomenon is called *paedogenesis*.

The development of familiar insect larvae, such as the caterpillars, maggots, and hellgramites, is caenogenetic. Of these the caterpillar at least is familiar to everyone, and the complete lack of resemblance between it and the adult butterfly or moth into which it develops. Neoteny is illustrated by the axolotl, a salamander found in Mexico, which remains an aquatic form and retains its larval gills throughout life. Salamanders usually develop into terrestrial adults and this metamorphosis can be artificially induced in the axolotl. Excellent examples of paedogenesis have been reported in a few species of insects from the more primitive families of two-winged flies. Both larvae and pupae have been observed to produce young in these families.

The Environment. The environment to which organisms are adapted is complex. We recognize that every organism has an association with the surrounding world, from which it receives stimuli of a chemical and physical nature, as well as the materials of which its body is composed. To this environment it responds by more or less complex reactions. By its intricate responses and by the ultimate return of the substances which it has used during its life, every organism contributes to the complexity of the environment of every other organism. The trees shade other plants, and so modify their relations to sunlight. They also give a home to arboreal animals, and when they die, furnish food for insects, bacteria, and fungi. The green plants make other forms of life possible. Thus the organism as a whole is related to an inorganic and an organic environment, but within itself there are related parts which show that we must consider further an internal environment of any organ. Many organs, indeed, respond only to stimuli from this internal environment, and are reached directly from without only by accident, if at all. To all of these phases of environment the organ or organism must be adjusted if it is to live successfully. Lines of demarcation are not necessarily sharp for an organ may be effective in more than one way. It is possible, however, to note definite adaptations to definite conditions in all organisms.

Non-Adaptive Characters. Following the publication of Darwin's *Origin of Species* there was a marked tendency among scien-

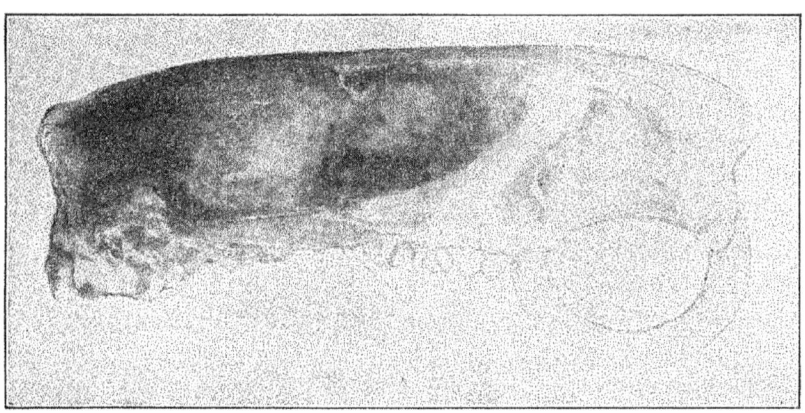

Fig. 123.—Skull of woodchuck, showing an upper incisor that had grown in an arc of a circle until it entered the roof of the mouth, after the opposing lower incisor had been broken off. (From Reese's *Economic Zoology*, with the permission of P. Blakiston's Son and Company.)

tists to seek and describe marvellous adaptations. It is not surprising that this should have occurred, since Darwin showed how wonderfully species are associated with their environments and how important the usefulness of adaptations may be, but the tendency to regard any character as adaptive to the external environment must be regarded as extreme.

We now recognize that an organism may possess many characters which cannot be construed as having any value in meeting the conditions of adaptation. Such characters have been called non-adaptive, and it is evident that organisms are made up of both this type and distinctly adaptive characters, both of which must be explained by any theory of evolution if the theory is to be adequate.

We cannot, however, avoid the belief that there is stimulus and response in all organic conditions. Non-adaptive characters therefore serve to emphasize the importance of the internal environment, in response to which parts of an organism may attain even harmful development with regard to external conditions. But even in these cases, inherent powers of development can be realized only if the proper conditions surround the part.

The chisel-like incisors of rodents are an evidence of these relationships. Ordinarily those of the upper and lower jaws are exactly opposed to each other so that they wear away equally and maintain a constant length and position. Several cases are on record of the serious effects of loss of one incisor; one of these is illustrated in Figure 123. The lower incisor in this case was broken and the opposed upper incisor, continuing its normal growth without any compensating wear, finally penetrated the brain and caused the death of the animal. This case involves definitely adaptive structures but the fate of the animal was due entirely to the power of growth inherent in its own body and to the removal of an influence normally supplied by its body. The abnormal growth of the upper incisor was in no way adaptive, but it was due to definite responses no less than truly adaptive structures.

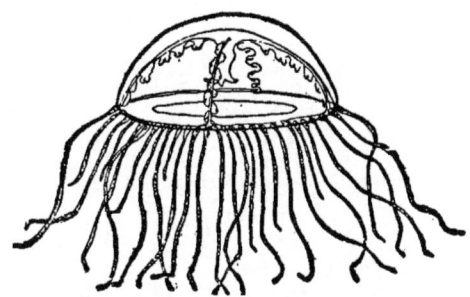

FIG. 124.—*Gonionemus*, a hydrozoan jelly-fish. (From Hegner, after Hargitt.)

Adaptation to the Physical Environment. Such adaptation is closely linked with the three major habitats, water, earth and air. An organism may be adapted to two or to all three, but in any case it shows its fitness by structural characters. The **loon**, for example, is highly developed for aquatic life. Its feet are effective for swimming, but they also enable it to move about on land as a terrestrial organism. Like most birds, it is also able to fly, and the volant adaptations of birds surpass those of all other animals.

Aquatic Adaptation. Purely aquatic organisms are safe from dessication, consequently they have no need of a moisture conserving integument. They are buoyed up by the water in which they live so effectively that they have no need of rigid structure. The result is that the most delicate organisms are aquatic. Jellyfishes are made up mostly of water (Fig. 124). Their beautiful, filmy bodies, if removed from the water, fall into a shapeless heap and quickly dry into a small organic residue.

Benthos. Some of the aquatic animals remain on a solid substratum, and are either attached to immersed objects or move from place to place over the bottom (Fig. 125). Some rigidity is

Fig. 125.—Sea anemones. (From Hegner, after Coleman.)

obviously necessary in these animals as protection, not only against their enemies, but also against the motion of the water in which they live. It is also an advantage for the sedentary forms to be able to reach out in all directions, since their food must come within reach instead of being sought. Consequently radial symmetry is a common character. Radially symmetrical animals are made up of similar parts arranged about a common center, in contrast to the more common bilaterally symmetrical forms, which are made up of similar halves flanking the longitudinal axis. Regardless of their powers of motion, these bottom forms are called the benthos. The shallow seas are rich in benthonic forms, such as the sponges, sea anemones, corals, barnacles and many others.

ADAPTATION

Plankton. In contrast to benthonic animals are those which merely float in the water, drifting about with its movements. This group is called plankton, and comprises innumerable species both of animals and plants. Single-celled organisms are very numerous. The Coelenterata are also well represented, and molluscs of some classes. The plankton includes the most extreme aquatic organisms, such as the jelly-fishes. Radial symmetry is common in this group as in the sedentary benthos, and for the same reason, lack of locomotion. Transparency is also common in the plankton.

Nekton. The transition from plankton to the third division of aquatic life, the nekton, is gradual. The essential characters of organisms belonging to the nekton are correlated with their ability to move freely through the water, resisting all motion of the medium in which they live. This necessitates a non-resistant body form, well illustrated by the spindle-shaped bodies of common

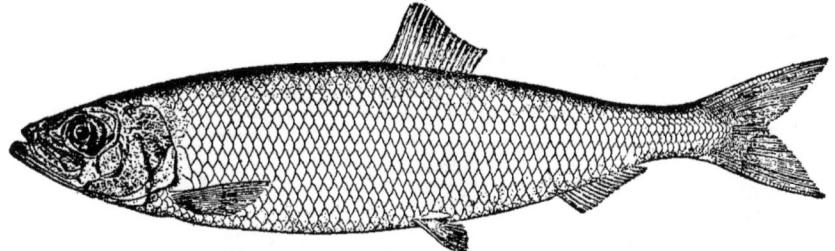

FIG. 126.—The herring, *Clupea harengus.* (From Hegner, after Jordan and Evermann.)

fishes (Fig. 126), characteristic organs of locomotion, broadened to offer the necessary resistance to a fluid medium, and bilateral symmetry, a common corollary of well-developed powers of locomotion. Such invertebrates as aquatic insects and some mollusca, and most aquatic vertebrates are included here. The fishes are excellent examples.

The Abyssal Realm. The ocean affords still other examples of adjustment to the physical environment in the peculiar fauna of the deep sea. This environment includes depths below 100 fathoms, where light does not penetrate. The pressure is great, increasing at the rate of one ton per square inch with every thousand fathoms, and the temperature is low, near the freezing point in the open ocean. Because of the absence of light plants do not grow, and animals must subsist by eating each other and the things which settle from above. It seems only natural that conditions so different from those in which we live should produce organisms which are, from our point of view, bizarre (Fig. 127). Strange

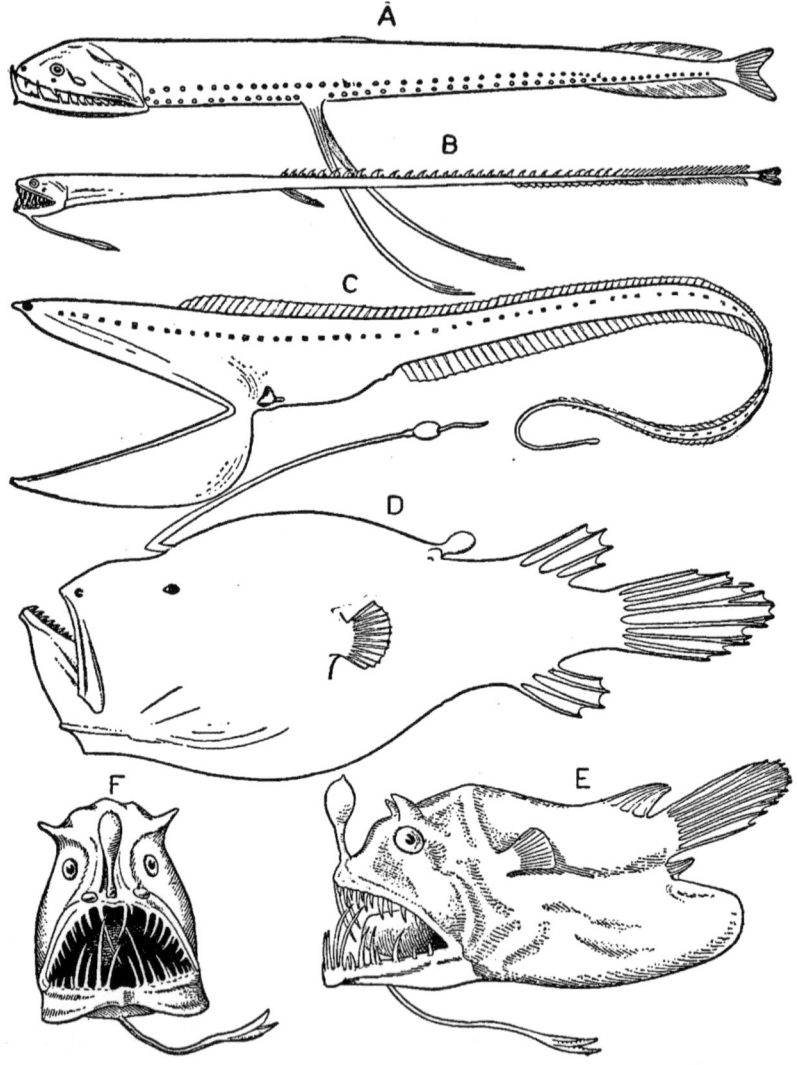

Fig. 127.—Deep-sea fishes. A, *Photostomias guernei*, length 1.5 inches, taken at a depth of 3500 feet; B, *Idiacanthus ferox*, 8 inches, 16,500 feet; C, *Gastrostomus bairdii*, 18 inches, 2300–8800 feet; D, *Cryptopsaras couesii*, 2.25 inches, 10,000 feet; E, F, *Linophryne lucifer*, 2 inches. (From Lull, after Goode and Bean.)

body forms prevail, and luminescence is common. The purpose of luminous organs seems usually to be the attraction of prey or the provision of light for their possessor, but Beebe, in *The Arcturus*

Adventure, has recently reported a striking variation in a deep-sea prawn. The cephalopods living in shallow waters conceal themselves when molested by discharging a cloud of brown secretion into the water. The prawn behaved likewise, but the darkness of the abysses could hardly be darkened, so the discharge of this animal was a luminous cloud, which in the general absence of light would conceal its movements and often obtain its escape.

Terrestrial Adaptation: Locomotion. Lull very logically applies the classification of aquatic organisms to those living in the air, so that volant forms may be looked upon as aërial nekton, and ordinary terrestrial forms as aërial benthos. Aërial plankton is very limited. Bacteria are known to float in the air, but they probably remain there only temporarily. The same is true of the spores and pollen of plants. Consequently, while there is an aërial plankton, no organisms can be said to belong to it permanently.

The differences between terrestrial and aquatic life have already been considered under the emergence of the terrestrial vertebrates. A conspicuous feature of the adaptation of terrestrial forms is the modification of supporting and locomotor organs. Since the air by which the animal is surrounded does not support it like the water, its points of contact with the rigid substratum must serve both for support and locomotion. The pentadactyl appendage (Fig. 54) in vertebrates and the jointed appendage in Arthropoda are the outcome of this need while in other forms the body lies on the ground and locomotion is accomplished by creeping, aided sometimes, as in the annelids, by setae or other projections to increase the hold of the organism on the surface which supports it.

While creeping is a very simple process, in general the same in all groups which move about in this way, the development of jointed appendages of either type paves the way for a variety of modifications. Thus we find the vertebrate limb modified for walking, running, jumping, burrowing, and climbing and the invertebrate appendage for most of these functions.

Ambulatory Adaptations. The ambulatory, or walking type obviously involves the least change of form. Such locomotion requires in addition to supporting function no further power than successive shifting of position of the several limbs in relation to the ground. Without sufficient development of muscles for this purpose, the structures could not function even as supports for the body, consequently the ambulatory condition is the most primitive stage of the terrestrial appendages.

Cursorial Adaptation. Cursorial animals must move their limbs rapidly, and for a maximum rate of speed must also be able to move by long strides. Rapidity of movement is in part a physiological adaptation, but it is aided by the structural modifications which result in lengthened stride. These include lengthening of the limb and slenderness, so that the greatest reach is attained with minimum bulk. The limb is a lever of the third order in which the point of articulation to the body is the fulcrum and the insertion of a muscle the point of application of power. By lengthening of the entire limb, particularly its distal

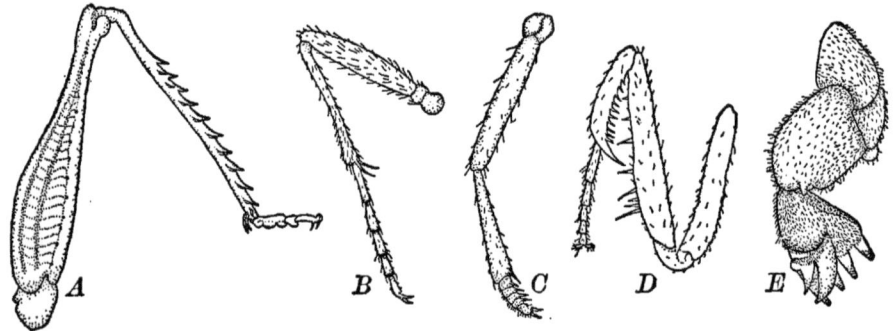

Fig. 128.—Types of insect legs. *A*, grasshopper, a jumping leg; *B*, tiger beetle, a running leg; *C*, gyrinid beetle, a swimming leg; *D*, mantis, a raptorial leg; *E*, mole cricket, a burrowing leg. (From Sanderson and Jackson's *Elementary Entomology*, with the permission of Ginn and Company.)

segments, the ratio of work arm to power arm is increased, consequently greater range of movement is acquired at the expense of power. In cursorial Arthropoda this lengthening and slenderness is general (Fig. 128A). In vertebrates it involves the distal segments of the appendages (Fig. 56), while the proximal muscles remain short and bunched, making for quick and powerful contractions.

Correlation with flight lessens the necessity for cursorial adaptation among the Arthropoda, but such insects as the tiger beetles (Cicindelidae) are able to run very rapidly for their size. Their legs are long and slender.

In the vertebrates the first step in cursorial adaptation is change of posture. The animal rises to its toes, becoming digitigrade, instead of plantigrade, and thus adds to the free length of the limb the length of the foot. Later the foot and lower limb elongate and the change of posture becomes more extreme, giving rise to the unguligrade habit, in which only the extreme tips of the digits touch the ground. With high specialization in this direction the final step, reduction of the digits, occurs. Such a change becomes possible through emancipation of the appendages from other functions than locomotion; it is an advantage in reducing the weight of the appendage with the reduction of its efficiency as a lever.

Vertebrates of the classes Reptilia, Aves and Mammalia have developed cursorial powers, but none show any higher specialization than the horse, which we have already considered in detail. Some existing lizards are cursorial, but the extinct dinosaurs attained the highest speed adaptations in this class. Their adaptations included also bipedality. The flightless birds, such as the ostriches, and some of our common birds including the quail, are excellent runners, although the power of flight, when not completely lost, is always a last resort when speed is necessary. Such adaptations as are shown by the horse are developed in lesser degrees in many mammals, consequently we are most familiar with cursorial animals in this class. The ungulates including dogs and cats are highly developed cursorial animals.

Saltatory Adaptations. Saltatory, or jumping animals, are more conspicuously different in the two groups. In the vertebrates the hind legs are always predominant appendages for propulsion, while the front legs act as supports for the anterior end of the body, serve to catch the body at the termination of a leap, and in a minor degree aid in propulsion. This predominance of the hind legs is well illustrated by their earlier specialization in fossil horses. Locomotion in cursorial vertebrates is therefore associated with saltatory power, since a gallop is essentially a succession of jumps. In the rabbits and the kangaroos and wallabies the development of the hind legs is carried to such a point that the animals may be called truly saltatory (Fig. 129). This is especially true of the kangaroos, since they are able to move by a series of jumps, without the aid of the fore-limbs.

The insects show saltatory adaptations of a very different type. The basis for such development is probably twofold. Structurally, the presence of six legs affords an opportunity for walking, even though one pair be modified to such an extent that they are practically useless for that purpose. In addition, the size of insects is so small in relation to objects about them that a single powerful leap is often a guarantee of safety, since it may carry them readily into concealment. The suddenness of this type of locomotion is its most valuable feature.

The structural modification of a saltatory insect leg includes great enlargement of the femur, which contains large extensor

Fig. 129.—The Rock Wallaby *Petrogale xanthopus*. (From Parker and Haswell, after Vogt and Specht.)

muscles (Fig. 128B). Elongation is common, but not essential, for among the beetles and the fleas are included very powerful jumpers with hind legs as short as is normally the case in other insects. The Orthoptera include more familiar jumping insects, however, and in this order the long hind legs of the grasshoppers and crickets are familiar to everyone. In these legs another specialization is evident. The foot is not a dependable support, but heavy spines are developed at the tip of the tibia, analogous

ADAPTATION

to the lower limb of a vertebrate, and these guarantee a non-skid take-off from any surface providing the slightest of holds. For obvious reasons only the hind legs of jumping insects are specialized.

Fossorial Adaptations. In contrast to cursorial and saltatory adaptations, fossorial species have the front legs most highly developed, since they must open a way for the body through the earth. Among the invertebrates without appendages burrowing is accomplished by the simple means of forcing the slender, tapering body through relatively loose soil, or by passing earth through the alimentary tract as the animal progresses. Arthropoda, however, make use of the mouth parts and of the legs in digging. An extreme specialization of this type is seen in the mole-cricket,

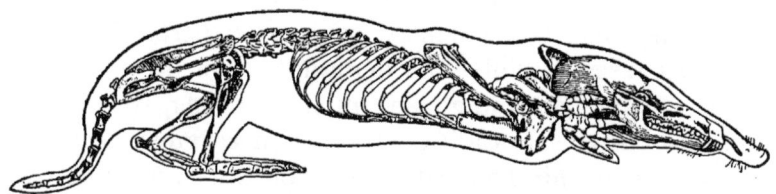

Fig. 130.—Common mole, *Talpa europaea*, showing skeleton and outline of body. (From Lull, after Pander and D'Alton.)

whose front legs are strong, notched, shovel-like appendages with which the insect digs rapidly and effectively (Fig. 128C). These strange little creatures are also covered with moisture-resisting down.

The mole is the most highly specialized fossorial vertebrate since it lives entirely underground (Fig. 130). Its pectoral girdle and fore-limbs are massive, and the forefeet are very broad and provided with strong claws. In addition to this elaborate mechanism it has an elongate pointed snout which aids it in forcing its way rapidly through soft earth. Its progress may be a combination of digging and spreading of the earth before it. The mole, like its namesake, the mole-cricket, has extremely fine vestiture, resistant to moisture. The eyes are vestigial in the mole, and the external ears are lacking. Either organ would be liable to injury in burrowing, and neither could be of use to an animal whose life is spent in darkness surrounded by solid earth.

Many other vertebrates representing all classes above the fishes are fossorial, though they do not remain altogether underground.

Such animals, as might be expected, show adaptations similar to those of the mole but less extreme. The fore-limbs and claws are well developed, but not extremely. The eyes are, of course, useful above ground, and remain functional, but they are reduced in proportion to the amount of time spent in burrows. The same is true of the ears. Tapering of the body may occur, but is no more extreme than in many animals which do not burrow.

In addition to the animals which form burrows for concealment there are some that dig for food. The elephant uses his tusks for this purpose. The snout of hogs and of some snakes is turned up at the tip, forming an effective organ for burrowing to slight depths in soft ground.

Scansorial Adaptations. Such adaptations are chiefly related to the organic environment, but since they are for the purpose of locomotion they may conveniently be treated here. Whether an animal climbs trees or cliffs, the demands upon its body are the same. The appendages and girdles must be strong, to support the weight of the body. In addition to this there must be provided some means of maintaining a safe grip upon the supporting object. Finally the proximal segments of the limbs are seen to be elongated in some arboreal vertebrates. This reversal of the condition noted in cursorial forms is due, no doubt, to the fact that intrinsic strength is necessary in the distal parts which are in immediate contact with the support, so that reach must be gained elsewhere if at all.

Appendages are modified in several ways to enable animals to cling to branches. The claws are usually involved, but in some cases the appendages themselves are prehensile. Animals which climb by clinging to the bark of tree trunks, and run along the upper surface of branches have sharp claws which give them an adequate grip on the surfaces which support them. Squirrels are perhaps the most familiar example, but many birds and some reptiles are similar in habits. The sloths normally suspend themselves from branches, and are provided with great hook-like claws as an aid to this habit (Fig. 131). The development of the appendages is so extreme that they are scarcely able to move about on the ground. Prehensile appendages are best developed in the primates, where opposability of the thumb and locomotion of the type known as brachiation are found. Adaptations of the last kind have already been discussed under the evolution of man.

Adhesive organs are found in insects and in such vertebrates as the tree frogs. Adhesion may be accomplished by the secretion of fluids or by the vacuum-cup principle, and is especially effective for climbing on smooth surfaces. The ability of a fly to walk up a window pane or across the ceiling is due to such organs.

Some animals use other organs than the limbs for climbing. The true chameleon of Africa and some monkeys have prehensile

FIG. 131.—The two-toed sloth, *Choloepus didactylus*. (From Parker and Haswell, after Vogt and Specht.)

tails which are used to grasp branches, and parrots use their powerful beaks for the same purpose.

Adaptations to Light. In addition to adaptations of the appendages terrestrial animals are adapted to conditions of light and moisture, although to a lesser degree than the green plants. Since sunlight plays so large a part in the metabolism of the latter organisms, they are nicely adjusted to it. Shade-loving species are of more delicate texture, for various reasons, and have broader leaves than those which live in the open. Violets afford a familiar illustration; most species have entire leaves, but in those which **are** found **on** the dry, brightly lighted prairies the leaves are finely divided. Animal adaptations concern visual functions chiefly, and in animals which have no eyes, the sensitiveness of the skin to light rays. Burrowing animals may retain light sensitiveness, as is true of the earthworm, but vertebrates like the mole tend to lose their eyes. Such elaborate organs are obviously of no use in darkness, and consequently they disappear or lose their functions in cave-inhabiting animals as well as fossorial species. Salamanders and insects without eyes have been recorded from caves, and some blind fishes from subterranean waters (Fig. 132).

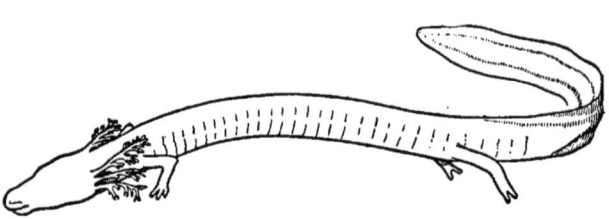

FIG. 132.—Blind salamander, *Proteus anguinus*, from underground waters. A European species. (From Lull, after Gadow.)

Another character which is directly correlated with absence of light is reduction or loss of pigmentation. The presence of pigment in the skin acts as a protection against the rays of the sun, as everyone has experienced through sunburn followed by tanning. It is also of value in the development of color and pattern for interrelations with other animals. Where light does not penetrate, obviously neither use exists.

Adaptation to Aridity. *Animals.* Water is essential to all life, and is almost universally abundant. Only in desert regions is it scanty, and animal adaptations to meet this lack are invariably correlated with adaptation to other conditions of desert life. Adaptations for the conservation of water in animals are of three types: (1) storage reservoirs, such as are found in the stomach of the camel; these have already been mentioned; (2) lack of the power to perspire; (3) ability to absorb water through the skin.

Plants. Plants require water as well as sunlight for photosynthesis, and since they cannot move about in search of it, but must depend upon rainfall, their adaptation to lack of moisture is not limited to desert species. If one month of the plant's life must be passed without rain, it must be able to withstand this lack of moisture or perish, no matter how great the abundance of moisture at other times. Plants so adapted are known as xerophytes. They meet conditions of dryness in several ways, viz., (1) by extensive root systems, which draw moisture from a large volume of ground; (2) by reduction of leaf surface through small size or subdivision of leaves, or in extreme cases by complete loss of leaves and shifting of their functions to the branches; (3) by the development of thick cuticle; and (4) by scaly, hairy and spiny surfaces (Fig. 133). The last three serve to prevent rapid transpiration, and so conserve the moisture which the plant contains. Desert plants, such as the cacti, are extreme adaptations. They are without leaves and are provided with a moisture conserving cuticle. Some contain large quantities of water. A desert fern, *Notholaena*, has wiry stipes and waxy fronds, densely scaly below. More common illustrations are found on the western prairies, where harsh grasses and other plants with harsh, hairy and finely divided leaves are common. The roots of many of these plants penetrate far into the ground, although their branches may extend less than a yard above the surface.

Flight. Volant animals, since the air is too light a medium to buoy them up, must combine their flight adaptations with others. They may be entirely aquatic and still possess some slight power of flight, but ordinarily they are in some degree terrestrial. Thus a combination of terrestrial, aquatic and volant adaptations in one individual is common, and no animals are exclusively volant.

Structure. The fundamental requirements for flight are correlated with the fluid quality and lightness of the air. They include lightness of structure, broad planes for support, steering, maintenance of equilibrium and propulsion, great muscular power and power of endurance. Newman's treatment of "The Bird as an Automatic Aëroplane" is a graphic account of these adaptations, since the birds are the most highly developed flying animals.

Lightness is secured by the development of hollow bones. Rigidity and strength are maintained, in spite of the relatively fragile bone structure, by their form. Many of the bones are formed like the T and I beams used in structural steel work. The sternum, which bears the great stresses of flight, is an especially fine example of T beam. Fusion and overlapping of bones also add to rigidity of the bird skeleton, while the presence of air at the relatively high body temperature adds to the buoyancy of the entire animal. The feathers form the lightest broad supporting surface known in the organic world, and enclose air about the body; "nearly half of the contour volume of a bird is air-filled" (Newman).

Organs for supporting, steering, propelling and balancing are

FIG. 133.—A group of Xerophytes. The cactus is *Cereus giganteus;* absence of leaves and the development of spines are conspicuous. At the right are an agave with broad fleshy leaves and a yucca with harsh, narrow leaves; both have moisture-conserving epidermis. The small plants are *Mesembryanthemum* and *Sedum*, both of which have fleshy leaves for the storage of water and some protective covering, such as hairs or thick epidermis, to aid in its conservation. (From Campbell, after a photograph by Dr. F. M. MacFarland.)

ADAPTATION

formed of feathers in birds. Both tail and wings are included, the latter alone as propelling organs. In other flying animals, such as the bats, similar organs are formed of folds of skin extending from limb to limb along the sides of the body or stretched between elongated bones (Fig. 134). The fins of flying fishes express the same adaptive tendency as fins in general, but carried to an extreme correlated with the lightness of the air (Fig. 135). Insect wings are entirely different; since the exoskeleton provides rigidity

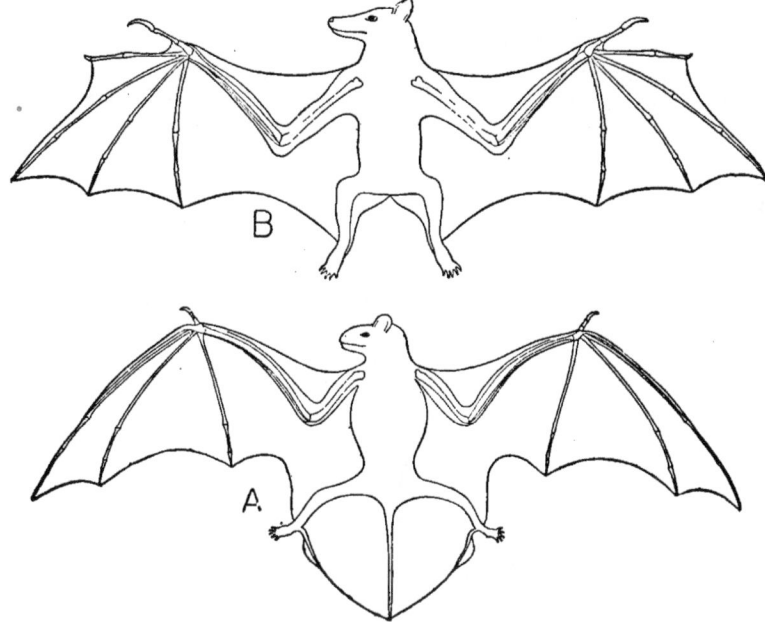

FIG. 134.—Bats. A, *Vespertilio noctula*, an insectivorous species; B, *Pteropus* sp., frugivorous. (From Lull.)

throughout the body, the wings are simple broad flat evaginations of the body wall, stiffened by local thickenings, the veins, between which the cuticula is extremely thin and light (Fig. 136).

Power and endurance are closely correlated. Enlargement of the muscles that move the organs of flight is necessary, and is extreme in birds. The large pectoral muscles, attached along the keel of the sternum, make up the masses of breast meat with which everyone is familiar. The rhythmic contraction of these muscles during long flights demands an abundant supply of food and oxygen, and rapid removal of wastes. These needs are met by modifications of the alimentary tract, providing for the storage of a supply of undigested food in the crop, and for rapid digestion. The respiratory system is so modified that air enters the alveoli of the lungs through one system of tubules and passes out through another, so that fresh air passes constantly over the respiratory epithelium. These things result in rapid metabolism, which is correlated with high body temperature. Many birds maintain a

Fig. 135.—Flying fish, *Dactylopterus volitans*. (From Lull.)

temperature of 105° F., and the best fliers reach 110° to 112°. High temperature is of value not only in promoting rapid metabolism but also as a protection against the low temperature of the upper air.

The extent to which animals are adapted for flight varies greatly however. Some forms are able to move through the air to a limited degree only, by gliding, while others are capable of true flight.

Gliding is a self-explanatory term. If an animal is provided with sufficiently extensive membranes, it is buoyed up by the air pressure induced by gravitation, and is able to coast from higher to lower levels on the air. The same principle enables birds to soar on air currents without moving the wings, but this demands a nice adjustment of equilibrium of which less highly adapted organ-

ADAPTATION 229

isms are incapable. Gliding animals include the remarkable flying dragon, a reptile of the Indo-Malayan region (Fig. 137), the flying squirrel (Fig. 138), and a tree frog, *Rhacophorus*. In the flying dragon the supporting planes are developed as membranes on the sides of the body, covering elongated ribs. The flying frog has broad webbed feet which serve as gliding planes, and the flying

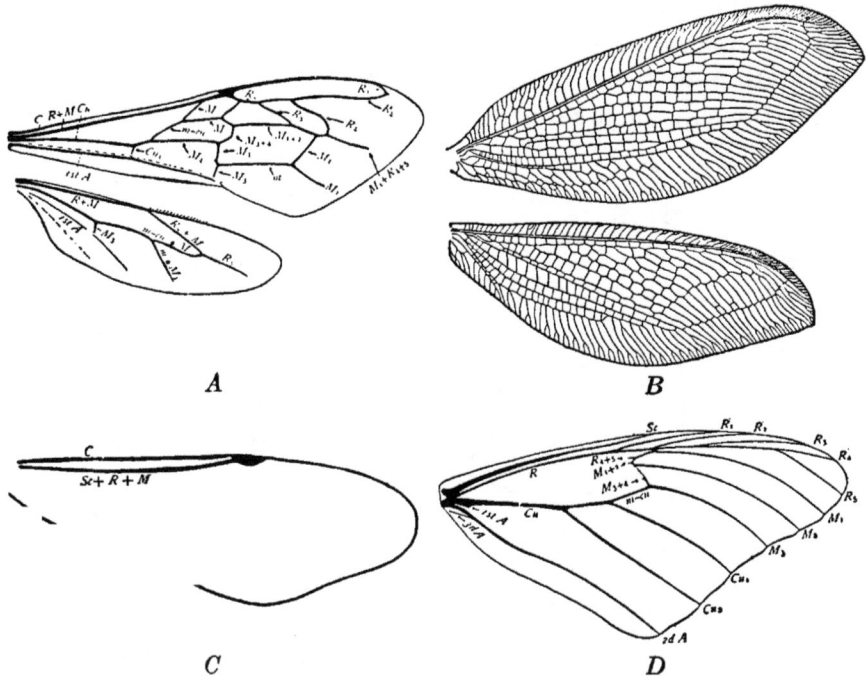

Fig. 136.—Wings of insects, showing the supporting veins. *A*, honey-bee; *B*, *Osmylus*; *C*, *Evaniellus*; *D*, fore-wing of *Anosia*, a butterfly. (From Constock's *Wings of Insects* with the permission of the Comstock Publishing Company.)

squirrel has folds of skin extending along the sides of the body from limb to limb.

True flight of a very limited degree is possibly found in the flying fishes. These animals have greatly enlarged pectoral fins, which are flapped during their long jumps through the air. However, some biologists construe their progress as gliding.

The power of sustained flight has been developed independently in three classes of vertebrates. Among extinct forms the reptiles are represented by the pterodactyls. The birds as a whole are flying animals or derived from flying ancestors. The mammals are represented by *Galeopithecus*, the "flying lemur," which is intermediate between the insectivores and bats, and by the entire order Chiroptera, the bats.

The Pterosaurs had wings formed of skin folds stretched between the hind-limb and one greatly elongated digit of each fore-limb (Fig. 139). In *Galeopithecus* the membrane stretches from limb to limb, and to the tail, and in the bats a similar condition prevails, but the elongation of several digits of the fore-limb extends the membrane into a wing-like form.

Adaptation to the Organic Environment. *The Web of Life.* The relationship of organisms with each other is infinitely complex. All organisms depend directly or indirectly upon photosynthesis for food. The green plants carry on the process for themselves, and some animals feed upon the green plants, but other animals secure their food by eating animals. Some organisms depend upon others for protection or concealment, for tempering the intensity of light, for shelter from the weather and for many other things. The resulting complexity has given rise to the conception of the web of life, in which organic associations are likened to a woven fabric. The disturbing of one thread affects the relationships of others. The more intimate the association with the source of disturbance, the greater the change, but step by step the effect may be transmitted throughout the whole fabric.

FIG. 137.—The flying dragon, *Draco volans*, a lizard. (From Lull.)

Darwin called attention to a striking illustration of such interaction. He noted that red clover was fertilized by bumble bees, and that protected flower heads did not produce seed. Since field mice destroy the nests of bumble bees, they are a check on the production of red clover, and since cats destroy mice they are beneficial to its production.

Modern economic biology is full of such illustrations. We spray plants to get rid of aphids, but it is often necessary to destroy a

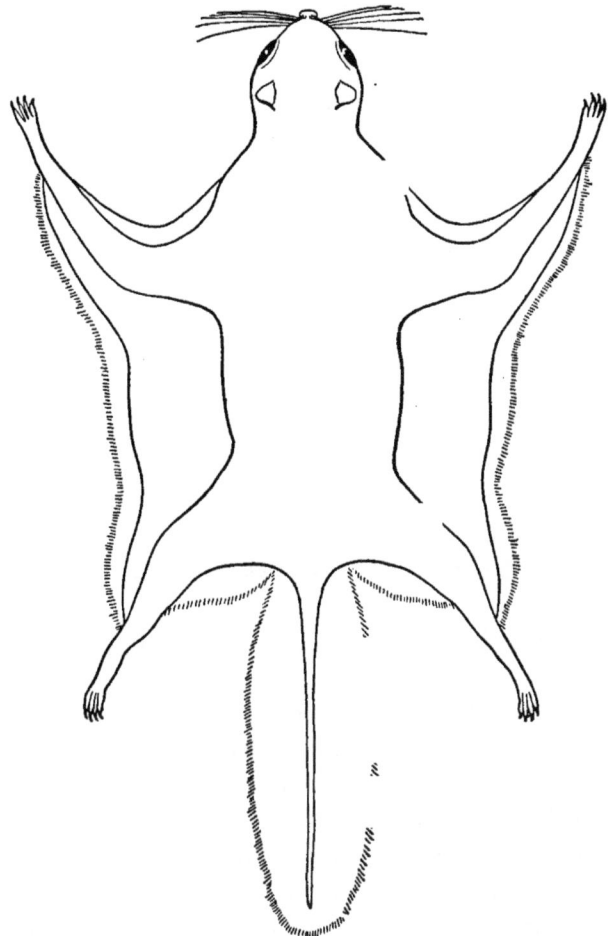

Fig. 138.—Flying squirrel, *Sciuropterus volucella*. (From Lull.)

neighboring ants' nest before a cure can be accomplished, for the ants may bring other aphids to replace those killed by spraying. We also poison, trap and swat the fly, but if horse manure is piled in the neighborhood, we can destroy more flies with less effort by having it removed. We raise fruit, and many orchardists have found it to their advantage to rent bees to be placed in their orchards during blossom time, in order that cross fertilization may be more certainly accomplished. Pages could be filled with a mere listing of these known associations.

The Nature of Organic Relations. A primary reason for the association of organisms is the securing of food, but since this involves the destruction or injury of other organisms except in the case of the green plants, there must always be adaptations for the protection of those subject to attack. Finally there are associations of convenience, wherein animals of the same or different

FIG. 139.—Pterodactyl, *Rhamphorhynchus phyllurus*. (From Lull.)

species find it possible to meet the requirements of life better through the assistance of others.

Food-Securing. *The Structures Involved.* In securing food the adaptation of the mouth is necessarily important. We have already noted a striking example in the evolution of the teeth in elephants and horses, and another in the heterodont dentition of the primates in connection with omnivorous habits. The lips of animals also show adaptation. Grazing species have a prehensile upper lip which is useful in gathering tufts of grass into the mouth, while rabbits nibble at leaves and fruits without the necessity for securing a quantity of pieces at one mouthful. Whatever may be the food, the teeth at least are formed so that it may be effectively chewed. The appendages are often modified as accessory structures.

Anteaters. Such animals as the anteaters are extreme adaptations to a limited diet. The great ant-bear has strong, hooked claws with which it tears open the nests of ants (Fig. 140). Its snout is slender and elongate, its tongue long and sticky, and it is entirely toothless. The tongue serves to gather the multitude of tiny insects which are necessary to nourish its seven-foot body, and since ants are so small and soft, teeth are quite unnecessary. The claws are incidentally effective weapons.

Carnivorous Animals. Such animals, while they have specialized teeth, are no less dependent for food upon adaptations of the appendages. The ability to stalk prey silently, to pursue it rapidly, and to spring upon it quickly must be brought into play before the teeth are necessary to hold and kill, and the shearing

Fig. 140.—Ant-bear, *Myrmecophaga jubata.* (Through the courtesy of the New York Zoological Society.)

molars to cut through the relatively tough flesh as the prey is eaten.

Insects. In all forms of animals such adaptations are found. Insects have mandibles whose strength is in proportion to the harshness of their food. Species which live on fluids have the primitive mandibulate mouth highly modified to form suctorial structures, and in some cases piercing structures to enable them to reach their food (Figs. 141 and 78). Both types of mouth are found in both phytophagous and carnivorous insects. In the latter some type of powerful grasping leg is also present.

Birds. Birds likewise have powerful hooked beaks and strong claws if carnivorous and hooked beaks but weak claws if they eat carrion. Seed eaters have thick, powerful beaks, but in insectivorous species the beak is more slender (Fig. 142).

Protective Adaptation. *Reproduction.* A simple protection for species which are preyed upon by others is the production of enough offspring to maintain the species in spite of its constant loss. Among many species this adaptation of reproduction takes the place of more active defenses. The oyster, for example, lays approximately 16,000,000 eggs each year, and many fishes are said to lay millions. In other animals this enormous rate of reproduction is supplanted by the production of less young, with paren-

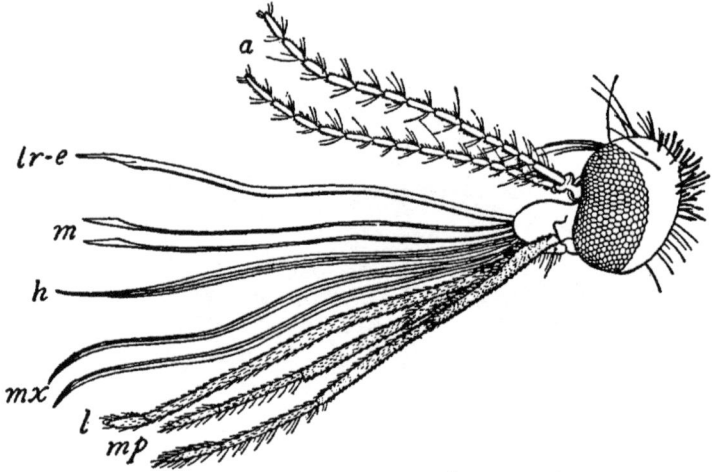

Fig. 141.—The head and mouthparts of a mosquito, *Anopheles* sp. a, antennae; lr-e, labrum-epipharynx; h, hypopharynx; m, mandibles; mx, maxillae; l, labium; mp, maxillary palpi. (After Nuttall and Shipley, from Comstock's *Introduction to Entomology*, with the permission of the Comstock Publishing Company.)

tal care to aid them in reaching maturity. Such adaptations often involve structural modifications of the organism. They may result in viviparity, i.e., the production of living young, so that the inert egg and tender embryo are not subjected to the vicissitudes of independent existence. In extreme degrees viviparous species bear young when sufficiently developed to care for themselves to a marked degree, while in others little more than the embryonic period is eliminated from the independent existence of the individual. Mammals are the best known viviparous animals, and in this group parental care is also highly developed. Normally oviparous groups, such as the fishes and insects, may also include viviparous species. Viviparity is not essential to parental care, however; the birds are oviparous, but their care of their young is a matter of common knowledge. In all cases, of course, the rate of reproduction must be great enough to offset destruction or the species must ultimately become extinct.

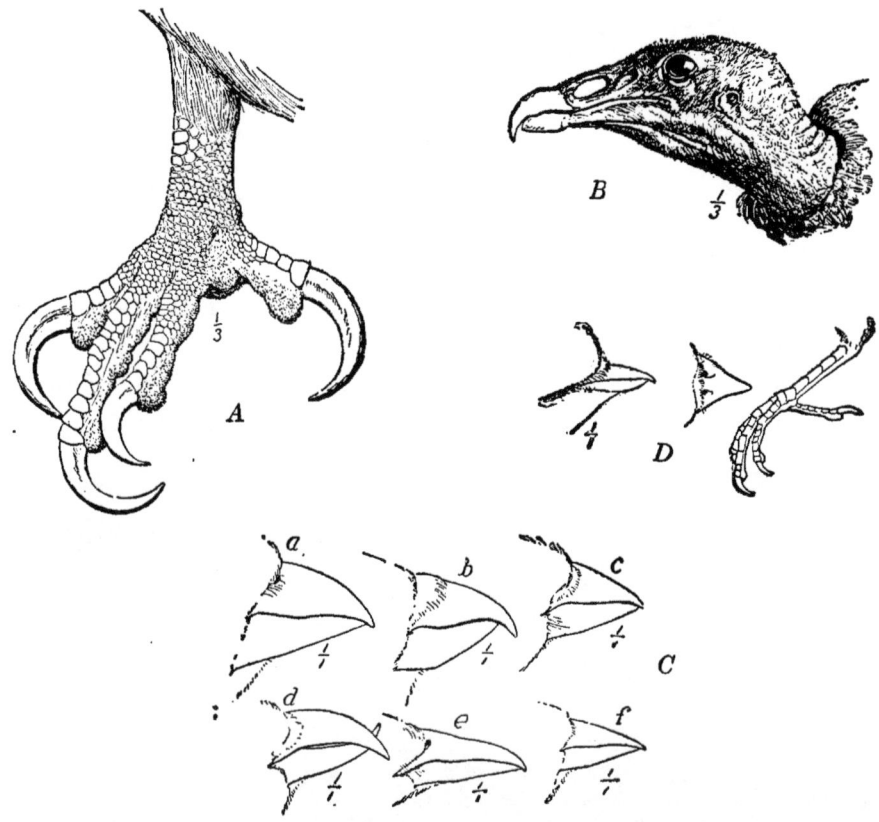

Fig. 142.—Beaks and feet of birds. *A*, foot of bald eagle; *B*, head of turkey vulture, a carrion eater; *C*, beaks of seed-eating birds; *D*, beak and foot of swallow, an insect eater. (From Chapman's *Handbook of Birds of Eastern North America*, with the permission of D. Appleton and Company.)

During the independent life of the individual it escapes various dangers by running away, by resisting them through combat, or by various concealing or repellent means. Some animals are armored, and are therefore almost immune from attack by others, and still others are protected by such peculiar powers as **autotomy**.

Armor. Armor was highly developed in some of the extinct reptiles, such as *Stegosaurus* and *Triceratops* (Fig. 143). These animals were sluggish creatures, and were well protected by their bony plates in such encounters as they must have had. Such

FIG. 143.—Restoration of the dinosaur, *Triceratops*. Length 20-25 feet. Upper Cretaceous of western North America. (From Lull.)

modern species as the alligators are no mean illustrations of protective armor, but the armadillo is among the best (Fig. 144).

FIG. 144.—Nine-banded armadillo, *Tatu novemcinctus*. (Through the courtesy of the New York Zoological Society.)

These little mammals are so completely covered with bony dermal plates that when they roll into a ball little else is exposed.

Speed. This quality as an adaptation is too well known to need discussion. It is conspicuous in such animals as the horse and deer as a primary protective adaptation.

Defensive Weapons. These may be the same as aggressive in vertebrates. The teeth and claws of many herbivorous animals are excellent weapons. They also include structures in no way correlated with securing food. The tail of *Stegosaurus*, for example, with its enormous bony spines, must have been a terrible weapon, and the tails of modern Crocodilia are effective clubs. Horns of ungulates should be mentioned here, although they are also organs of aggression in individual combat within the species. The stings of insects are among the most highly developed defensive weapons (Fig. 145).

Glandular Secretions. Some animals are provided with glands which secrete poisonous substances or repulsive scents. Many insects have such glands. The stink-bugs derive their popular name from their unpleasant secretions, and many other true bugs secrete more or less unpleasant substances. The larvae of the common swallow-tail butterflies have an eversible scent organ just behind the head which has been shown to be repulsive to birds. The scent produced by the skunks requires no detailed discussion, nor does the venom of poisonous snakes.

Electrical Organs. A few animals are able to defend themselves by discharging electricity from special organs. The electric rays and eels are capable of giving severe shocks.

Concealing Discharges. Such discharges have already been mentioned under adaptations to deep-sea life. The discharges of Cephalopoda and of the deep-sea prawn are the only familiar examples of this adaptation.

Autotomy. One of the most striking of all defensive adaptations is autotomy. In some lizards the tail is brightly colored and so is most likely to be the part seized by a carnivorous species, but if grasped it breaks away from the body and the lizard is able to scamper away to raise a new appendage. Crabs are said to drop off claws if seized by them. Such powers are usually correlated with ability to regenerate the lost appendages. The term autotomy is sometimes a misnomer in this connection since the loss of parts may apparently be due more to fragility of structure than to the animal's power to drop appendages at will, although the latter is sometimes present. The term is also, and more accurately, applied to the process of fission.

Animal Associations. *Gregariousness.* Many animals obtain by association with others of the same species or with different

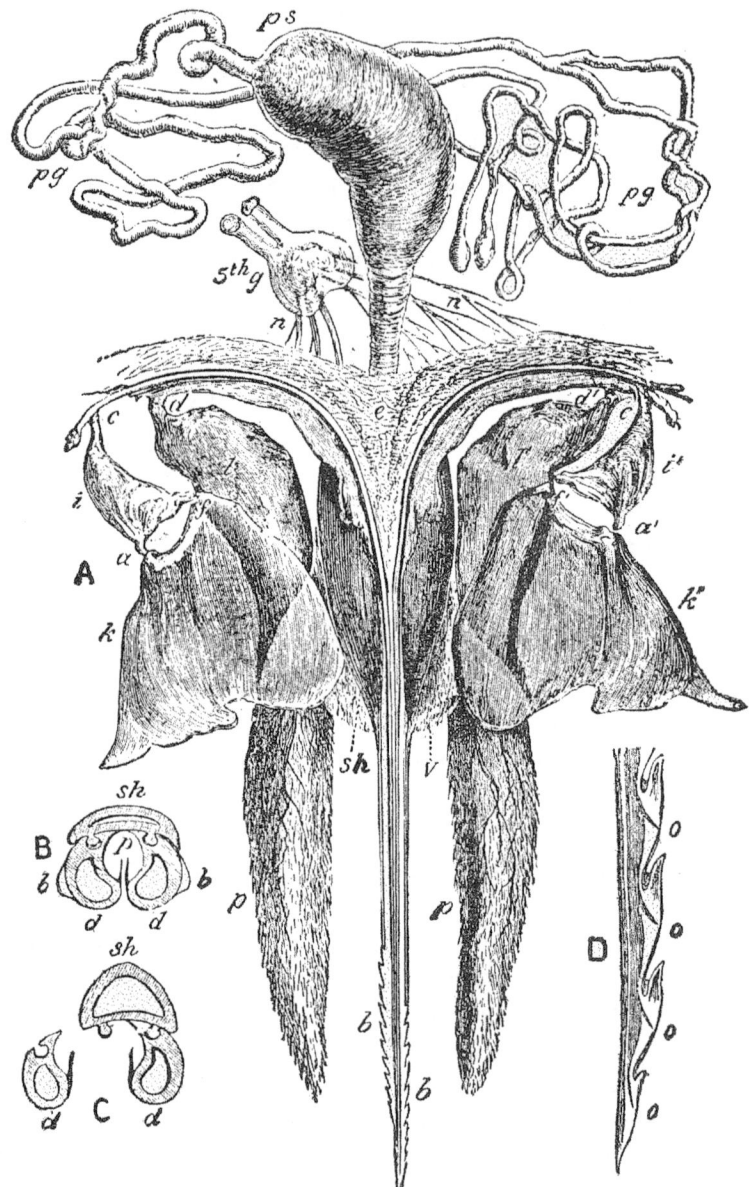

Fig. 145.—Sting of honey-bee, x30. A, sting separated from its muscles; *ps*, poison sac; *pg*, poison gland; *5th g*, fifth abdominal ganglion; *n, n,* nerves; *e*, external thin membrane joining sting to last abdominal segment; *i, k, l* and *i′, k′, l′*, levers to move the darts; *sh*, sheath; *v*, vulva; *p*, sting palpus or feeler, with tactile hairs and nerves. B and C, sections through the darts and sheath, x300; *sh*, sheath; *d*, darts; *b*, barbs; *p*, poison channel. D, end of a dart, x200; *o, o*, openings for escape of poison. (From Packard, after Cheshire.)

species the benefits which free-living individuals must secure for themselves. Association within the species may be limited to the mere banding together of individuals as in herds of grazing animals; such association does not reduce the potential independence of the individual but gives him the protection of numbers. Carnivorous species may run in packs, like wolves. Lull characterizes this arrangement as "a sort of armed truce, the idea of mutual aid for defense evidently being foreign to their code of ethics, for they will at once turn upon, destroy, and devour one of their own band who happens to be wounded, even though it delay the chase." Crows flock together apparently for no other end than companionship, although they probably derive some measure of benefit in the possibility of more certain warning of danger.

Communal Associations. Such associations are accompanied by a division of labor and in most animals by some structural specialization of the different castes. Among the insects the termites (Order Isoptera) and ants, bees and wasps (Order Hymenoptera) include many highly developed communal forms. The condition found in the honey-bee colony is a relatively simple illustration of communal association, since the number of forms involved is only three (Fig. 5), many less than in the ants and termites, although the division of labor is intricate and no individual can long exist independently.

The honey-bee colony is made up of a single queen, many workers, and during the breeding season some drones. The queen is a perfect female. She alone mates and lays eggs under normal conditions, and she must even be fed by the workers. The worker caste in a strong colony of Italian bees may include 100,000 or more individuals, although a wild colony is likely to be much smaller. During the winter under natural conditions there are less workers present. These insects are females, but they are imperfectly developed and never mate. They are capable of laying eggs under abnormal conditions, but these eggs can develop only into drones, which always come from unfertilized eggs. The drones are males, and are tolerated only during the breeding season. In the fall they are killed by the workers or thrust out of the colony to die. There is definite structural difference between each of the three forms and in addition a division of labor not wholly correlated with structural differences. Workers, when newly emerged from the pupa, are idle for a few days. They then care for the young brood and build comb for a longer period, and finally fly out to gather nectar and pollen for food, and propolis to stop up crevices in the hive. When there are no young workers available, as in the spring, the older ones take up the work of the hive in addition to their own duties afield.

Communal life is nowhere more complex than in the human species, but here it depends on differences of training, and not on structural adaptation. This is made possible by the high development of intelligence.

Commensalism and Symbiosis. Different species are sometimes associated for mutual benefit. If the association is not indispensable, it is known as commensalism. A classic example is that of the hermit crab and the hydroid which grows on its shell. The crab is more or less protected by the disguise afforded by the hydroid, and probably by the latter's stinging cells, while the hydroid is carried from place to place and is able to benefit by the scraps of food discarded by the crab (Fig. 146). If the association is indispensable, it is known as symbiosis. A familiar example is the association of fungi and algae to form lichens. Among animals the white ants and their intestinal fauna of protozoa have recently been proved equally necessary to each other. The termites eat wood, but cannot digest the cellulose of which it is composed; the protozoa digest the wood for the termites, and in turn receive a favorable environment and plenty of food.

Parasitism. Not all associations are mutually beneficial, however. Some organisms live upon or within others, and entirely at their expense. Such organisms are called parasites, and those upon which they live are known as hosts. Parasitism is extremely varied and complex. It cannot be treated adequately in such an account as this, but in all cases certain tendencies are to be noted. In proportion to the dependence of the parasite, it degenerates structurally and becomes incapable of independent life. Obligate parasites are thus our most degenerate organisms. In proportion to the risk involved in changing from host to host during reproduction or during the life of the individual, compensating adaptations must be acquired. These are usually reproductive and are either in the form of a very high rate of increase to meet enormous destruction, or the elimination of the more vulnerable stages. In proportion to the constancy of the association, the powers of locomotion may be reduced.

Internal parasites include many protozoa, flatworms and roundworms, crustacea and insecta. Such forms are, of course, the most extremely modified. External parasites are chiefly arachnida (mites and ticks), and insects. The latter are often adapted in texture, vestiture and structure to aid locomotion on the body

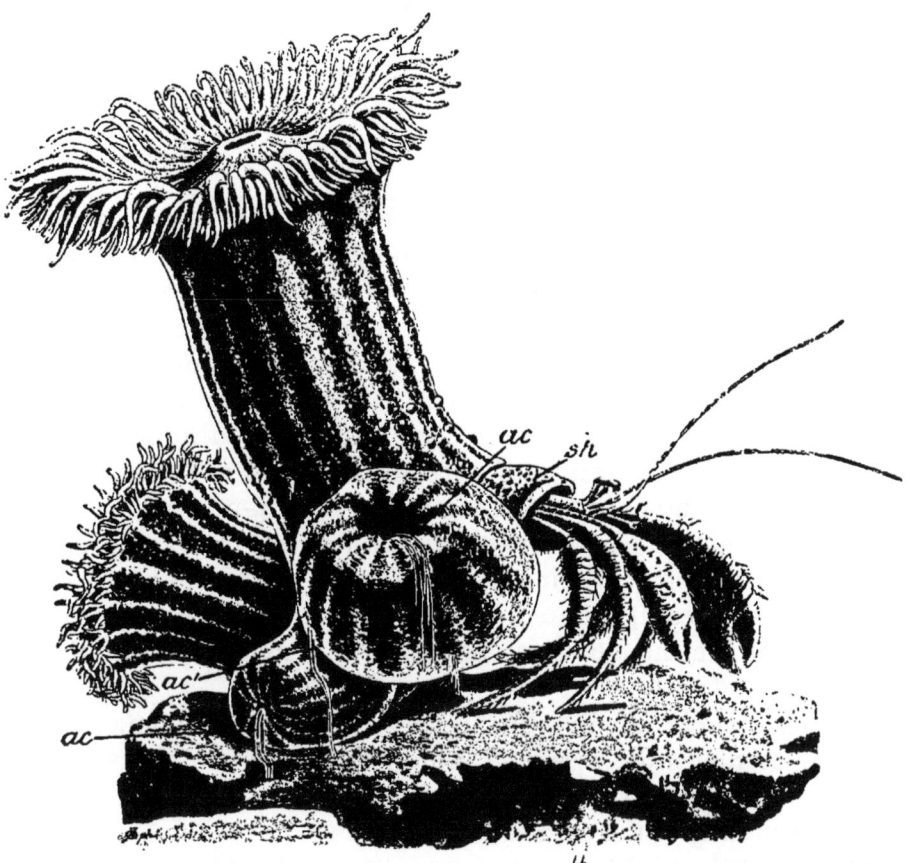

Fig. 146.—A hermit crab occupying the shell of a Gasteropod mollusc which is almost concealed by four sea anemones. ac, ac' are acontia of the anemones; sh, shell of Gasteropod. (From Parker and Haswell, after Andres.)

of the host and to resist capture, but they retain all of the structures of independent organisms and must be looked upon as highly specialized rather than as degenerate creatures.

Coloration and Mimicry. One of the most remarkable phases of adaptation to the organic environment is coloration and mimicry. The complexity of such adaptations is well indicated by the fact that they can be classified under the following diverse heads:

1. Resemblance to surroundings, involving both color and form, for the concealment of both hunter and hunted. This has been called protective and aggressive coloration.
2. Alluring colors to attract prey.
3. Warning colors indicating noxious qualities.
4. Signal and recognition marks.
5. Confusing colors.
6. Sexual colors.
7. Mimicry of other organisms.

Protective and Aggressive Colors. The efficacy of these adaptations has been questioned, and we may well imagine that animals often perish in spite of them. However we cannot doubt that the white winter coat of a prairie hare is more of a protection when the ground is covered with snow than its brown summer coat would be. Nor can we doubt that the similar resemblance of the arctic fox to its environment enables it to approach its prey with greater certainty of success than if it were red. The writer and a friend once took a picture of a whip-poor-will resting on dead leaves, and could never decide which object in the picture represented the bird! As in other cases, the effectiveness of these adaptations, even if limited, is sufficient to account for their continuation, although not necessarily responsible for their origin.

Alluring Colors. Such colors involve a degree of mimicry, in that the animal must resemble something attractive to its prey. In approaching the supposedly desirable object, the prey comes within the reach of the aggressor.

Warning Colors. Many poisonous or unpalatable animals are brightly colored, and are usually avoided by other animals. Among these are the Gila Monster of southwestern deserts, the red-banded coral snake, and the brightly colored or conspicuously marked insects such as the wasps and bees and many unpalatable species. Marshall records definite evidence that a few trials of such insects result in their refusal by birds and monkeys. There is, of course, nothing intrinsically repellent about their colors, but they serve to impress upon the memory of a bird or animal, that the insect which wears them is not good food, and to bring about ready recognition upon later encounters.

ADAPTATION

Signal Marks. Signal marks include the white tails of deer and rabbits, which are shown only when the animal runs or when the tail is raised.

Recognition Marks. The red lateral spots of trout have been interpreted as recognition marks, since they are visible to fishes

FIG. 147.—*Catocala ilia* Cramer. A, on the bark of a tree with wings folded; B, the same moth with wings spread. The hind wings are orange with black bands.

at their own level, while above and below trout have the characteristic dark and light protective colors so common in fishes. The colors of moths and butterflies have been added to this category, but there is abundant evidence pointing toward the sense of smell as the chief means of recognition among such insects.

Confusing Colors. Colors of this nature are undoubtedly effective. The Catocala moths and many grasshoppers have brightly colored underwings which are conspicuous during flight (Fig. 147). Their other colors are dull. When the insect flies, its bright colors attract attention, but when it settles again they are so suddenly concealed that it seems to disappear, and it is very difficult to determine its exact location. Another effect of these bright parts is that they are likely to be seized by a bird if it overtakes the insect in flight, so that the insect escapes with no more injury than a torn wing.

Sexual Colors. Sexual colors are again complex. Many birds have brilliant colors in the male sex, and dull in the female. To what extent the colors are attractive to the female, we cannot say. It is almost certain that they afford some degree of protection to the nesting birds, since marauders would be more easily attracted to the conspicuous male than to the brooding female. We can judge no better the value of sexual coloration than by Beebe's story of the tinamou (*Crypturus variegatus*), in which sexual behaviour is completely reversed. In this species the male raises the young, and the female does the courting. Through song and dance, and the display of such color as she possessed, he observed one of these birds carrying on her courtship in the jungle of British Guiana. The bright colors are usually concealed beneath the rudimentary tail, and then the bird blends wonderfully into its jungle background.

FIG. 148.—Larva of a geometrid moth, resting extended from a twig. (From Woodruff, after Jordan and Kellogg.)

Mimicry. This characteristic is common in insects. Some caterpillars can scarcely be distinguished from a dead twig; in form, color, and behaviour when they are disturbed the imitation is very nearly perfect (Fig. 148). Other insects resemble leaves, still others the bark of trees, and some, flowers or other insects. The effect is either concealment, for protection or aggression, or safety through the mimic's power to profit by some repulsive quality of the imitated.

Concealment through mimicry is often imperfect. The bilateral symmetry of many moths resting on tree trunks discloses their presence rather easily to the practiced eye, though it is impossible to know how many are passed by unnoticed. One species, *Stenoma schlaegeri* and probably other related species of the same genus, conceals its symmetry by lapping one front wing over the other; this moth resembles a small bird-dropping so closely that either may easily be mistaken for the other.

The explanations of adaptation are many. Whether we shall ever know how all of them have come about is very doubtful, for the details of the problem are infinite. They are universal, however, and afford us unlimited material for the study of evolutionary processes. From such studies theories have been formulated to explain the modification of organisms and in these theories we find the probable foundations of this varied and intricate association of organisms and environments. It is certain that some process or processes of change have given rise to the modern condition since evolution points to a common origin for all living things.

Summary. Adaptations are the characters of organisms which fit them for life in a given environment. In the individual we see both the process of adaptation and the adaptations resulting from it, while in species we see only the latter. It is the function of evolutionary theory to account for the wonderful adaptations which are characteristic of species. When we consider the environment with which they are associated we find that it involves external factors, both physical and organic, and in addition the internal environment which the individual body furnishes for any of its parts. Adaptations which are due to response to conditions of the internal environment have been called non-adaptive characters but they must be due to cause and effect no less than characters directly related to the external environment. The more evidently adaptive characters in the latter category have been classified in detail. They include adaptations for aquatic and terrestrial life and for flight, adaptations to varying degrees of moisture and light, and adaptations for association with other organisms. Organic relations necessitate special structures for securing food and for protection. Protective adaptations are exceedingly varied, including structural adaptations for speed and many other characters. Both ends are gained by more or less permanent associations of animals and by coloration and mimicry. The idea of the web of life expresses the complexity of organic relationships from which is derived much of the subject-matter of evolution.

REFERENCES

Evans, A. H., *Cambridge Natural History*, Vol. IX, *Birds*, 1900.
Beddard, F. E., *Cambridge Natural History*, Vol. X, *Mammals*, 1902.
Weismann, A., *The Evolution Theory*, 1904.
Banta, A. M., "The Fauna of Mayfield's Cave," *Carnegie Inst. of Washington*, Publication No. 67, 1907.
Gadow, H., *Cambridge Natural History*, Vol. VIII, *Amphibia and Reptiles*, 1909.
Thayer, G. H. and A. H., *Concealing Coloration in the Animal Kingdom*, 1909.
Roosevelt, T., *African Game Trails*, Appendix E, 1910.
Wheeler, W. M., *Ants*, 1910.
Mast, S. O., *Light and the Behaviour of Organisms*, 1911.
Henderson, L. J., *The Fitness of the Environment*, 1913.
Lull, R. S., *Organic Evolution*, 1917.
Newman, H. H., *Vertebrate Zoology*, 1920.
Chandler, A. C., *Animal Parasites and Human Disease*, 3rd edition, 1926.
Comstock, J. H., *An Introduction to Entomology*, 1924.
Beebe, Wm., *The Arcturus Adventure*, 1926.

CHAPTER XIII

THE BASIS OF ADAPTATION

Variation. We have already noted that organisms vary and that variation itself is the most invariable thing in nature. The possibility of change in organisms is directly associated with this phenomenon.

If an individual were incapable of varied responses change of environmental conditions beyond its power to respond could only result in death. In the latitude of response within the individual we find the foundation of individual success. Individuals differ in their power of response, however, even within the same species because of their variation, so that conditions which make life impossible for one may be met easily by another.

Since species are merely aggregations of individuals these things are also the basis for success of species in meeting varied conditions and in becoming fitted for life in a limited environment. Extinction may occur and has occurred in many known cases, but it can result only from a change so extreme that no individuals can meet it. Any change less extreme could be met by some; these survivors would continue to perpetuate the species which would differ from its former state through the loss of the unsuccessful individuals. The responses of the species as a whole would then be those of the surviving individuals; it would be changed to meet the changed environment.

Environment. Although change of external environmental relationships must always accompany the evolution of organisms, change of the external environment need not always be the inciting cause. Non-adaptive changes in organisms produce structures correlated wholly with the internal environment, although their establishment may later exact adjustments to the external environment as the price of existence. It is probable, at least, that any change in an organism will modify its ability to avail itself of surrounding conditions in some degree.

Whatever may be the source of change in organisms, it involves changed response to conditions in the complex environment.

Ultimately, as species give rise to other species, this change of response must find expression in change of structure and function as heritable properties.

Many recognizable forces may contribute to the initial shift of an organism from one environment to another. Change of climate due to geological factors might readily cause readjustment. Change of food supply, involving either climatic factors and soil conditions, or interspecific relationships probably occurs even more often than marked geological changes. Overpopulation is commonly regarded as a potent factor in evolution. In connection with all three forces, an animal's power of migration may determine the nature of its response, and in some cases it is probably a primary factor in itself.

Change of Climate. Geology and paleontology together show that from time to time in the remote past climatic conditions have varied greatly. Our own continent has passed through tropical and arctic conditions, as well as temperate periods with climates like that of the present. "Geologists now know of seven periods of decided temperature changes (earliest and latest Late Proterozoic, Silurian, Permian, Triassic, Cretaceous-Eocene, and Pleistocene) and of these at least four were glacial climates (both Proterozoic times, Permian and Pleistocene). The greatest intensity of these reduced temperatures varied between the hemispheres, for in the earliest Late Proterozoic and Pleistocene it lay in the northern, while in latest Proterozoic and Permian time it was more equatorial than boreal. Cooled climates occur when the lands are largest and most emergent, during the closing stages of periods and eras, and cold climates nearly always exist during or immediately following revolutions, when the earth is undergoing marked mountain making" (Schuchert) (Fig. 149).

Factors Which Influence Climate. The things which modify temperature may also be instrumental in determining the amount of precipitation, the relation of seasons, and minor fluctuations. A high range of mountains, for example, may cause abundant rainfall on one side and desert conditions on the other by cooling moisture laden currents of air as they blow toward it from the nearest ocean. This is true of mountains of the western United States. The same effect may be responsible for localization of rainfall. High altitudes are also subject to fluctuations of many degrees in temperature between day and night, so that organisms

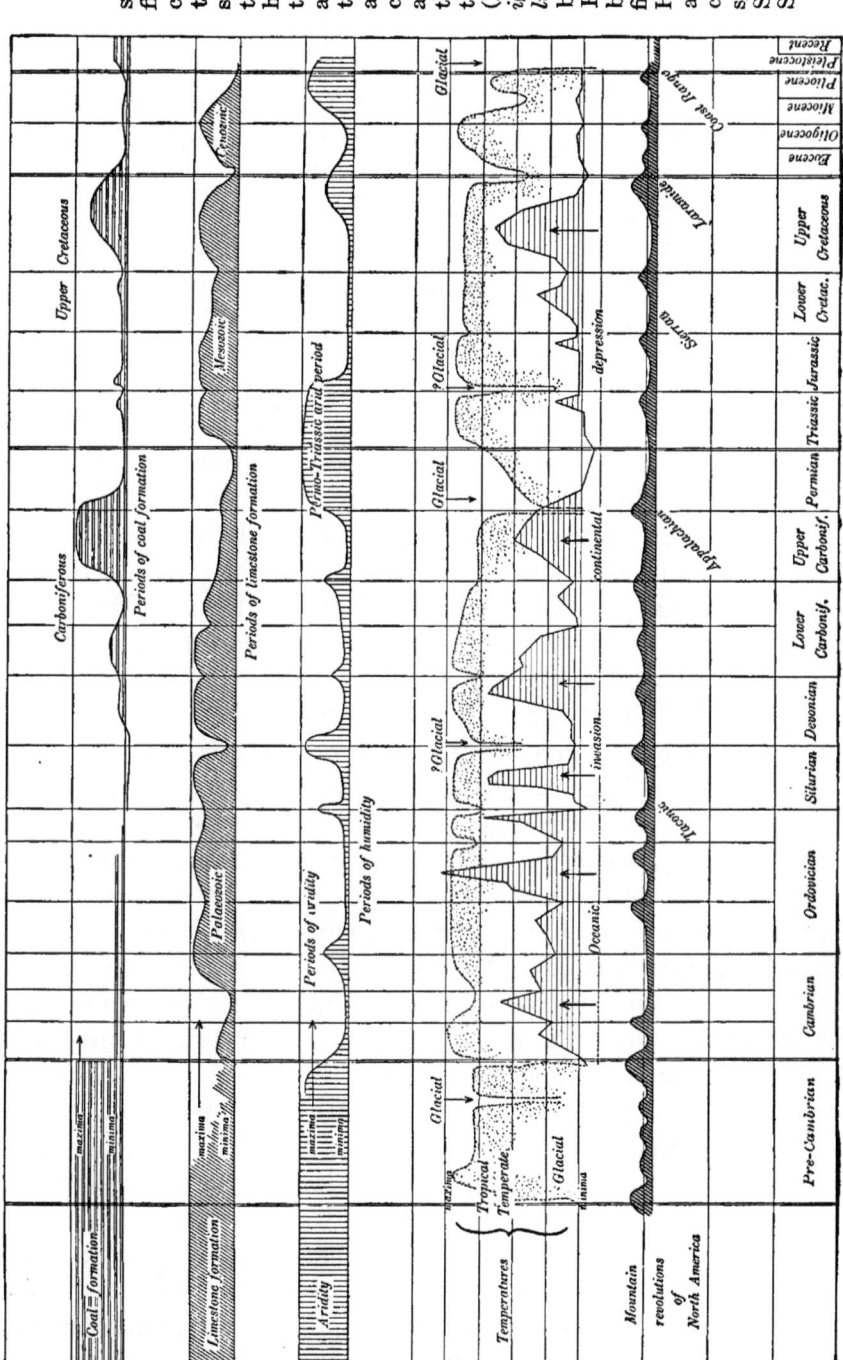

Fig. 149.—Changing environment during the last fifty million years or more.

must be adapted to both extremes. The same quality is expressed in short summers and long winters at high altitudes even in lower latitudes. Snow may be expected throughout the year at 15,000 feet in latitudes where it never falls at sea level.

A geological change such as the elevation of continents would change the climate of a region completely. If local, in the form of elevation of a mountain range, it would have a similar effect on the region involved, but would exert various influences on other parts of the same land mass. Its modification of precipitation might be extreme. Drainage might be affected in such a way as to change the distribution of fresh waters and the character of the streams. These changes could not fail to modify the environment of many organisms.

Climatic Factors and the Food Supply. As a result of geological changes and their indirect influences, the flora of a continent is likely to be modified. One of the most conspicuous indications of the climate of past geological periods is, in fact, the nature of the plant remains. This is a natural outcome of the intimate association of plants with the physical environment. Change of climatic factors to them means direct modification of the food supply, and their response, in turn, means modification of the food supply of all other organisms, directly or indirectly. Through the web of life a change may be transmitted to unsuspected lengths.

Food supply is one of the most important contributions of the environment to individual life. Its modification is therefore a matter of deep concern, perhaps more vital than the modification of other factors, but it is easily seen to be intimately linked with climatic factors on one hand, as a causative factor in evolution.

Rate of Reproduction. No less intimately associated with modification of the food supply is the factor of overproduction, although in this case it is not the actual supply that is modified, but its availability to individuals. The principle of overproduction was emphasized by Malthus in his work on population, which furnished the inspiration for Darwin's and Wallace's theories of natural selection. In man it seems less conspicuous than in animals, although man was Malthus' subject and the same process was later recognized in animals. It is obvious, however, that all organisms produce more individuals than can survive, and that only constant destruction prevents overpopulation. Interruption of the natural balance by transportation of animals to foreign

THE BASIS OF ADAPTATION

countries has given us illustrations in numerous cases. Rabbits introduced into Australia years ago have become a nuisance. The English sparrow has been equally successful in North America. Such insects as the San Jose scale, gipsy moth and Japanese beetle have been even more serious. The first named threatened to wipe out the citrus fruit industry of the west until it was traced to its oriental home, whence natural enemies were secured to hold it in check. The others have cost millions in the eastern United States, and are still a problem.

The production of an excessive number of offspring is essential to the maintenance of the food supply of all organisms, since those which are destroyed are either killed for food or become food incidentally. It is also closely linked with the food supply of the species concerned, for the persistence of an abnormal number of individuals might well carry the needs of the species beyond the limits of the available food supply and necessitate adjustments of some kind.

Response of Organisms. The response of organisms to changes in any of the above factors might be accomplished by change of location. The power of locomotion is developed in some animals to such a high degree, however, that it may well be a primary factor in evolution itself, since a species may gradually extend its distribution from an original center into regions which make different demands upon it.

The Interaction of Factors. The entire series of factors interact in a complex way, for the operation of one may well bring another into action. Geological changes alone are beyond the influence of the others.

Migration. Whenever environmental conditions change, two courses are open to the organism. It may either migrate from the region in which it lives in search of more favorable surroundings, or it may remain and carry on its existence by taking advantage of conditions which were formerly of no importance to it. Either of these courses depends to some extent upon its latitude of response, but the latter especially. To migrate may carry the animal into an environment but little different from that to which it is accustomed. To remain in a changing environment demands change inevitably. The two responses are in some degree associated, for the former is certain to contribute to the adaptive modification of some organisms, and the latter is equally certain to result in interchange between divisions of the region concerned which amount to a limited migration.

The opportunities available to a species within a limited region depend partly upon its established habitat. A terrestrial form is intermediate between fossorial and arboreal habitats, and between aërial and aquatic. Any terrestrial form may avail itself of the advantages of partial occupation of one or more of these neighboring habitats without relinquishing its primary terrestrial associations. A species living in the shallow sea is in a position to enter the abysses or fresh water streams, or to move into the intertidal zone of the littoral fauna. Such species are also subject to transition between the three stages, benthos, nekton and plankton. The more specialized forms, such as fossorial, abyssal and benthonic animals, have more limited contacts and less ability to respond to varied conditions. Specialization is fundamentally a reduction of possibilities, since it demands particular fitness of the organisms' equipment for some limited mode of life. In spite of this limitation, however, even specialized forms retain some latitude of response and diversity of contacts. It is only in the most extreme forms, such as the endoparasites, that complete inability to modify the mode of life in some slight degree is likely to be found, and even some of these develop immunity to drugs used for their destruction.

Adaptive Radiation. The results of animal response to the various possibilities of a limited environment have been expressed in Osborn's law of adaptive radiation (Fig. 150). This law was formulated in connection with studies of mammalian evolution, but is generally applicable with minor modifications. As Lull notes, "adaptive branching" is a more satisfactory term, since the process is not always one of radiation from a common central point. The principle involved is, in any case, much as it was formulated by Osborn in his original law. It may be stated in terminology as near as possible to the original as follows: Each isolated region, if large and sufficiently varied in its topography, soil, climate, and vegetation, will give rise to a diversified fauna in any group of animals. The larger the region and the more diverse the conditions, the greater the resulting variety of forms. From a primitive stem form, new lines of adaptation will go out into associated habitats. One result is divergence of form in related animals, which has already been mentioned.

Although it is generally applicable, there is probably no better illustration of the process than the mammals. Originally terrestrial, they have radiated into four habitats. On the basis of limb modifications, from the original ambulatory stem fossorial, volant, cursorial and aquatic species have arisen. Such animals as the shrews represent the primitive type. Scansorial animals like the squirrels may well be transitional to volant types. The bats are the most nearly perfect volant mammals (Fig. 134). The mole is the most nearly perfect fossorial form (Fig. 130).

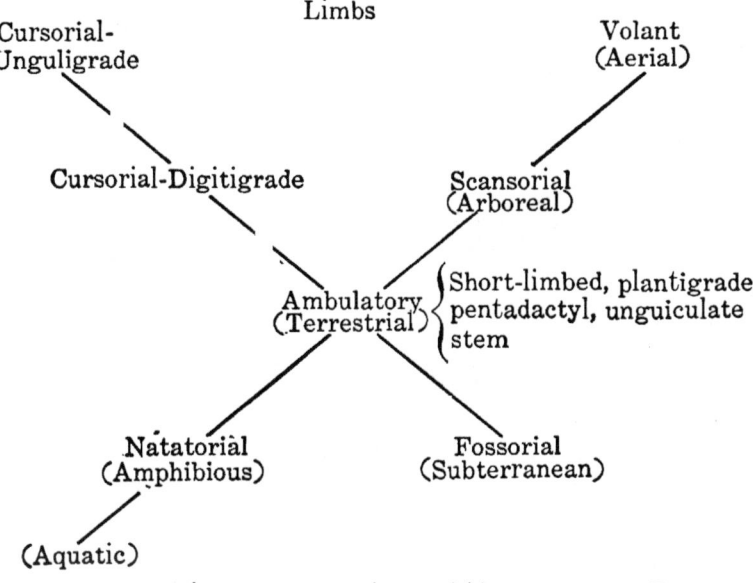

FIG. 150.—Diagram of adaptive radiation. (After Lull.)

From ambulatory forms digitigrade runners like the carnivores arose, and through similar limb posture came the highly specialized unguligrade appendages of animals like the horse. Among aquatic mammals the whale is extreme; many amphibious forms suggest the possibility of gradual change in this direction. The teeth also show adaptive radiation from the primitive insectivorous stem to herbivorous, carnivorous, omnivorous and ant-eating types, the last with greatly reduced dentition.

Adaptive Branching. Among insects the bugs (Order Hemiptera) show many evidences of adaptive branching (Fig. 151). They were originally, no doubt, terrestrial ambulatory species with volant powers. We now have highly aquatic species, which still retain the power of flight (*Belostomatidae, Corixidae, Notonectidae, Nepidae*). These insects have the legs highly modified as swimming organs. Other species (*Gelastocoridae*) are found along shores, while still others (*Veliidae, Gerridae*) skate about on the surface film of fresh waters. One genus of the *Gerridae* is marine. This genus, *Halobates*, may be found far from land, and Beebe has recently confirmed the belief that its eggs are deposited on floating feathers. Some are fossorial (*Gelastocoridae, Saldidae, Cydninae*). In food habits they vary greatly. Since they have

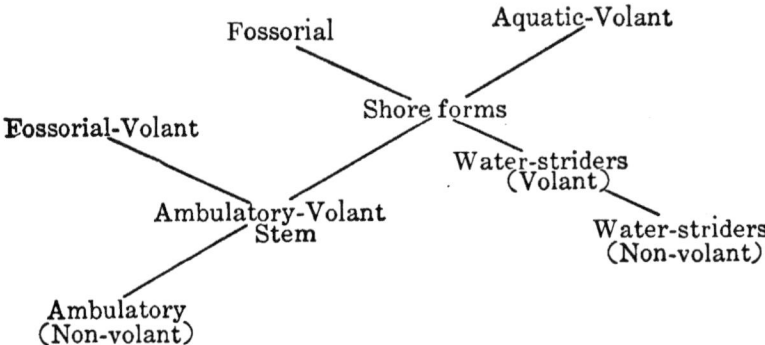

Fig. 151.—Diagram of adaptive branching in the Hemiptera.

suctorial mouths, they are limited to liquid foods, but they include both phytophagous and carnivorous forms. Some are highly predacious, some parasitic, and some probably scavengers.

Geographical Distribution. Animals that migrate, whether in response to the pressure of necessity or to the inherent ability and disposition to roam, have the world and its varying conditions at their disposal. Within certain limits of time and space they are restricted by the physical characteristics of the earth's surface, but these features have changed repeatedly during the past, so that what is now impossible may once have been easy. Migration is likely to carry animals unchanged into favorable regions or to carry them by gradual steps into regions which would demand new adaptations. It has resulted in the population of the earth with a unified fauna and flora. In similar habitats, however remote, we may expect similar organisms, and only those land masses whose isolation extends back beyond the origin of modern species are peopled with organisms fundamentally different from those of other similar regions.

THE BASIS OF ADAPTATION

Zoögeographical Realms. The earth is now divided into several zoögeographical regions, indicative of the surficial distribution of animals (Fig. 152). Lydekker proposed three chief divisions, each including territory which has been sufficiently isolated in the past to produce a characteristic fauna. These are the Arctogaeic realm, including America north of the Mexican plateau, Europe, Asia, Africa, and all adjacent islands; the Neogaeic realm, including South America and adjacent islands, Central America north to the Arctogaeic, and the West Indies; the Notogaeic includes Australia and the islands of the Pacific. The Notogaeic realm has been isolated since the beginning of the Tertiary, and has consequently developed a peculiar fauna including a number of primitive forms. The Neogaeic also enjoyed a long period of isolation during the Tertiary, and has a characteristic fauna as a result, but the greater area of the Arctogaeic has been open more or less constantly to intermigrations, and is populated by a remarkably homogeneous fauna.

Within these realms various subdivisions are recognized in which the faunae are distinguished by minor characteristics. The subdivisions commonly used are:

1. Palaearctic, including Europe, Asia north of Persia, the Himalayan Mountains and the Nan-ling mountains of China, the adjacent islands, and Africa north of the Sahara desert.

2. The Nearctic, including the American portion of the Arctogaeic realm.

3. The Neotropical region is the same as the Neogaeic realm.

4. The Ethiopian region includes Africa, Arabia and adjacent islands, south of the Palaearctic region.

5. The Oriental region includes India, the Malay peninsula and other parts of Asia south of the Palaearctic boundary, and in addition the Malayan archipelago.

6. The Australian region is the same as the Notogaeic realm. There is some transition between it and the last in the Malayan archipelago.

Few places on the earth are completely devoid of life, although such limited areas as the Great Salt Lake desert appear to be so. In this region the earth is a glistening mass of salt for miles. Not a plant is visible within the outer borders of the desert, where a few stunted shrubs are able to live. Excepting such relatively small areas, every region has its living forms. No matter how

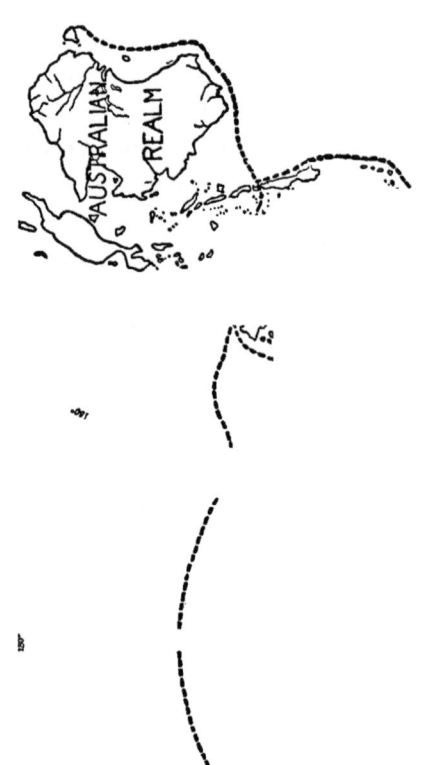

Fig. 152.—Map of Zoögeographical realms; polar projection. (From Lull.)

cold, how hot, how wet or dry, or how rocky, some organisms can be found clinging precariously to life, unless some deleterious substance or influence prevents for a time.

Barriers. When any of the factors mentioned at the beginning of the chapter make it necessary or desirable for animals to seek new homes, migration is apt to be one of the chief means of adjustment. It may be accomplished through the animal's own powers of locomotion with considerable rapidity, but in some parts of the world there are insuperable barriers. The oceans are impassable to terrestrial species, high mountain ranges are often so, and climatic conditions in some regions may hinder the passage of animals not adapted to them. In addition new contacts with other species may hinder migration.

Mountain ranges are effective barriers when parallel to the equator. The effect of a few thousand feet of altitude on such a range is equivalent to a change in latitude of many degrees. "This is notably true," says Lull, "of the great Himalayan Range in northern India, which rears its mighty summits far beyond the limits of perpetual snow. On the south we have the hot, moist plains of India, with a very distinct tropical fauna which in many respects resembles that of Africa. North of the barrier, conditions of climate, both in temperature and degree of moisture, are entirely changed, and with them appear animals, with some notable exceptions, of a totally different sort, more nearly comparable to those of Europe."

The difference between these ranges and mountains that run north and south, like the Sierra Nevada and Rocky Mountains, is easily understood. The temperature fluctuations, and resulting climatic conditions, at an altitude of eight to ten thousand feet in the latitude of central Colorado are much like those at normal levels far north of the United States. Consequently there is a zone of similar climatic conditions extending gradually upward from north to south in these mountain ranges, and migration may be accomplished along this zone without sudden change. The northern hemisphere contains several species of insects which illustrate the condition admirably. Two of these, related to the butterflies, are found in northern Europe, Asia, and North America. In the central part of Canada they range southward nearly to the international boundary, but in the mountains of the east and west they extend far to the south. One has been taken at an altitude of 13,000 feet in southern Colorado and in the Appalachian range as far south as North Carolina. The other is found at lower altitudes in the western mountains and extends as far south as Arizona. Such irregularity of distribution shows clearly how transition from one side of a range to the other may be accomplished through the compensating influences of latitude and altitude.

Oceans are effective barriers to the migration of terrestrial animals, as also are the climatic conditions of extensive desert areas and to some extent, perhaps, other climatic extremes. To volant species, however, such areas are not obstacles. Butterflies have been observed three hundred miles from land and even the normal seasonal migration of birds may carry them a much longer distance over unbroken ocean.

Aids to Dispersal. The passage of such areas is not impossible to other organisms when aided by favorable circumstances. In the geological past oceans have been bridged by narrow isthmuses such as the Isthmus of Panama which now connects North and South America. There was no such connection during most of the Tertiary, consequently the relationships between the Nearctic and Neotropical species are either older or more recent than that time. On the other hand we know that Alaska was once joined to Asia by an isthmus extending from the Alaskan peninsula to Kamchatka which allowed free communication between the Nearctic and Palearctic regions (Fig. 153). All that now remains above the ocean is the chain of the Aleutian Islands, but the water between them is nowhere very deep. The faunae are very similar on the two continents because of the early association. Apparently the remarkable Galápagos Islands were also once connected to the Americas, but their isolation must have been remote, and even the early occurrence of a connection is disputed.

The land bridge between North America and Asia furnished a convenient passage for many mammals during the Tertiary. Extinct proboscideans and horses, as well as more obscure forms which we have not considered, passed from continent to continent over this route, and it is very probable that the ancestors of the American Indian entered the continent from Asia in the same way.

Drifting debris and ice floes may also carry terrestrial animals many miles over the ocean. Floating logs and masses of tropical

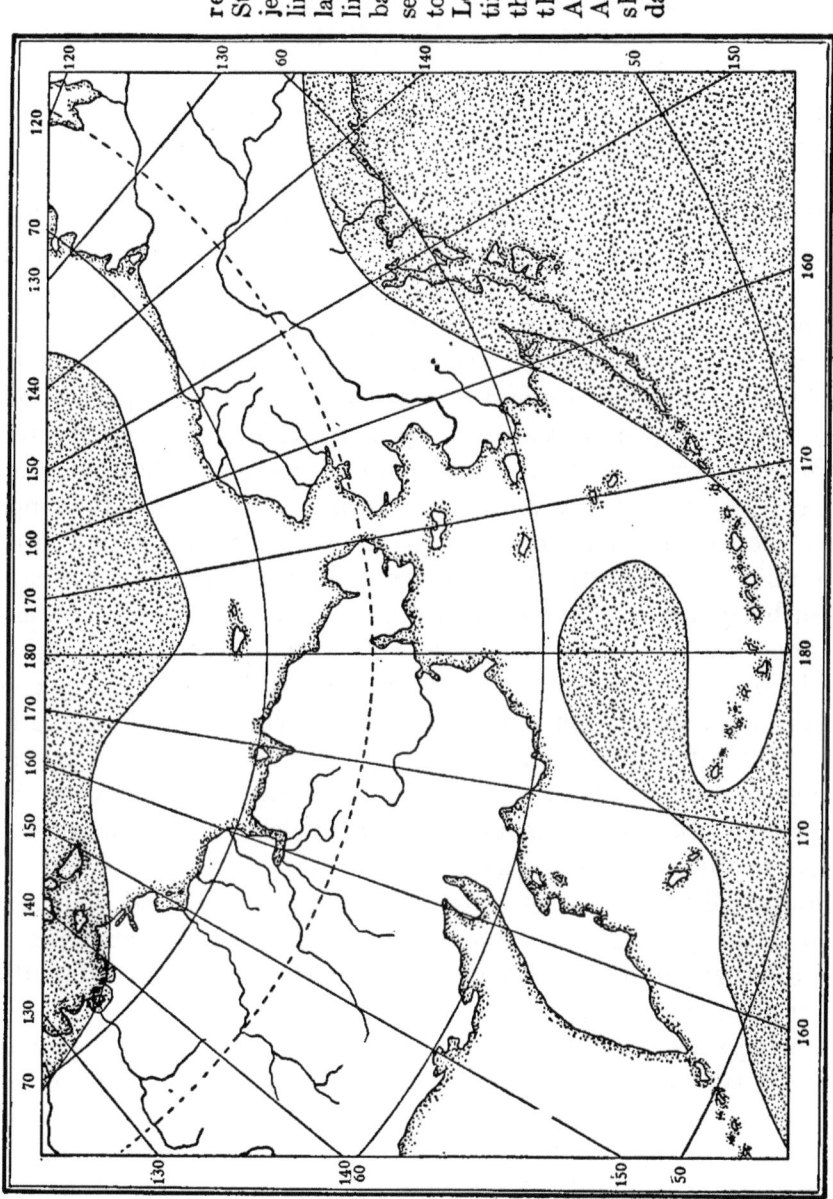

FIG. 153.—The region of Bering Strait, polar projection. The coast line of the modern land masses is outlined with a stippled band. The ancient sea as it is supposed to have existed in Lower Cretaceous time is stippled and the land mass which then connected Asia and North America is left unshaded. (From data by Schuchert.)

vegetation have been observed far from shore. Beebe records in *The Arcturus Adventure* a population of fifty-four species on a single log, and in his interesting chapter *An Island of Water*, in the same work, lists four species of plants, five of birds, eight of shore fishes, four of shore crabs, and eight of insects observed within a space of ten days which might have become established had his "island" been a new-born bit of land. If this could occur in ten days, what might not pass by in a year? With such evidence before us we can hardly imagine land remaining uninhabited.

Elsewhere the transportation of polar bears on ice floes has been recorded, and the observation of animals of considerable size on drifting masses of vegetation, which are apparently of common occurrence in tropical waters.

Wind is an important factor in the dispersal of animals of small size, and an aid to the dispersal of volant forms in particular, whatever their size. Insects are especially likely to be transported in this way, for their bodies are so small and light that they are unlikely to fly strongly enough to resist high winds. It is not uncommon to find battered specimens of butterflies a hundred or more miles to the north of their normal range, often after a period of southerly winds. These individuals, because of highly specialized food habits, do not usually establish the species in the region to which they are carried. In some cases, however, a species appears occasionally on cultivated plants which is not normally seen. Its intermittent occurrence may well be due to gravid females carried by flight and wind. Such forces have no doubt been responsible for the peopling of oceanic islands with organisms similar to those of the nearest mainland.

Plants are in many cases specially adapted for wind dispersal through the production of buoyant seeds. Wing-like expansions and tufted appendages, such as those of maple and milkweed seeds, are familiar examples.

Flowing waters, either ocean currents or streams, act in a similar way, but they can act only on aquatic organisms or in connection with some buoyant object. The seeds of plants may be transmitted in this way usually without special adaptation, since many of them are incidentally light enough to float. Some are aided, however, by buoyant accessory structures, such as the fibrous husk of the cocoanut and the receptacle of water lilies.

THE BASIS OF ADAPTATION

The agency of other organisms is probably of greater importance than we realize in dispersal. The transportation of plant seeds by birds and other animals is familiar. It may come about through use of fruits as food and failure to digest the seeds, or through accidental means. The transportation of such specialized fruits as the cockleburs, beggar's-ticks and Spanish needles by animals is furthered by their special adaptations for adhesion to fur or skin. Small animals may also be transported to some degree in this way. Parasitic, commensal and symbiotic species are so distributed, beyond doubt, but the activities of men are probably of greater importance in the dispersal of free living animals than the incidental associations of lower forms. The railroads and other vehicular traffic are known to have carried organisms accidentally for long distances, and some of our most troublesome pests, the San Jose scale, gypsy moth, Japanese beetle, European corn borer, English sparrow and starling have been imported, intentionally or accidentally, from other continents.

Agencies of this kind may well transport animals from one land mass to another accidentally, but the results would be similar to those arising from forced migration. Accidental transportation, however, would be much less likely to carry organisms into regions favorable for their development than intentional migration.

For these reasons no part of the world is able to remain long without a population of living organisms if it provides the bare necessities of life. Sooner or later, through accident or the pressure of competition in other regions it receives pioneers which may or may not be able to maintain themselves. The repetition of such occurrences is bound in time to result in the establishment of some species and with every addition the possibility of adaptive branching increases. Ultimately the region is itself a source of migrants to other places.

The Constancy of Change. It must not be supposed that these are abrupt transitions. Only such limited land masses as coral and volcanic islands, lava flows, and inundated and glaciated regions can have been utterly devoid of life at any time since life began. The transitions of most of these have been gradual, and with their gradual changes must have come gradual development of flora and fauna as opportunity arose.

In most regions life has been continuous, as it is now, and as now, fluctuating in details. Emigration and immigration go on constantly in these regions, and minor adjustments are as constantly taking place. In an area of less than a square mile the writer once watched within a decade a transition from a beautifully wooded valley to a weedy patch of dead and dying trees, of prairie land to a field of sweet clover, and of a grassy meadow to a slough, overgrown with coarse sedges and with water standing in its lower spots, all without the interference of man. In the same area it was often possible to see walnut trees defoliated by larvae of a moth. The overpopulation of a tree resulted in the destruction of the food supply, and the caterpillars were forced to leave the tree in search for another, some perhaps successfully, some perhaps not. Such cases are examples on a small scale of the factors which bring about adaptation.

Adaptations are the result of change, change in the organism through influence of its internal environment, change brought about by factors in the external inorganic environment, or change due to competition within the species or with other organisms, but always change. Static conditions do not demand fluctuating response, and static organisms, if such could exist, would be incapable of responding in more than one way. The organism varies and everything about it varies. In varied responses, whether migrations into new regions or reassociations within the same limited region, lies the beginning of the many adaptations which characterize the various creatures of the earth. Adaptation is a process and that process is evolution.

This much it is easy to say, but the fact remains that adaptation as a process is visible to us only in the individual. In order to accomplish the evolution of species it must be a process in the entire aggregates of individuals that constitute species, and here we face the complex problem of chronological as well as immediate association of individuals. The unity of successive generations is the field of heredity. It became evident many years ago that the linking of generation with generation was an important point of attack for the solution of the problems of evolution and the attention of many biologists has been turned upon it. Out of their efforts has grown the science of genetics, whose importance, both practical and purely scientific, is inestimable.

Summary. The components of individual existence, viz., heredity, environment and response, are variable. Change of environment in the broad sense necessitates change of response which is possible within certain hereditary limits of variation, and so furnishes a basis for the modification of the organism. Differences of heredity make it necessary for organisms to seek different conditions of environment. Since the individuals making up a species are not all the same, changing environment may favor some and destroy others, so that the limits of variation of the species may be narrowed. Response of organisms to changing conditions may be accomplished in various ways, but one important result is interchange between various habitats and regions. The result is a constant dispersal of organisms from their centers of development. This dispersal is hindered and aided by various factors, and in turn causes changes which may further influence the behaviour of other living things. Dispersal has brought about a definite geographic distribution of organisms and has been accompanied by adaptive radiation or branching within limited groups. The complexity of these interactions is the immediate explanation of adaptation. The transfer of changes from the individual to the species as a permanent component of the heritage is yet to be explained.

REFERENCES

LYDEKKER, R., *A Geographical History of Mammals*, 1896.
PAGENSTECHER, A., *Die Geographische Verbreitung der Schmetterlinge*, 1909.
CLARKE, W. E., *Studies in Bird Migration*, 1912.
MEEK, A., *The Migrations of Fish*, 1916.
LULL, R. S., *Organic Evolution*, 1917.
DAHL, FR., *Ökologische Tiergeographie*, 1921.
COMSTOCK, J. H., *Introduction to Entomology*, 1924.
HOLDHAUS, K., *Schröder's Handbuch der Entomologie*, II (7), 1927.

CHAPTER XIV

THE FOUNDATIONS OF GENETICS

The transmission of definite characters from parents to offspring is an obvious phenomenon, but it was not until early in the twentieth century that scientific knowledge of the subject was sufficient to be regarded as a division of science. Under the name genetics, coined by the English scientist, Bateson, it has since taken its place among the important biological sciences.

Genetics deals with the origin of individuals, and in this is closely connected with embryology. They differ in that the latter science is concerned solely with the gradual differentiation of a complete individual from the germ cells, while genetics attempts to explain the appearance in every individual of the characters previously found in its progenitors, and to account for the differences which occur. The correlation of these two fields of study is a problem whose solution will be very valuable to biology.

Heredity and Adaptation. The facts of adaptation considered in the last chapter are too orderly for explanation on any other basis than that of correlation of the organism and its environment. Correlation involving change in any factor cannot fail to bring about different results, which in the organism would necessitate modification of structure and function. In order to affect a species, these modifications must not only affect individuals belonging to it, but must reappear generation after generation. The constancy of the species, such as it is, must be preserved, and yet somewhere in the succession of generations change must be introduced if evolution is a reality.

It is the difficult task of genetics to unravel this paradoxical situation and to determine the limits of permanence of hereditary characters. Fortunately there has been no lack of available material. Although working hypotheses still constitute a large part of the fabric of evolution and genetics, they are supported by an ever increasing mass of facts. Evolutionary change has not yet been brought under control so that it can be produced in the laboratory, but more and more evidences of its existence become known as time advances and genetics in the meanwhile has given us a convincing account of the underlying mechanism of the entire process.

Variation, one of the fundamental factors in adaptation, is no less important in the science of genetics. Were it not for variation we could have no knowledge of the methods of inheritance. The fact of heredity would be no less evident; indeed, absolute likeness of all individuals of a species through successive generations would be even more definite evidence of its occurrence than the partial resemblance which is known to occur. It is in the variability of individuals, however, that the behaviour of individual characters can be traced. When differences are mixed in one generation and reassorted in the next they furnish a contrast in which the course of any one character can be seen and traced.

Kinds of Variations: Nature. In variation lies the range of possibility of change within the organism. Extensive studies have shown how universal variation is, and have given rise to a classification based on the nature, degree, heritability and evolutionary tendency of variations.

According to their nature variations are of three kinds:

Morphological. Variations in structure may involve differences in either the form or the size of parts, or in the case of duplicated organs they may involve differences in the extent of duplication. Differences in number are very common in plants. The petals of flowers, lobes of leaves, leaflets, and other structures which are usually duplicated may vary widely in number. In animals such variations are less commonly available, but the radially symmetrical forms often have more or less than the normal number of parts. *Hydra* with six or even seven tentacles in place of the usual five and *Asterias forbesii* with four or six rays instead of five are frequently encountered in the laboratory. In man extra fingers or toes sometimes occur, a condition known as polydactyly (Fig. 154). Differences in form and size of parts are evident to all of us in everyday contacts. It is more difficult to find human beings with approximately the same appearance than with very different appearances.

Physiological. Variations in function are a necessary corollary of morphological variations, since functions are merely the activities of structures. In the varied capacity of human beings to resist the same disease, to digest the same food, or to respond to the same stimulant we have common examples of physiological

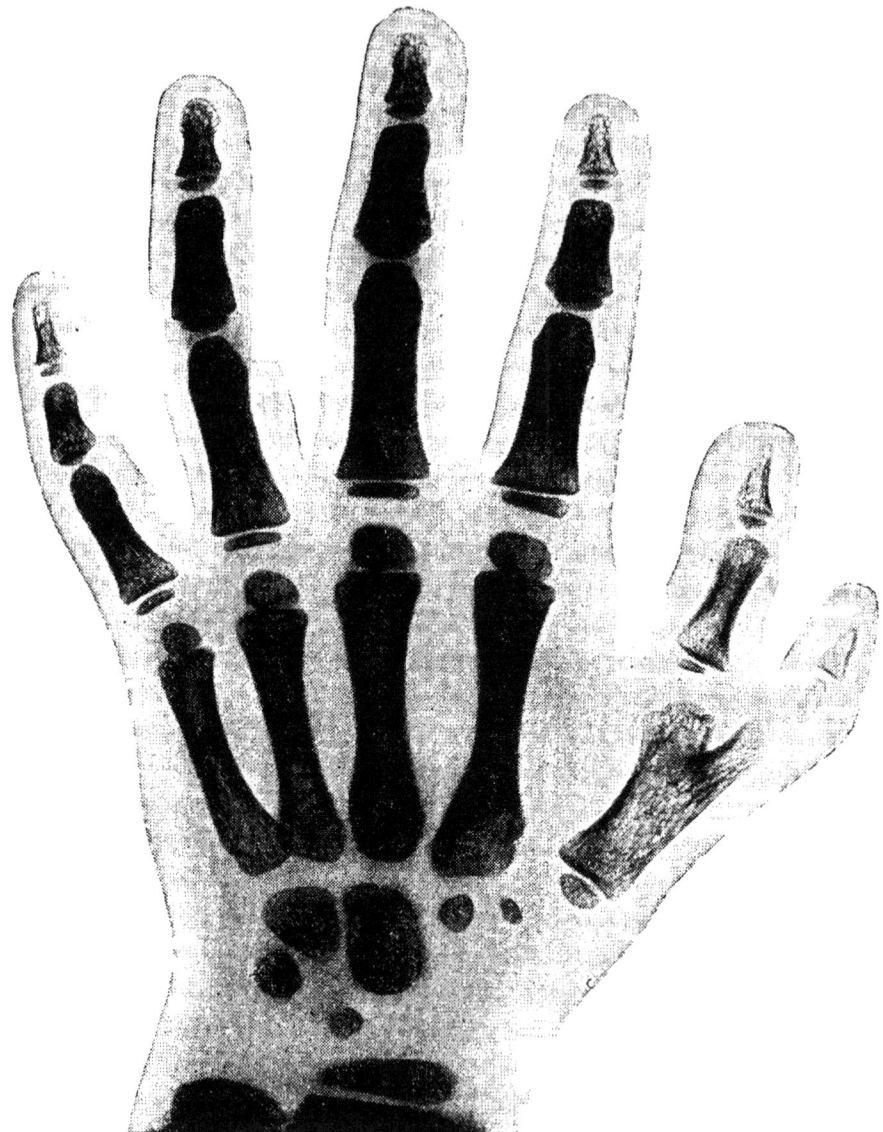

Fig. 154.—Radiograph showing polydactyly in a child's hand. (From Guyer's *Being Well Born*, copyright 1916–1927, after Dr. W. B. Helm; used by special permission of the publishers, The Bobbs-Merrill Company.)

variation. Everyone is familiar with the difference in individual response to caffeine. Some people cannot drink a demi-tasse in the evening without enduring a sleepless night, while others drink as much coffee as they desire without unpleasant effects. Walter cites an interesting case in the kea parrot of New Zealand. This bird was herbivorous but became carnivorous after the introduction of sheep into its native home. Thus physiological variations may occur in individuals as well as among the different individuals making up a species.

Psychological. Variations in mental qualities and processes are evident in all of our associates. They too may be witnessed in the individual from day to day as well as among the various individuals of a group or species.

To what extent they are separable from a morphological and physiological basis is less evident than the association of those types of variation; there are still many persons who look upon mind as an independent phenomenon. Biologically they are the variable expression of brain functions. That they are intimately associated with other bodily processes is amply attested by the effects of indigestion or any other slight illness upon mental efficiency, which most human beings have unwillingly experienced. In spite of our relatively meager knowledge of nervous functions it is therefore logical to believe that our mental processes and their variations are as definitely and completely associated with them as are any other phenomena of vital activity with underlying organic causes.

Variations of Degree. According to degree two kinds of variations are commonly recognized:

Continuous. Such variations grade through long series of individuals without an apparent break or abrupt transition. The dimensions of any individuals of a given species, for example, usually vary gradually from minimum to maximum. The curliness of human hair also varies by minute degrees from the straight hair of the Mongolian races to the kinky hair of the negro. Curliness is based on the form of hair; straight hair is cylindrical while kinky hair is distinctly flattened, and various intergrades occur.

Discontinuous. Discontinuous variations, on the other hand, exhibit abrupt transitions. Variations in the number of parts of an organism are always of this type, for there can be no gradation between six and seven petals or four and five rays. The term has also been applied in an entirely different sense to mutations; these are mentioned in the following pages under a different category.

Evolutionary Tendency. The effect of variations in evolution becomes evident only through the study of other related phenomena. It is impossible to determine this quality of variations by the examination of an existing species alone, but complete information about a phylogenetic series often discloses that variations in the past have followed either of two tendencies through successive generations. They have been either fortuitous or orthogenetic.

Fortuitous Variations. These are the ordinary fluctuating variations which appear generation after generation apparently always within the same limits and about the same mean. There is no available evidence to prove that they contribute to the evolution of species, although positive proof that they take no part in evolution is equally lacking.

Orthogenetic Variations. Such variations are evident in the field of paleontology. Their trend is in a definite direction through a phylogenetic series, toward the ultimate modification of the species in which they occur. A striking case of orthogenetic variation is the gradual succession of changes in the foot of the horse from *Eohippus* to the modern genus *Equus* as described in chapter X.

Heritability. A more fundamental classification of variations from the point of view of genetics is based on their heritability. They are divided into three groups.

Modifications. Modifications are changes which appear in the individual during the ordinary course of its life. They are usually looked upon as a product of environment, but should be regarded, with very few exceptions, as the product of inherited powers responding to environmental conditions. Such are tanning of the skin and muscular development. It is evident that these characters are not inherited as they develop in the individual, but must develop anew with each generation if the proper conditions are encountered. For the purposes of genetics they may therefore be looked upon as not heritable.[1]

Combinations. Since no two individuals are exactly alike, it follows that biparental reproduction will mix in the second genera-

---

[1] Modifications are usually regarded as the effects of environment on organisms. Few characters can be caused wholly by the environment, however, and most of the so-called modifications are as described above. For this reason it seems desirable to retain the term as it is used here. The question is considered further in chapter XXIV.

tion some of the different qualities of the first. A guinea pig may have the black color of its father and the angora coat of its mother. Such variations are based on the rearrangement of heritable components and are consequently heritable. They are an important source of change in organisms, but are necessarily limited by the range of characters present in the species. They have been widely exemplified in the development of domestic races.

Mutations. When an individual appears with characters distinctly different from those of preceding generations it is said to be a sport if its offspring return to the parental condition. In some cases, however, differences which appear thus suddenly are heritable and constitute permanent characters of the succeeding generations. The characters in question are then said to be mutations, and the individual possessing them is a mutant (Fig. 155). Such variations are due to an abrupt qualitative change in the organism and because of their heritability they play an important part in the modification of species. Their value as a cause of evolution is treated elsewhere in this volume.

Source of Variations. Walter summarizes this question

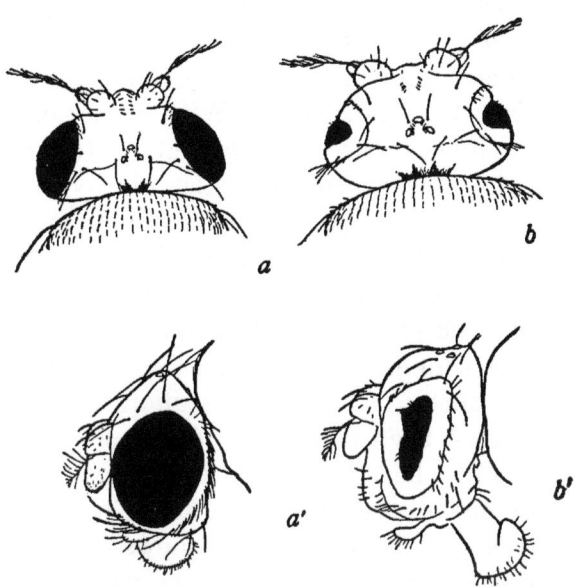

FIG. 155.—A mutation in *Drosophila melanogaster* compared with the normal character. a, a', normal eyes; b, b', bar eyes. (From Morgan et al., *Mechanism of Mendelian Heredity*, with the permission of Henry Holt and Company.)

with four opinions: (1) Darwin considered variations as axiomatic; (2) Lamarck and his followers looked upon them as either produced by the environment or by the organism in response to environmental conditions; (3) Weismann regarded them as purely an intrinsic product of the body; (4) Bateson regards inquiry into the causes of variation as premature.

Modern emphasis upon the inseparability of heritage and environment in the existence of organisms seems to dispose of the question rather effectively. Although it does not explain in absolute terms, it at least makes possible a logical interpretation in place of pure opinion. If this emphasis is well placed, the coöperating factors in existence can hardly fail to be jointly responsible for the resulting expression of organic characters. Any organism is infinitely complex in its inherited qualities, and its environment is no less complex and variable. Either through changes emanating from the organism, such as locomotion, or through such factors as climatic fluctuations, the balance between organism and surroundings is likely to be in a state of very delicate and constantly shifting adjustment. It seems reasonable to suppose that the result would be variation.

To this extent the interpretation is a combination of the opinions of Lamarck and Darwin. With variation before us in its complexity, however, we are forced to the belief of Weismann as well, who saw in the mingling of diverse parental qualities in each generation a provision for the diversification of the species. It is necessary to bear in mind that an organism as complex as a vertebrate or an arthropod is independent of the outer world to the same degree that it possesses within itself the proper mechanism for the maintenance of normal living conditions. In such a state, with a constant shifting of stimuli through any of the changes which have been considered as causes of adaptation, differences may readily be expected to arise from conditions wholly within the organism or partly without. There is sound reason in both Lamarck's and Weismann's views, and Darwin expressed the gist of the matter in his terse analysis. If we seek exact relations between cause and effect in this field, however, we can only follow Bateson's view; we do not know, and at present cannot expect to know, exactly what condition will cause a given variation, nor is it necessary for the purposes of genetics.

Importance of Heritability. Since the science of genetics deals with the resemblance of different generations, it is obvious that variations in any of the above categories must be heritable if they are to furnish material for the study of problems of genetics. Only combinations and mutations, therefore, are available sources of information in this science. The latter, when they appear, can be manipulated for the production of new combinations so as to indicate the fundamental methods of transmission of heritable characters.

Methods of Study. The superficial facts of inheritance were known long before there was a scientific foundation for their interpretation. It requires no profound knowledge to establish the fact that children have the traits of their parents, that sometimes they resemble more remote generations, and that individual peculiarities are likely either to be transmitted through many generations or to appear only intermittently. In a practical way man has taken advantage of this knowledge for the production of various breeds of domestic animals and many varieties of plants. Hound and dachshund, mastiff and pekingese, are products of the same wolf-ancestors of many generations ago. Sweet corn, flint, and dent varieties, as well as the primitive maize of the American Indian, have been traced back to the wild teosinte grass which now resembles them so slightly. Selection of animals and plants showing the desired qualities and propagation exclusively from these individuals have been responsible at least in part for such changes. In other cases man has taken advantage of the useful qualities of two species or varieties and combined them by hybridization. The mule, a cross between the horse and the ass, is such a hybrid.

Galton's Laws. Before these principles became a recognized part of scientific procedure attempts were made to formulate laws of heredity, with partial success. Best known are the two laws of Sir Francis Galton, who worked extensively with statistical data on human characters. They are:

The Law of Ancestral Inheritance. Each parent of an organism contributes one quarter of its inherited qualities, each of its four grandparents one sixteenth, and so on. In other words, the resemblance of an individual to all of its ancestors of any generation is inversely proportional to the remoteness of the relationship.

The Law of Filial Regression. Variation of parents from the racial mean is transmitted to offspring in a lessened degree (Fig. 156). Tall parents tend to produce tall offspring and short parents short offspring, but the normal expectation is that these offspring will be nearer average height than their parents. This law is of practical value, but is subject to modification according to facts discovered since Galton's time.

Influence of the Discovery of Cells. Although the cellular structure of organisms was recognized long before any important contributions to the study of heredity appeared, detailed knowledge of cell structure and behaviour was not available for many years. Mendel's important discoveries described in the next chapter were made without such knowledge, but they have much greater significance with the background of modern cytology than when they stand alone. The importance of cell behaviour

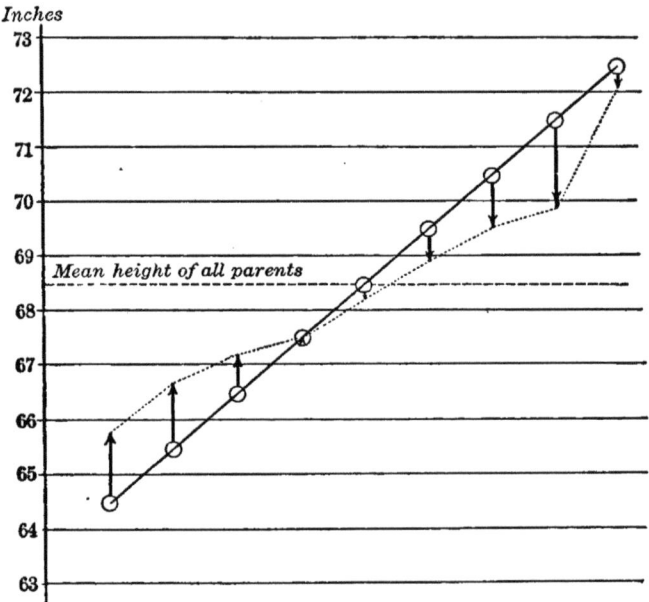

Fig. 156.—Diagram to illustrate Galton's law of filial regression. The circles represent graded parental heights while the arrow points indicate the average height of offspring descended from each group. The offspring of short parents are taller and of tall parents shorter than their respective parents. (From Walter.)

is strongly suggested by the methods of reproduction in organisms of various degrees of development.

Reproduction of Unicellular Organisms. Single-celled plants and animals reproduce in many cases by an equal subdivision which gives rise to two new individuals with complete loss of individuality on the part of the parent (Fig. 157). Such offspring are apparently as nearly as possible identical with each other and with the parent; there is abundant reason to believe them at least potentially the same.

Occasionally these single-celled forms undergo a complex process of nuclear reorganization accompanied by an exchange of nuclear material between individuals. Such a process is akin to sexual reproduction of higher forms and suggests the importance of the nucleus in the vital processes of the cell. It also indicates the possibility of combination of parts of two organisms in a single individual even in the lowest groups.

Reproduction of Multicellular Organisms. Above the single-celled organisms a process of budding or fission occurs which can

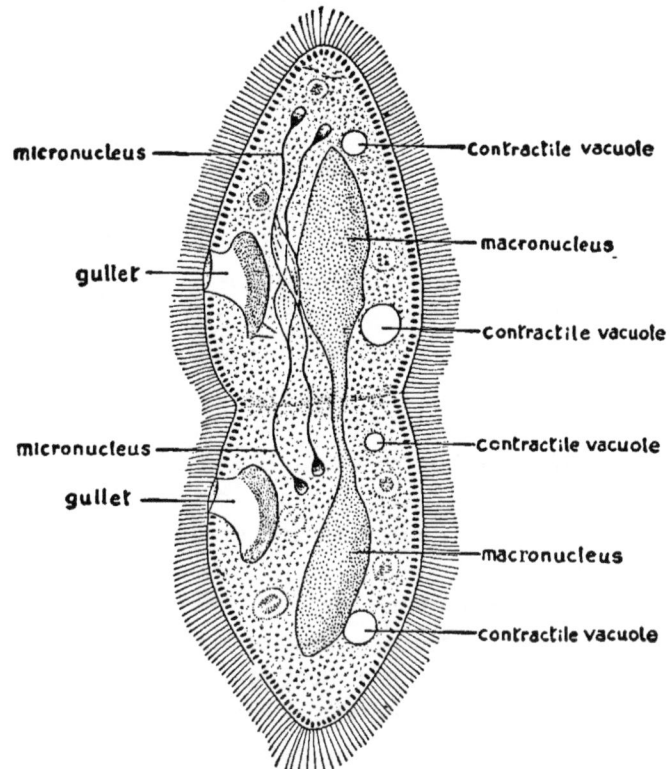

Fig. 157.—*Paramecium aurelia* dividing by binary fission. (From Newman, after Lang.)

also be expected to produce as nearly as possible identical individuals. *Hydra*, a common fresh-water coelenterate, frequently reproduces in this way (Fig. 158). Among the more highly organized forms, however, sexual reproduction is by far the most common process, although sporulation still persists in such highly developed plants as the ferns and parthenogenesis is not rare among animals. All of these cases differ from the more primitive asexual processes in the minuteness of the matter from which the new individual develops. This consists of a single cell, sometimes formed by the union of two independent germ cells derived from opposite sexes.

The Hereditary Bridge. Whatever the source of the single cell which gives rise to a new individual, it constitutes a link between generations which is of the greatest importance to the geneticist. All hereditary parts of an organism must necessarily come from its parents. In many cases the production of the original germ cells is the sole connection of the parents with the new generation. Obviously then everything which appears in the developing individual must be based upon something present in the cell from which it arises. It is the most compact expression of the complete organism and well deserves Walter's appellation, the hereditary bridge.

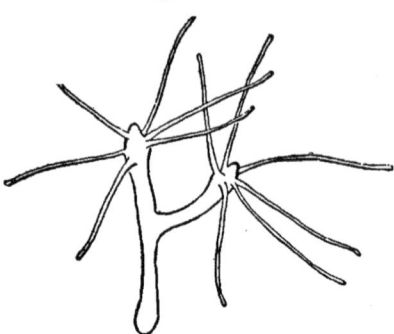

FIG. 158.—*Hydra* reproducing asexually by dividing lengthwise. (From Woodruff, after Koelitz.)

The value of studies of the behaviour of germ cells was recognized long ago and cytological investigations correlated with observations of the behaviour of characters in heredity are now a fundamental part of genetics.

Hybridization. Unscientific breeders have not been alone in utilizing this process. Crossbreeding of different strains within a species is now the most important source of material for the study of inheritance, for in the sharp contrast of characters thus obtained the behaviour of any one is usually conspicuous.

Hybridization demands first of all, known strains or varieties, which are best secured by careful breeding under observation to determine the normal course of their heredity. Once isolated, strains with contrasting characters can be crossed at will, and in the phenomena of reproduction through successive hybrid generations is found the source of important laws of heredity.

Selection. In the establishment of pure strains for hybridization selection is a necessary part of the methods of genetics. It has been important in other ways, but in this science it is chiefly an accessory of hybridization.

THE FOUNDATIONS OF GENETICS 275

Biometry. Many phenomena are of interest to geneticists which are not readily available for laboratory study. Such phenomena can be studied by the collection of statistical data. Still other phenomena involve variations which demand much more refined methods of analysis than the mere recording of visible differences, whether studied in the laboratory or in the field. These conditions have given rise to methods of exact mathematical measurement and analysis which constitute the science of biometry. A knowledge of biometrical methods is not necessary to an understanding of the principles of genetics and the subject will not be treated here. An excellent introduction may be found in Chapter III of Babcock and Clausen's *Genetics in Relation to Agriculture*.

An important application of biometry is the determination of the amount of evolutionary change which may have occurred in breeding experiments. Examples may be found in the work cited.

Summary. The phenomena of adaptation, in order to be a property of species and not merely of individuals, must be handed down from generation to generation. The variations on which they are based are also the materials from which our knowledge of hereditary processes is derived. For the purposes of genetics variations have been classified according to their nature, degree, evolutionary tendency and heritability. Their source is not definitely known, but their immediate origin seems adequately explained by the complexity of the various factors of organic existence. An understanding of the superficial facts of inheritance has long been put to practical use; Galton formulated laws based on such knowledge. The discovery of cells and their behaviour in reproduction has since made possible the correlation of the superficial phenomena of heredity with the morphological basis. Modern genetics also makes use of the established principles of hybridization and selection to provide materials for accurate study. Out of these methods have come definite laws of heredity.

REFERENCES

BATESON, Wm., *Materials for the Study of Variation*, 1894.
BABCOCK, E. B. and CLAUSEN, R. E., *Genetics in Relation to Agriculture*, 1918.
CASTLE, W. E., *Genetics and Eugenics*, 1921.
WALTER, H. E., *Genetics*, revised edition, 1923.
SHULL, A. F., *Heredity*, 1926.

CHAPTER XV

MENDELIAN HEREDITY

Johann Gregor Mendel, an Austrian monk, is the greatest figure in the history of genetics. With less formal training in biology than many college graduates of today he was destined through his keen powers of analysis and his painstaking experimental methods to give the world the first valuable laws of transmission of characters through heredity. These laws have stood the test of years and have been proved over and over again by the researches of the twentieth century. Cytological studies have established with reasonable certainty the physical foundation for them, and they now form an accepted basis for the usual procedure of genetic investigation.

Mendel's work was carried on in the peaceful atmosphere of his monastery garden at Brünn, where experiments covering many years led at last to the formulation of the laws that bear his name. His results were presented to the Natural History Society of Brünn and in 1866 appeared in the transactions of that society. They were then lost to the scientific world until 1900, perhaps because of the obscurity of the publication and perhaps because the world was not ready to receive them. Both causes have been suggested and both are probably true.

In 1900 three scientists are said to have revived Mendel's paper independently. They were De Vries of Holland, Correns of Germany and Tschermak of Austria. Suffice it to say that discoveries in which these men participated disclosed the sound value of Mendel's pioneer work and it became an accepted part of science. The succeeding quarter of a century has witnessed a constant increase in contributions to genetics, of which practically all are based on Mendel's laws. Mendel himself, like so many distinguished men, died long before his work was recognized, in 1884.

Mendel's Materials. The variable characters of garden peas constituted the material used by Mendel. These plants lent themselves equally well to crossbreeding and inbreeding, they were easy to raise under cultivation without forcing a change of environment upon them, as must happen when wild plants are used for experiment, and they were sufficiently variable to afford abundant confirmation of his conclusions. He could hardly have made a happier choice for pioneer genetic research. The characters with which he dealt were:

1. *Shape of seeds.* Some peas are round or nearly so and smooth when ripe. Others have their seeds shrunken into deep wrinkles.

2. *Color of cotyledons.* These seed leaves make up the greater part of the pea seed, and may be yellow or green.

3. *Color of seed coat.* Some seeds have a white coat and are produced by plants with white flowers, while others have gray to brown coats and are developed from purplish flowers.

4. *The form of ripe pods.* These may be either inflated or wrinkled and deeply constricted between the seeds.

5. *Color of unripe pods.* Green or yellow.

6. *Position of flowers.* Axillary or terminal.

7. *Length of stem.* Plants vary considerably according to environment, but under similar conditions peas fall into two classes. There are tall or climbing peas which may be several feet in height and dwarf peas which attain a maximum length of about eighteen inches.

Methods. By carefully controlled cross-fertilization Mendel produced hybrids of these varieties. For example peas bearing smooth seeds were crossed with those bearing wrinkled seeds. The result was the production of smooth seeds only, and in order to determine the fate of the other character these were planted, the plants self-fertilized, and a second generation of seeds produced. In other experiments plants bearing yellow seeds were crossed with those bearing green seeds, tall plants with short plants, plants with yellow cotyledons and those with green, and a number of others. The behaviour of these characters in the hybrid was noted, and the nature of their reappearance in successive generations.

Monohybrids. In the hybrids which Mendel produced only one of the two characters involved was evident. The hybrid peas of the smooth-wrinkled cross, for example, were all smooth. When the plants developed from these smooth seeds were self-fertilized, thus combining only the types of gametes which a hybrid could produce, the missing character appeared again. Approximately one quarter of the resulting seeds were wrinkled. On raising yet another generation the three quarters which were smooth were found to include two kinds of peas. One of these three quarters produced only smooth peas while the remainder produced a three to one ratio like their parents. The wrinkled peas bred true.

Mendel used a simple algebraic expression of his results. If we let S represent the character of smoothness and s the alternative, wrinkled, then in the original smooth and wrinkled strains only S and s respectively could be handed on to the next generation. In the hybrid there would be a possibility of either character being handed down. The total possibilities to be derived from either hybrid parent may therefore be represented by S+s, and all possible combinations in the second hybrid generation by the following computation:

$$\begin{array}{r} S+s \\ \underline{S+s} \\ SS+Ss \\ \underline{Ss+ss} \\ SS+2Ss+ss \end{array}$$

Later Punnett suggested a simple checker-board plan commonly called the Punnett square which expresses the same result (Fig. 159).

In the case of height all offspring of the hybrid seeds were tall. Here of course the development of the plant was necessary for the expression of the character. The tall peas proved to be mixed, however, for when self-fertilized they produced three tall offspring to one short. The short plants bred true, one third of the tall bred true, and the remainder again produced the three to one proportion. All of the seven characters behaved in the same way. Tables of Mendel's results have been prepared by several writers. In the one which follows the ratios are computed on the basis of unity for the smaller group. The numbers cited are those actually secured in breeding experiments.

	Character	Dominants	Recessives	Ratio
1	Form of seed	5,474 smooth	1,850 wrinkled	2.95+ : 1
2	Length of stem	787 tall	277 dwarf	2.84+ : 1
3	Color of cotyledons	6,022 yellow	2,001 green	3.00+ : 1
4	Color of seed coat	705 dark	224 white	3.14+ : 1
5	Form of pod	882 inflated	299 constricted	2.94+ : 1
6	Color of pod	428 green	152 yellow	2.81+ : 1
7	Position of flowers	651 axillary	207 terminal	3.14+ : 1
	Totals	14,949	5,010	2.98+ : 1

MENDELIAN HEREDITY

Terminology. The examples given above represent in all cases hybridization for a single pair of characters and are called *monohybrids*. *Dihybrids* and *trihybrids* for two and three pairs of characters respectively will be treated later. Any single character is called a *unit character* and the two alternative characters of a pair are called *allelomorphs*. It is customary to refer to the original parents as the parental or P generation, to the hybrid as the first filial generation or F_1, to the offspring of the hybrid as the F_2 generation or second filial and so on. In any example certain differences among individuals are found which are expressed in Figure 159. Here it will be seen that of the three similar F_2 individuals, in two the character wrinkled is represented even though it does not appear. Such a character is said to be *recessive*, while the allelomorph which conceals it is *dominant*. The

	Male gametes	
	S	s
S	Homozygous dominant S S Smooth	Heterozygous dominant S s Smooth, with latent determiner for wrinkled
s	Heterozygous dominant S s Smooth, with latent determiner for wrinkled	Homozygous recessive s s Wrinkled

(Female gametes)

FIG. 159.—Diagram to illustrate the F_2 generation of Mendel's hybridized smooth and wrinkled peas. Ratio of phenotypes 3:1; ratio of genotypes 1:2:1.

three individuals look the same, and are therefore said to belong to the same *phenotype*, while the characters actually represented within them determine their *genotypes*. When pure, like the SS individual, they are *homozygous*, when mixed like the two Ss individuals, *heterozygous*. Obviously where dominance is perfect only the test of breeding will disclose the genotype, except in the case of a homozygous recessive.

Fundamental Principles. The results of these experiments and the many others conducted since demonstrate three cardinal principles of inheritance, viz:

Unit Characters. Organisms are composed of many characters which may be combined in the same individual or distributed among different strains without losing their distinctive properties. These are commonly known as unit characters.

Segregation. The reappearance of the original unit characters in the progeny of a hybrid indicates that association does not modify characters but that during the process of reproduction they may be separated or segregated anew.

Dominance. The fact that of the two characters present in a hybrid one may completely conceal the presence of the other illustrates the principle of dominance.

Behaviour of Allelomorphic Characters. It is now known that many allelomorphic characters are not completely dominant and recessive to each other, but that both may be expressed in a hybrid. Three kinds of inheritance of allelomorphs are recognized, viz., alternative, mosaic and blending.

Alternative Inheritance. This is the type described above, in which one character completely dominates the other and the recessive makes its appearance only in homozygous recessive individuals.

Mosaic Inheritance. Mosaic inheritance differs from alternative inheritance in that both of the allelomorphic characters may be fully expressed, but in different parts of the body. Black and white spotted offspring of self-colored parents may be of this type, although they are sometimes explained in another way (see Chapter XVII).

Blending Inheritance. This type of inheritance differs from both of the preceding types in that neither character is fully expressed in any part of the body when both are represented. Such inheritance produces a character in hybrids which is intermediate between the parental characters. Many flowers inherit color in this way. The offspring of a red and white cross, for example, may have pink flowers. In the F_2 generation derived from the pink flowers, however, red, pink and white individuals appear.

Regardless of the manner of inheritance the integrity of the unit characters is preserved and their segregation may occur anew with every generation.

Effects of Allelomorphic Relationship. We have seen that in Mendelian monohybrids the ratio 3:1 appears in the F_2 generation when one character dominates the other, although breeding tests indicate that this is based on a 1:2:1 genotypic ratio. When characters are inherited in a blended or mosaic conditions, however, heterozygous individuals belong to a different phenotype as well as to a different genotype from the homozygous, and consequently the 1:2:1 ratio is evident without further breeding.

MENDELIAN HEREDITY

In the four-o'clock white flowers crossed with red produce in the F_1 generation only pink flowers. Inbreeding of this generation produces all three colors in the ratio of 1 white: 2 pink: 1 red (Fig. 160). Animal characters are well exemplified by the classic case of the blue Andalusian fowl. These chickens are produced by crossing black and splashed white Andalusian types, and never

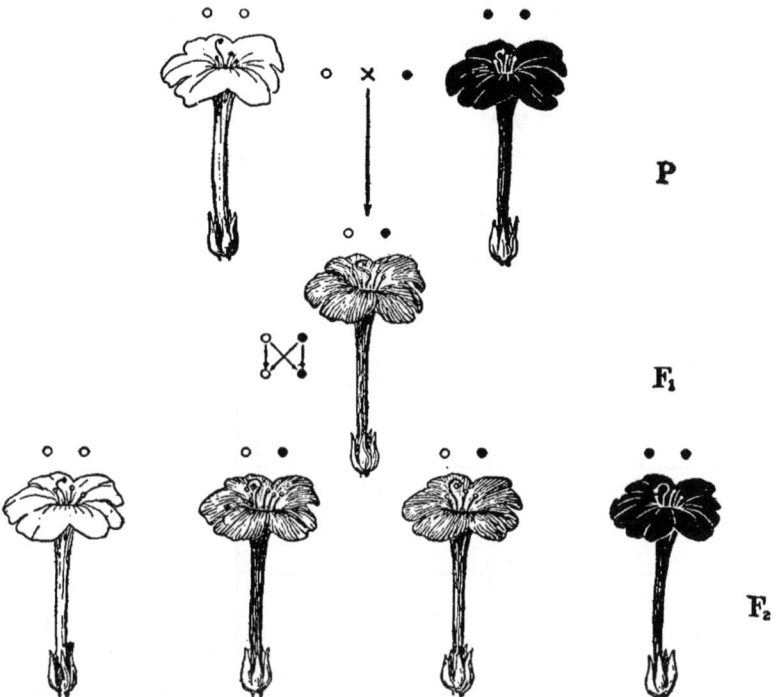

Fig. 160.—Diagram to illustrate the results of crossing white and red flowered races of four o'clocks (*Mirabilis jalapa.*) The somatic condition or phenotype is shown graphically; the small circles represent the genes involved. (From Woodruff.)

breed true. They are the heterozygous individuals of any generation and the race can be maintained only by constant hybridizing or by the elimination of the homozygous individuals of every generation. The condition of the blue Andalusian has been interpreted as a mosaic of finely divided colors.

Dihybrid Ratios. When two pairs of characters are associated in a hybrid the possibilities of reassociation in the F_2 generation are greatly increased, since not only the expression of each is variable but the association of the unrelated characters as well. There are four possibilities of distribution of the characters as

282   EVOLUTION AND GENETICS

they are handed down to the next generation from each hybrid parent, and the number of possible combinations received from the two parents is therefore sixteen. Out of the sixteen combinations some are duplicated. The phenotypic ratio is $9:3:3:1$. Nine are dominant for both characters, three for one, three for the other and one for neither (Fig. 161). In this diagram it is evident that the number of genotypes in the dihybrid is greater than the number of phenotypes, as was the case in the monohybrid.

In the F_2 generation many individuals are homozygous for one character and heterozygous for the other. Such individuals, when inbred, are capable of producing offspring in which the homozygous character always appears while the heterozygous character behaves as in monohybrids. An SsYY pea vine, for example, can produce only yellow seeds but they will appear in the ratio of three smooth to one wrinkled.

	Male gametes			
Female gametes	S Y	S y	s Y	s y
S Y	SY SY (1) 1	Sy SY (1) 2	sY SY (1) 3	sy SY (1) 4
S y	SY Sy (1) 2	Sy Sy (2) 5	sY Sy (1) 4	sy Sy (2) 6
s Y	SY sY (1) 3	Sy sY (1) 4	sY sY (3) 7	sy sY (3) 8
s y	SY sy (1) 4	Sy sy (2) 6	sY sy (3) 8	sy sy (4) 9

FIG. 161.—Diagram to illustrate the F_2 generation of Mendel's hybridization of smooth yellow and green wrinkled peas. Ratio of phenotypes $9:3:3:1$; ratio of genotypes $1:2:2:4:1:2:1:2:1$. The phenotypes are numbered in the upper right hand corners of the squares, the genotypes in the lower corners.

In experiments with animals similar results have been obtained. It has been found that a guinea-pig with short black hair crossed with a long-haired albino produces a short-haired black F_1 generation. When these F_1 guinea-pigs are inbred they produce an F_2 generation with four types of animals, viz., short-haired black, short-haired white, long-haired black and long-haired white. Of these the short-haired black animals may belong to four different genotypes, the long-haired albinos to only one and the others to two each. Figure 162 illustrates the kinds of guinea-pigs based on the reassortment of three different characters.

Trihybrids. As shown in Figure 163, a hybrid between individuals bearing the allelomorphs of three different characters

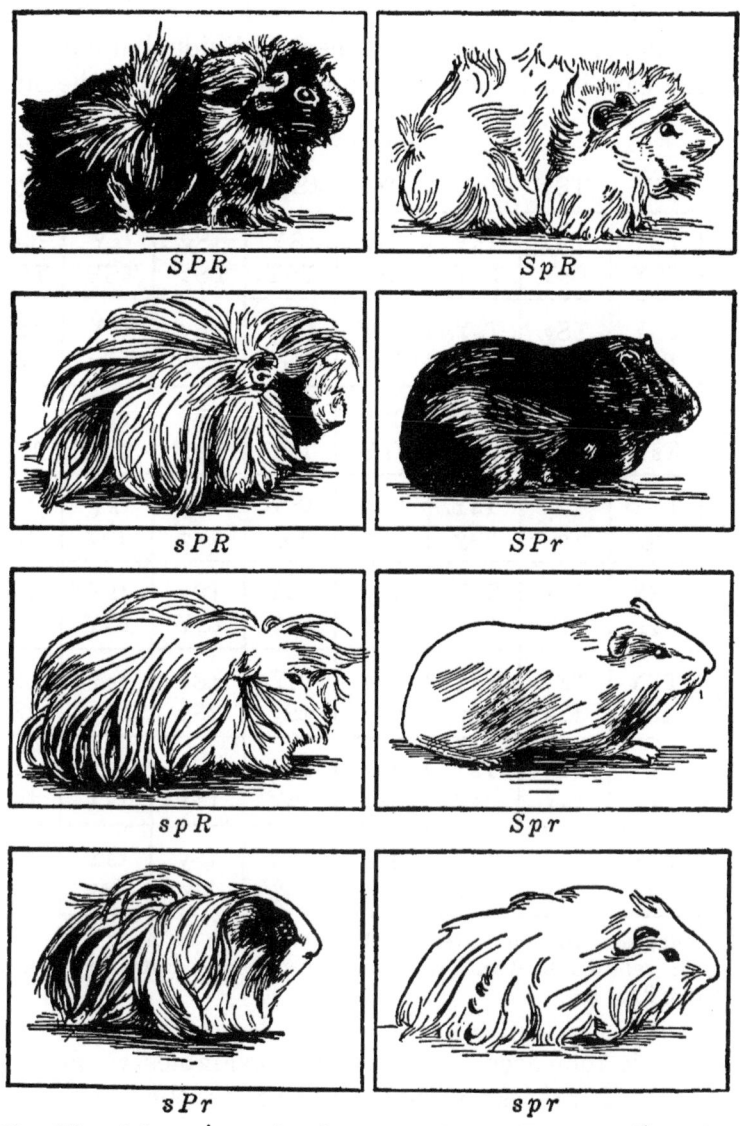

FIG. 162.—The eight guinea-pig phenotypes in the F_2 generation of a trihybrid. S, short hair; s, long hair or angora; P, pigmented coat; p, albino; R, rough or rosetted coat; r, smooth coat. (From Walter, after drawings from Castle's photographs by C. J. Fish.)

affords possibilities for eight different combinations in the gametes of the F_1 generation. The same number of combinations is present in the gametes of each parent (Fig. 163), so that sixty-four combinations from the two parents result in the individuals of the F₂ generation. Among these sixty-four the eight phenotypes express the trihybrid ratio 27:9:9:9:3:3:3:1. The diagram shows both the phenotype and genotype of each individual and illustrates the rapid increase in complexity of the F₂ generation as the number

Male gametes

	TSY	TSy	TsY	Tsy	tSY	tSy	tsY	tsy
TSY	TSY/TSY	TSy/TSY	TsY/TSY	Tsy/TSY	tSY/TSY	tSy/TSY	tsY/TSY	tsy/TSY
TSy	TSY/TSy	TSy/TSy	TsY/TSy	Tsy/TSy	tSY/TSy	tSy/TSy	tsY/TSy	tsy/TSy
TsY	TSY/TsY	TSy/TsY	TsY/TsY	Tsy/TsY	tSY/TsY	tSy/TsY	tsY/TsY	tsy/TsY
Tsy	TSY/Tsy	TSy/Tsy	TsY/Tsy	Tsy/Tsy	tSY/Tsy	tSy/Tsy	tsY/Tsy	tsy/Tsy
tSY	TSY/tSY	TSy/tSY	TsY/tSY	Tsy/tSY	tSY/tSY	tSy/tSY	tsY/tSY	tsy/tSY
tSy	TSY/tSy	TSy/tSy	TsY/tSy	Tsy/tSy	tSY/tSy	tSy/tSy	tsY/tSy	tsy/tSy
tsY	TSY/tsY	TSy/tsY	TsY/tsY	Tsy/tsY	tSY/tsY	tSy/tsY	tsY/tsY	tsy/tsY
tsy	TSY/tsy	TSy/tsy	TsY/tsy	Tsy/tsy	tSY/tsy	tSy/tsy	tsY/tsy	tsy/tsy

(Female gametes on left axis)

FIG. 163.—Diagram to illustrate the F₂ generation of a hybrid between tall smooth yellow and dwarf wrinkled green peas. Ratio of phenotypes 27:9:9:9:3:3:3:1.

of characters is increased. Any further examples would merely repeat the same process without adding to the facts already disclosed.

Theoretical Calculations. We have seen that a single pair of allelomorphs make possible four combinations according to the law of chance, that two make possible sixteen combinations, and three sixty-four combinations. It is therefore apparent that the number of possible combinations is equal to four raised to a power indicated by the number of different characters involved, and if we are concerned with ten characters, a modest portion of those present in any complex animal, the result is the astonishing total of 1,048,576. When we consider the number of unit characters which must be present in a human being it is easy to understand why no two individuals are alike.

The Back-Cross. In cases of complete dominance in multiple hybrids it would be an endless task to attempt the determination of genotypes by inbreeding. Self-fertilization would lighten the task for the plant breeder, since he could be certain of mating like with like, but the animal breeder would be at a loss to know which of his many dominants were genotypically the same, so that their crossing would be equivalent to self-fertilization. To meet the difficulty of this situation it is customary to test an individual by crossing it with a known recessive. Since recessive characters are always apparent in the phenotype, such individuals can be selected without difficulty.

As an example let us assume that we have the F_2 progeny of the hybrid represented in Figure 163. We know that each of the twenty-seven dominants contains the characters tall, smooth and yellow, but whether homozygous or heterozygous for these characters must be determined. If the individual in question is mated with a ttssyy, or homozygous recessive, and gives only tall smooth yellow and tall smooth green offspring we can be certain that it was homozygous for both of the first two characters and heterozygous for the third.

Kinds of gametes formed by F_2 dominant		
	TSY	TSy
Gametes of homozygous recessive } tsy	TSY tsy	TSy tsy

Fig. 164.—Diagram illustrating the back-cross of a tall-smooth-yellow pea of the F_2 generation with a short-wrinkled-green homozygous recessive.

For every heterozygous character it would give to one-half of its offspring the recessive and to one-half the dominant. Those which received the recessive, since they could receive only that character from the homozygous recessive parent, would necessarily express the recessive condition. The individual cited would therefore belong to the genotype TTSSYy. The results are expressed in Figure 164.

Linkage. Although these facts justify the belief that unit characters are independent of each other, it has been found that some always appear together under normal conditions. Such characters are said to be linked. Linkage does not imply that the characters are in any way dependent upon each other or that they are related in any way other than simple association in the same individuals. They may involve very different parts of the body, as in the case of bar eyes and small wings in *Drosophila*, the fruit fly. Characters may also be definitely associated with sex, and are then said to be sex-linked. Among these are color-blindness in man and various characters in birds, insects and other organisms.

The effect of linkage is a simplification of the ordinary hybrid ratios, for a group of linked characters give only a monohybrid ratio because of their normal inseparability.

Modern Investigations. During the twentieth century Mendel's discoveries have been repeatedly verified and considerably extended by breeding experiments. Both plants and animals have been used as materials, and in all cases satisfactory results have been secured, although accuracy is most nearly attained in plants and such animals as produce large numbers of young. Laboratory animals like the guinea-pig, rabbit and rat are sufficiently prolific to be useful, but not to give approximately accurate Mendelian ratios unless several litters of the same cross are used.

According to the law of chance, if a given phenomenon may occur in a number of ways and is allowed to occur a sufficient number of times, it will occur in all possible ways. The allelomorphs in a monohybrid can combine in only four ways, three of which are different. If animals produce only single young, obviously one litter cannot express the monohybrid ratio. There is a possibility that a litter of four may do so. A dihybrid ratio cannot be completely expressed by less than sixteen individuals, which is beyond the size of litters produced by most mammals, although the four phenotypes represented in this ratio may readily be produced in smaller litters.

Corn has come into use in the last few years as a laboratory illustration of Mendelian ratios, and because of the large number of grains on a single ear it usually approximates dihybrid ratios fairly well and monohybrid ratios very closely. Since each seed is a potential plant, the seed characters of a field of corn can be studied in a few cars. The characters available include purple and white aleurone and starchy and sugary endosperm, as well as other color characters. A single ear representing the F_2 generation of a purple-white starchy-sweet cross shows purple starchy, purple sweet, white starchy and white sweet grains. The sweet grains are shrunken and so stand out in sharp contrast to the smooth, plump starchy grains (Fig. 165).

The pendulum has swung strongly toward the use of fruit flies of the genus *Drosophila* in this work since the first important contributions of Morgan and his associates. These insects have several advantages for genetic research. They reproduce rapidly and in large numbers; they are easily reared in the laboratory; they are so organized as to facilitate cytological examination, and best of all they give rise to many mutations which furnish the

FIG. 165.—An ear of corn showing the F_2 dihybrid ratio. Four kinds of grains are present, viz., purple starchy, purple sweet, white starchy and white sweet, in the proportion of 282:86:74:27; this is very near to the expected ration of 9:3:3:1. The starchy grains are smooth and the sweet grains wrinkled.

character contrasts so necessary for accurate observation (Fig. 155). *Drosophila* has been so productive that almost any principle of genetics can be illustrated with it alone.

Practical Importance. From the foregoing account it is apparent that the transmission of characters, whatever their number, goes on according to definite laws. By taking advantage of these laws desired combinations of characters can be secured through the proper control of propagation without awaiting the accidental occurrence of the right individual. If a plant breeder has purple sweet corn and white field corn, for example, and wishes to produce white sweet corn, he can do so by hybridizing the two and selecting the homozygous recessives of the F_2 generation. Or if any other characters are desired the homozygous condition will be found in some individuals of the F_2 generation, and such individuals will breed true. One thing alone cannot be stabilized as an independent strain through ordinary methods of reproduction, and that is a variety based on the heterozygous condition, such as the blue Andalusian fowl. These must be maintained by repeated crossing or by breeding heterozygous individuals and selecting their heterozygous offspring.

Summary. Mendel established the fact that organisms are made up of unit characters which are segregated during reproduction. These characters may be alternative, blending or mosaic in the hybrid, but in the offspring of the hybrid they are reassociated in definite ratios, so that the characters of the original parents as well as of the first hybrid reappear. The ratios are fixed according to the law of chance for every number of characters. They are only approximated in breeding experiments unless large numbers of individuals are considered. Many organisms are now used for experimental studies in this field, but fruit flies of the genus *Drosophila* are the most important. The laws are of great practical importance to breeders of plants and animals.

REFERENCES

MORGAN, T. H., STURTEVANT, A. H., MULLER, H. J., and BRIDGES, C. B.,
The Mechanism of Mendelian Heredity, revised edition, 1922.
MORGAN, T. H., *The Physical Basis of Heredity*, 1919.

All general works on heredity contain more or less detailed accounts of Mendelian inheritance. Those cited above are especially valuable for detailed evidence of the various processes as illustrated in *Drosophila*.

CHAPTER XVI

THE CHROMOSOME THEORY OF HEREDITY

The remarkable behaviour of the chromatin during mitosis, described in Chapter VI, is fundamental evidence that this substance plays a peculiar and important part in the history of cells. During mitosis all other portions of the nucleus become merged with the cytoplasm, the centrosome plays a part which seems to be limited entirely to cell reproduction, and the cytoplasm and chromatin are equally divided between the daughter cells. The division of the cytoplasm is attended by no definite phenomena

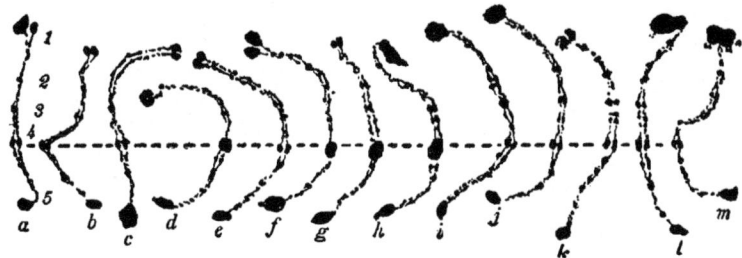

Fig. 166.—Examples of one of the chromosomes of *Phrynotettix* sp. taken from thirteen individuals. The dotted line passes through homologous granules. x 2200. (From McClung, in Cowdry's *General Cytology*, after Wenrich; with the permission of the University of Chicago Press.)

which would indicate both quantitative and qualitative equality of distribution. The chromosomes, however, are developed in many cases as slender structures showing longitudinal differentiation which is evident in well prepared specimens (Fig. 166) and the longitudinal splitting that occurs in the metaphase is apparently an equation division. Moreover in the many successive mitoses intervening between the union of the germ cells and the completion of embryological development extensive cytoplasmic differentiation occurs. Although it is accompanied by some nuclear differentiation evidence shows that the same chromosome complex persists even in highly specialized cells. This complex is characteristic of the different parts of the body of all individuals of the species, and in some cases of related species. McClung's

290 EVOLUTION AND GENETICS

data on the Orthoptera show striking chromosomal similarity among related species of these insects.

The entire body of facts relating to chromosomal behaviour is so significant that it has given rise to the chromosomal theory

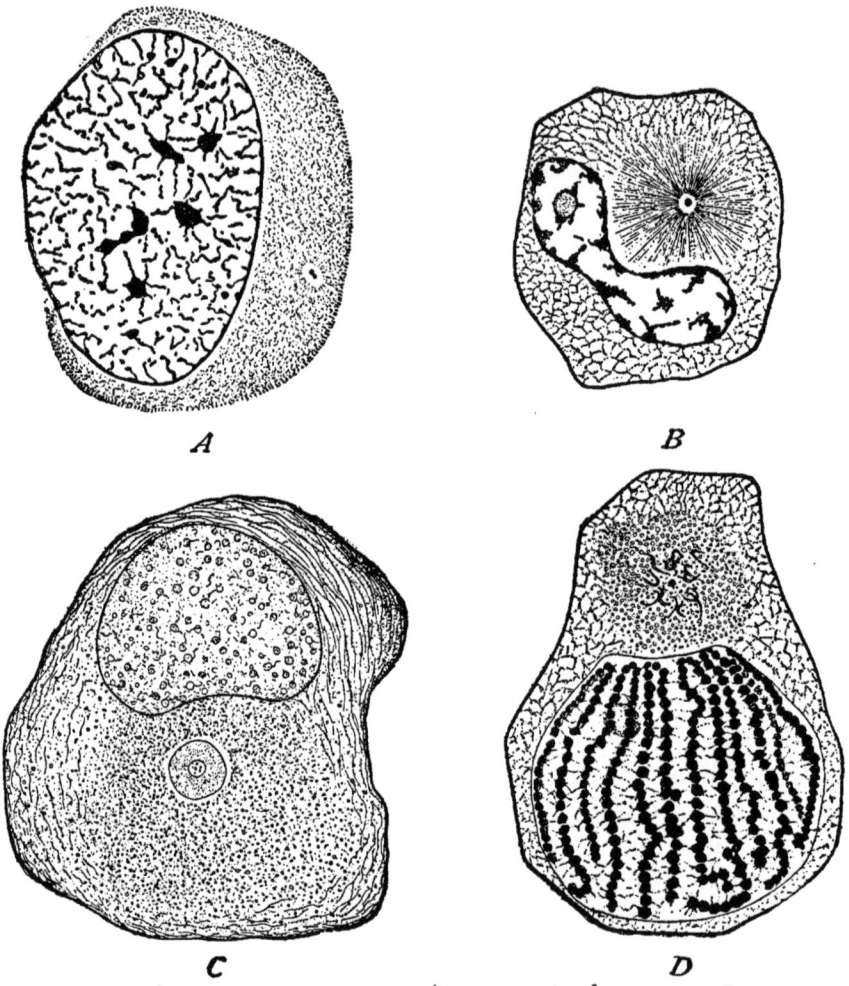

Fig. 167.—Various cells. *A*, from peritoneum of salamander; *B*, spermatogonium of frog; *C*, spinal ganglion cell of frog; *D*, spermatocyte of Pro*teus*, nucleus in spireme stage. (From Wilson, *A* after Flemming, *B* and *D* after Hermann, *C* after Lanhossek.)

of heredity which postulates that these minute bodies are the seat of the substances or particles which are handed down from one generation to the next as the foundation of hereditary resemblance.

THE CHROMOSOME THEORY OF HEREDITY   291

This theory was preceded by the hypothesis that the body contained some substance which acted in a similar capacity. This hypothetical substance was called the idioplasm. The term is reminiscent of Weismann's theories in which the terms id and idant were prominent, but before Weismann's contributions appeared, Roux (1883) had recognized the similarity of chromatin to idioplasm. Modern science has established a large body of facts which are little short of actual proof that the chromosomes are the material basis of heredity.

The Organization of Chromosomes. During the resting stage of a cell its chromatin is visible in stained preparations as a network of uneven texture or as an aggregation of granules (Fig. 167).

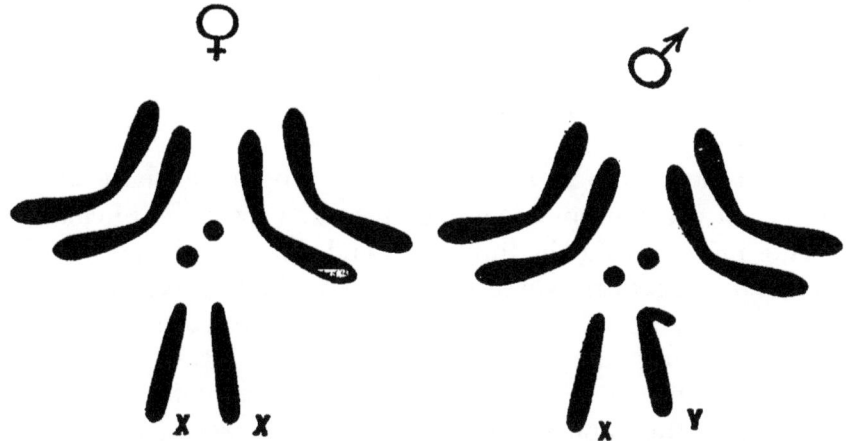

Fig. 168.—Diagram of female and male diploid groups of chromosomes of *Drosophila melanogaster*. The hook on the y chromosome is characteristic. (From Morgan et al., *Mechanism of Mendelian Heredity*, with the permission of Henry Holt and Company.)

This arrangement is lost when the nuclear membrane breaks down at the beginning of cell reproduction and the chromatin condenses, sometimes into a heavier thread which later breaks up into chromosomes and sometimes directly into chromosomes. In its condensed state the chromatin appears to be a homogeneous material with a marked affinity for many biological stains.

According to cytologists, however, the chromatin is actually composed of minute bodies called chromomeres which retain the same relationships in all stages where they are visible. Wenrich has shown that these chromatin units differ in size, so that their arrangement in chromosomes can sometimes be traced readily.

He has also made comparisons of many cells in which their arrangement appeared to be constant (Fig. 166).

Characteristics of Chromosomes. While the inner organization of the chromosomes is a subject for cytological studies of the utmost refinement, the bodies themselves are more readily observed and exhibit several phenomena of interest.

1. Chromosomes appear in the cells of a given species in the same number and form, with the exception of a common discrepancy between the sexes (Fig. 168).

2. Within the chromosome complex of a given species there are chromosomes of different sizes and shapes. In species which show chromosomal differences sufficient for accurate observation it is evident that these characteristics are constant for different individuals.

3. In the cells of the body the chromosomes are duplicated, i.e., there are two of each kind, with the sole exception of those associated with sex.

The Chromosomes in Reproduction. No single phase of the behaviour of chromosomes is as significant as the series of changes that take place during the formation and union of the germ cells. In the higher organisms these are of two distinct types, the female gamete or ovum and the male gamete or spermatozoön of animals. Both are fundamentally similar in that they possess one half of the total number of chromosomes characteristic of the species, usually called the haploid number, and in the process of gametogenesis by which they are produced. By the union of two germ cells the full or diploid number of chromosomes is restored.

Gametogenesis. The body of an organism during development contains many primordial germ cells which are directly descended from the original fertilized ovum with which its development began. These primordial germ cells multiply by mitosis, and by the completion of embryonic life have produced many other cells which lie in the gonads. Those of the female are called oögonia and those of the male spermatogonia (Fig. 169A). After a period of growth they become the primary oöcytes and primary spermatocytes respectively, and in this stage a significant step occurs in the behaviour of their chromosomes, known as synapsis (Fig. 169C).

Synapsis. In brief, synapsis consists of a pairing of similar chromosomes from the diploid complex. It is attended by a considerable degree of complexity in some animals, but in its

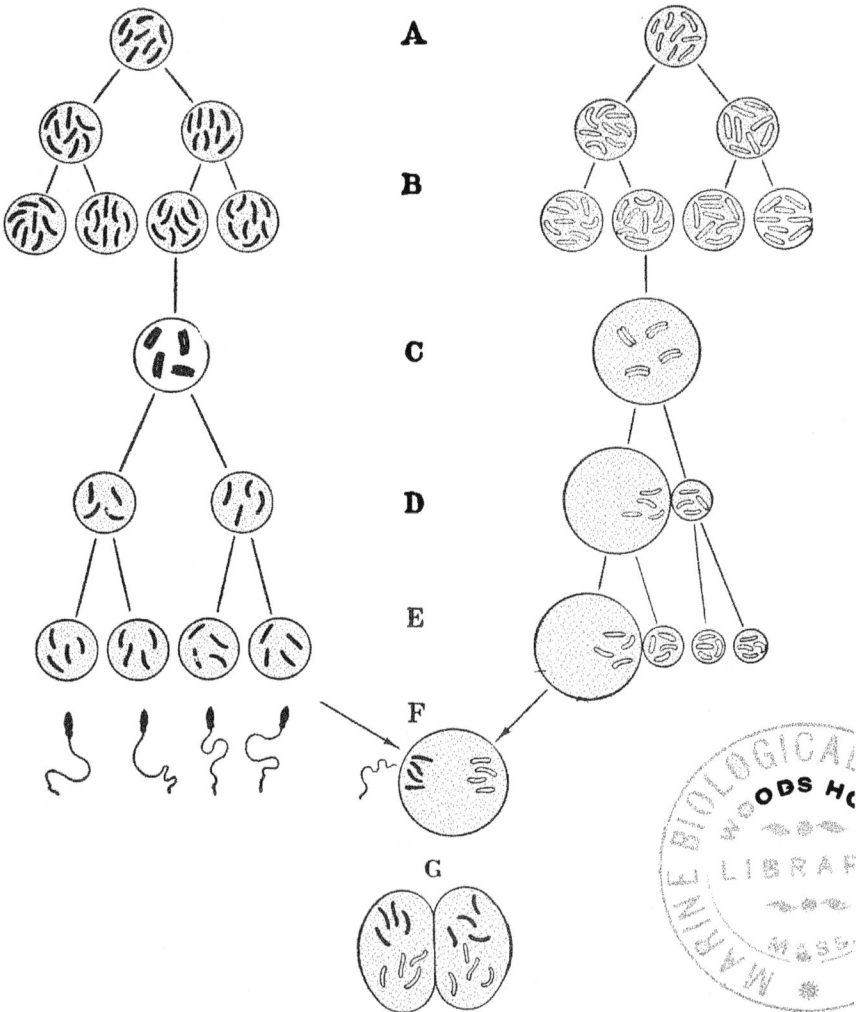

FIG. 169.—Diagram of the general plan of spermatogenesis and oögenesis in animals. The somatic, or diploid, number of chromosomes is assumed to be eight. Male to the left, female to the right. A, primordial germ cells; B, spermatogonia and oögonia, many of which arise during the period of multiplication; C, primary spermatocyte and oöcyte, after the growth period, with chromosomes in synapsis; D, secondary spermatocytes and oöcytes with haploid number of chromosomes. These cells are the product of the reduction division; E, spermatids, which develop into spermatozoa; in the female, one egg and three polar bodies. The cells of this stage are produced by the equation division. The order of the reduction and equation divisions is variable in different species; the two together are called the first and second maturation divisions. F, union of ovum and spermatozoön (fertilization); G, the diploid number of chromosomes in the daughter cells formed by the first cleavage of the zygote. This number persists in all cells of the body and in the germ cells until maturation. (From Woodruff.)

essential features is apparently always the same. Omitting consideration of the chromosomes associated with sex, we find that the fully developed primary spermatocyte or oöcyte contains as many pairs of chromosomes as there are kinds. In preparation for the two *maturation divisions* which follow close upon synapsis, each of the chromosomes has split longitudinally, so that each synaptic pair is in reality a group of four halves of chromosomes called a *tetrad*. Tetrads are not represented in Figure 169, which is designed to show in the simplest possible way the fundamental features of the process. In Figure 170E is shown a primary oöcyte of *Ascaris megalocephala* containing two tetrads. The diploid number of chromosomes in this species is four.

Maturation Divisions. Closely following the formation of tetrads each primary cell undergoes two successive divisions which distribute the four parts of each tetrad among the four resulting daughter cells. One of these divisions separates the two synaptic mates, or entire chromosomes, and produces daughter cells with only the haploid number. This is called the *reduction division*. The other merely separates halves of chromosomes, so that daughter cells and parent cell are similar. It is called the *equation division* (Fig. 169C–E). The order in which the two divisions occur may vary among different species.

The cells derived from primary spermatocytes are called secondary spermatocytes, and those of the next generation spermatids. Spermatids undergo a process of differentiation and become motile spermatozoa.

The maturation divisions in the female differ in that only one primary and one secondary oöcyte are produced. Each of the maturation divisions results in the concentration of cytoplasm in this one cell, while the necessary chromosome reduction is accomplished by the formation of abortive *polar bodies*. The first polar body (Fig. 169D) is similar in nuclear organization to the secondary oöcyte and the second is similar to the mature ovum, with the exception of such qualitative differences as may result from reduction. The first polar body sometimes divides as shown in Figure 169E.

The Mature Germ Cells. Although the mature germ cells of both sexes contain the haploid number of chromosomes they may be very different in other ways. The ovum is a large cell containing an abundance of cytoplasm and often a large quantity of stored food. It is developed to an extreme in the eggs of birds, in which the yolk is the true ovum but the active cytoplasm is

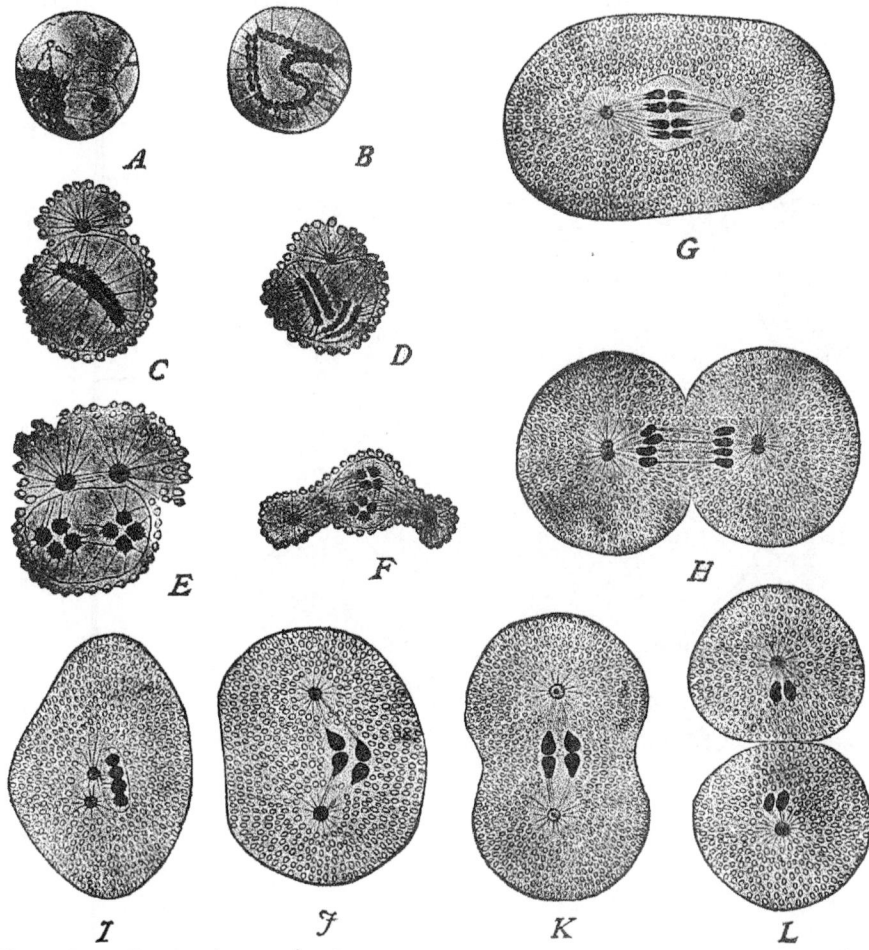

Fig. 170.—Spermatogenesis in *Ascaris megalocephala*. *A–G*, successive stages in the division of the primary spermatocyte. *D* shows the two tetrads in profile and *E* shows an end view of them; *F, G, H*, division of the primary spermatocyte to form two secondary spermatocytes, each containing two dyads; *I*, secondary spermatocyte; *J, K*, the same dividing; *L*, spermatids, each with two chromosomes and a centrosome. The normal diploid number of chromosomes in this species is four. (From Wilson, after Brauer.)

restricted to a small mass on one side. The remainder is yolk in the strict sense, and is food for the developing embryo. The white is an envelope of albumen of similar importance outside of the cell.

296 EVOLUTION AND GENETICS

The spermatozoön, on the other hand, is highly differentiated. It consists of several regions, particularly the head, neck and tail. The head is a very compact nucleus, and the tail

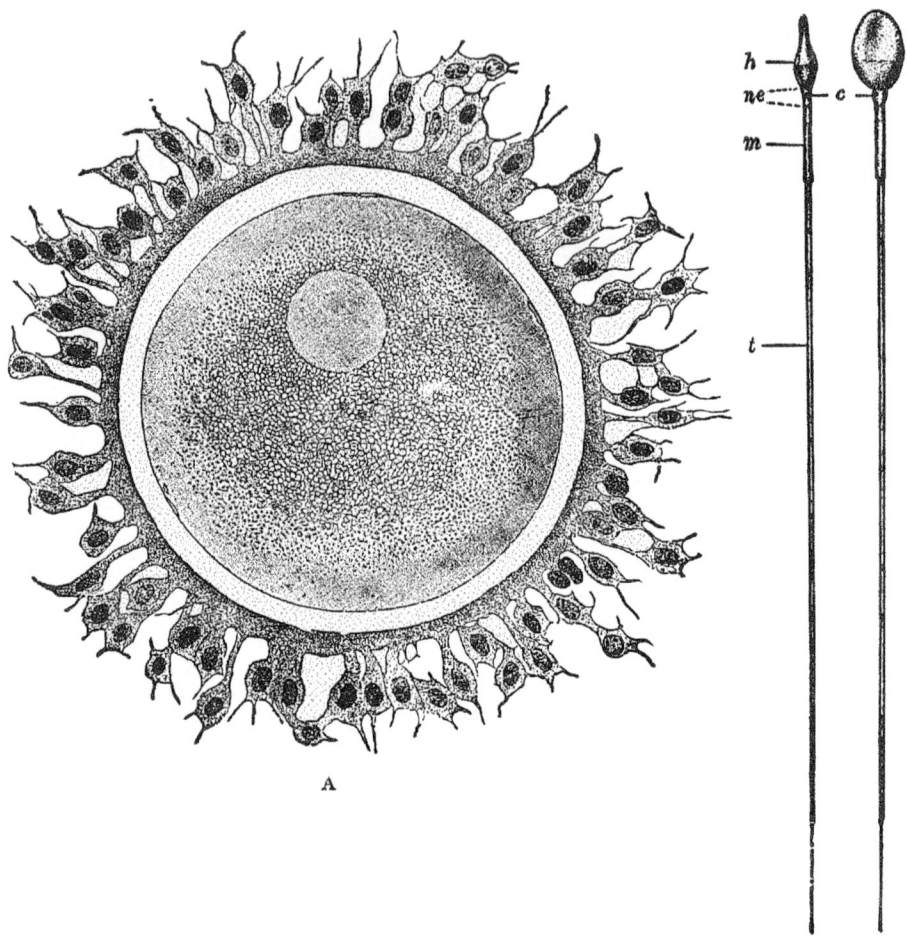

Fig. 171.—Human germ cells. A, ovum, x 415, surrounded by follicle cells from the ovary; B, sperm cell x 2000. (From Woodruff, after Retzius.)

is an organ of locomotion. The human germ cells are shown in Figure 171.

Fertilization. This is the process of union of the two kinds of germ cells. When a sperm cell comes into contact with an ovum which has not previously been fertilized it penetrates the membrane surrounding the ovum. The head, at least, enters the cytoplasm of the ovum, while the tail remains outside. The surface of the ovum then undergoes changes which normally prevent the entrance of other sperms.

The head of the spermatozoön then undergoes changes which transform it into a *male pronucleus*, resembling a resting nucleus, which is similar to the *female pronucleus* already present. A centrosome appears, divides, and forms a mitotic spindle. The source of this centrosome is uncertain. It was once thought that it was introduced by the male cell, and this may sometimes be the case, but there is evidence to show that it may arise from the cytoplasm of the ovum. After the formation of the mitotic spindle the pronuclei undergo changes similar to those which occur in the prophase of an ordinary mitotic division and the resulting chromosomes group themselves together in the equatorial plate. From this point the process is similar to the three final stages of mitosis, viz., the metaphase, anaphase and telophase, and its completion is the formation of two daughter cells. This is the first cleavage of embryonic development.

The significant fact of fertilization is the combination in one cell of two similar groups of chromosomes derived from different parents. The diploid complex is maintained in the new individual by mitosis. By referring to Figure 169F and G, in which the chromosomes derived from the female parent are outlined and those from the male are black, the possible results may be seen. There is apparently no definite association between the different kinds of chromosomes, hence the reduction division in the individual derived from this cell may place some maternal and some paternal chromosomes in each resulting germ cell. Such a reassortment of parental chromosomes is of the greatest importance. The number of combinations possible is obviously limited by the number of chromosomes present.

Sex Chromosomes. The foregoing account omits consideration of a common contrast in the chromosome complex of males and females of the same species. Sometimes a difference is discernible between two synaptic mates when the number is constant, and in other cases there is a numerical difference, one sex having an odd number of chromosomes and the other an even number due to the presence of a synaptic mate for the odd chromosome. These phenomena have given rise to a theory of chromosomal determination of sex which appears to be largely sound, although recent investigations have indicated that the chromosomes are not alone responsible.

In some species the male is characterized by an odd number of chromosomes and the female by an even number. The odd

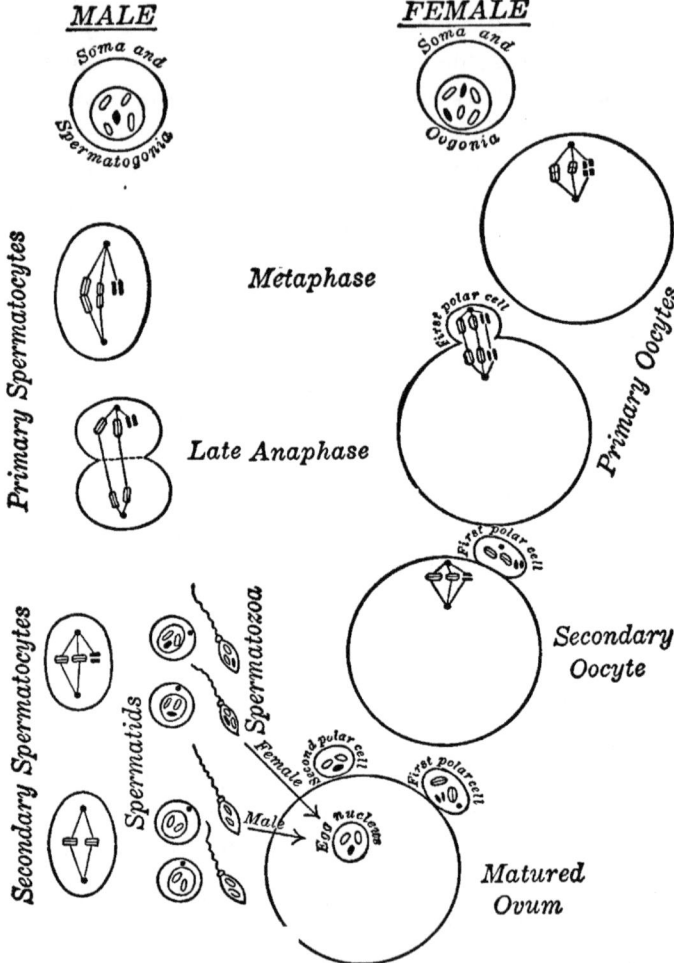

FIG. 172.—Sex determination in species having heterogametic males. (From Walter.)

chromosome in the male is without a synaptic mate and consequently the reduction division produces two types of gametes. Such species are said to have heterogametic males. The single chromosome is called the x chromosome. Since the females have an even number including two x chromosomes all female gametes are the same. The sex of the individual is therefore determined

THE CHROMOSOME THEORY OF HEREDITY

by the type of male gamete with which an ovum unites. Inheritance of sex in these species is graphically indicated in Figure 172.

Some heterogametic males have an even diploid number of chromosomes, but the synaptic mate of the x chromosome, called the y chromosome, is at least different in function. The chromosome complex of *Drosophila* is of this type (Fig. 168).

Other animals, including the birds and Lepidoptera, are known to have heterogametic females. The behaviour of the chromosomes is similar but two kinds of ova are produced and only one kind of spermatozoa. The sex chromosomes are called z chromosomes in such cases.

Sex chromosomes of all kinds are sometimes called allosomes, and the remaining chromosomes autosomes.

Abnormal Behaviour of Chromosomes. The phenomena of chromosome behaviour just described are the normal. In such a complex series of changes it would be surprising to find that abnormalities never occur. Several modifications have, in fact, been demonstrated, including (1) interchange of parts between synaptic mates and (2) changes in the number of chromosomes. The latter may result from multiple fertilization or from the failure of chromosomes to separate during the reduction division so that the chromosome complex of some of the daughter cells is increased and of others decreased.

Interchange of parts between synaptic mates cannot actually be seen, but during certain stages of synapsis the thread-like chromosomes have been found twisted together, and some of the phenomena of heredity indicate that the separation of these mates may be attended by rupture at one or more points of crossing. This would result in the formation of new chromosomes, each formed of parts of both of its predecessors (see Chapter XVII).

The multiplication of chromosome numbers does not change the nature of the chromosomes involved, and so may not change the nature of the individual. Haploid, triploid and tetraploid complexes are on record as well as the usual diploid complex and in some cases they have been produced experimentally. It is worthy of note that gigantism is a common result of the tetraploid condition as recorded by DeVries in the evening primrose, Marchals in mosses and Winkler in tomato and nightshade, yet in cases of abnormal numbers of sex chromosomes no abnormality is apparent except the modification of sex ratios.

300 EVOLUTION AND GENETICS

The inheritance of sex in *Drosophila* in a case involving non-disjunction of the x chromosomes in the female parent illustrates the effect of abnormal allosome number. Females of *Drosophila* normally produce only one kind of egg; non-disjunction results in

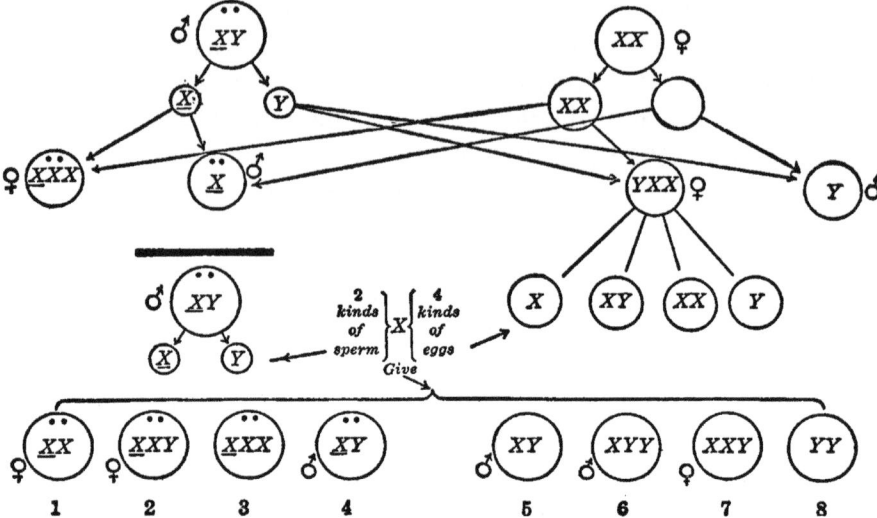

Fig. 173.—"Non-disjunction and its results in *Drosophila*. The two large circles in the first row represent male and female flies producing sperms and eggs respectively. Non-disjunction in the female gives two kinds of eggs, with xx and no sex chromosomes, instead of the normal single kind with one x. At fertilization there are possible four combinations rather than two, as shown in the large circles of the second row. Owing to the several ways in which her three sex chromosomes may be distributed at maturation, the female represented by the third circle produces four kinds of eggs. When mated to a normal male (below the horizontal line) with two kinds of sperms, eight combinations are possible (last row). Numbers 1, 4 and 5 are normal flies and give the usual type of progeny. Numbers 2, 6 and 7, owing to the presence of three sex chromosomes, give exceptional results when bred. Types numbers 3 and 8 do not appear in the cultures, probably because they die very early. The original male has red eyes and the original female white eyes. Red eyes (represented by the dots) appear in every fly bearing the x-chromosome of the original male." (From Walter, diagram by Sharp based on data from Bridges and Morgan.)

the production of two kinds, one with no x chromosome and one with two. In combination with normal spermatozoa they give the results shown in Figure 173.

Evidence for the Chromosome Theory

The Contributions of the Sexes. *Nuclear Components.* In the phenomena of Mendelian inheritance it is evident that the male and female parents contribute equally to the heritage of their issue with the sole exception of sex and associated characters.

Since the germ cells are actually a bridge between the generations and must therefore contain the potential equivalent of all that is to appear in the new individual it is reasonable to expect similarity of those which combine to the same extent that we find similarity of the characters which they transmit. This similarity of male and female germ cells we have but briefly considered.

The development of sexual reproduction begins with the union of obviously similar gametes. Differentiation such as that exhibited by the germ cells of mammals and birds is an extreme development. In the transition between the two stages it appears that anything which remains constant is fundamental, and the chromatin of the nucleus alone fails to undergo marked change. The highly specialized spermatozoön conveys little or no cytoplasm into the ovum which it fertilizes; the ovum contributes much. The nuclear material, especially the chromatin, is equivalent with the exception of sex chromosomes. If cytoplasm played a large part in the determination of the qualities of the new individual the preponderance of maternal characters would seem to be inevitable.

The Rôle of the Cytoplasm. The cytoplasm is apparently only a plastic material for the expression of the heritage. Lillie and Just make the significant statement that "the materials of the cytoplasm are . . . being constantly consumed in the metabolism, and the process of renewal and increase of such materials involves interaction of nucleus and cytoplasm; therefore the purely maternal cytoplasm soon disappears, and is replaced by cytoplasm formed under the influence of the biparental zygote nucleus." There is certainly some increase, if not replacement, of material in the chromosomes as development proceeds, but it is evidently only to meet the demands of growth and multiplication, while cytoplasmic interchanges are at the root of the intricate vital processes in the living body.

A most convincing experiment is that conducted by Boveri on sea urchins. Enucleate cytoplasmic fragments from the eggs of one species were fertilized by the sperms of another species. These fragments developed sufficiently to show definite characters, and the characters produced were those of the species from which the sperms were derived. The wholly cytoplasmic maternal contribution was apparently without effect.

The cytoplasm of some eggs shows marked differentiation. The eggs of *Styela partita*, for example, are reported by Conklin as having an orange pigment distributed over the surface before maturation, while during fertilization this pigment retreats to the vegetal pole. Above the orange area there appears a layer of clear cytoplasm and in the remainder of the egg gray yolk. Such regional differentiation is sometimes associated with the initial

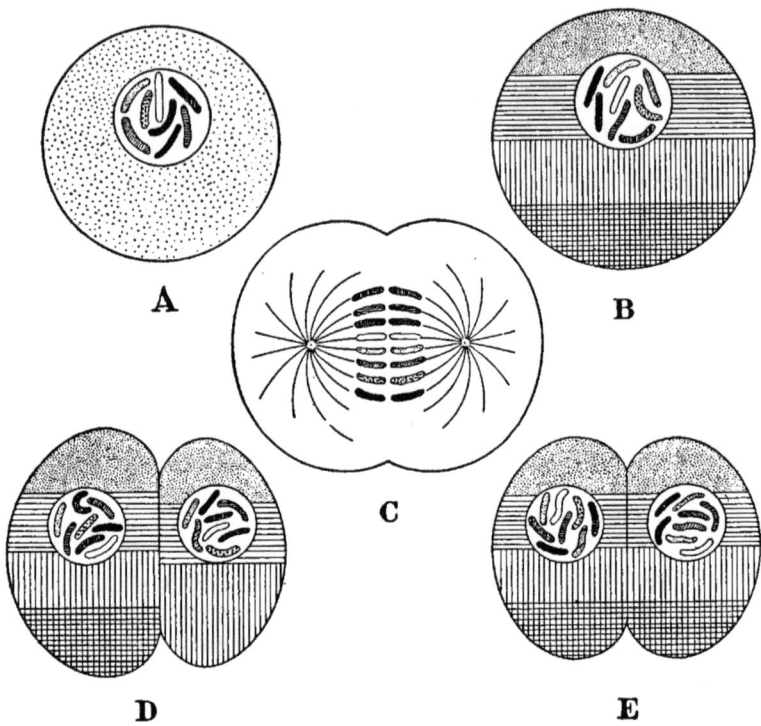

Fig. 174.—Diagram of zones of cytoplasmic differentiation and their distribution at the first cleavage. A, immature egg with no visible cytoplasmic differentiation; B, mature egg with four zones; C, division of the egg; D and E, two types of two-cell stages; D, *Dentalium* or *Styela*, in which one cytoplasmic zone passes completely into one cell; E, sea urchin or *Amphioxus* type, with equal distribution of cytoplasm. (From Woodruff, after Wilson.)

cellular differentiation of the developing embryo but it has not been shown to influence hereditary characters (Fig. 174).

It is necessary to draw the conclusion that cytoplasm is merely a material of construction. It is as essential in reproduction as in individual life but is apparently entirely under the control of the nucleus, and particularly of the chromosomes. The two are mutually dependent; cytoplasm alone can give expression to the chromosomes and the chromosomes alone can activate the cytoplasm.

Mitochondria. These are bodies included in the cytoplasm, to which functions in heredity have been attributed. A paragraph from Lillie and Just will serve as an ample refutation of this proposal: "Whatever may be the function of the mitochondria in cell physiology, it must be admitted that the study of fertilization has shown no reason for the assumption that their introduction into the egg by the sperm in certain species is concerned in the transmission of paternal characteristics. The variable quantity in different cases and the distribution to single blastomeres in certain cases exclude the hypothesis that they have any specific paternal hereditary effect. There is no reason to deny that sperm mitochondria function in the egg when present, but if so it is probable that they are not differentiated in their chemical composition or genetic behaviour from the mitochondria of the egg itself."

Mendelism and the Chromosomes. There is no stronger evidence for the chromosome theory of heredity than the close resemblance between the behaviour of Mendelian unit characters and that of the chromosomes. The points of resemblance may be summarized as follows:

1. Unit characters are duplicated in the organism, as is shown by the behaviour of allelomorphic units when they are combined in a hybrid. The monohybrid ratio indicates that there are two and no more than two of these units present, since it expresses almost exactly the mathematical possibilities of such a number. The same duplication prevails in the chromosomes of the body, where the diploid condition is normal.

2. Segregation of the determiners for unit characters occurs during reproduction. The reappearance of recessive characters in the F_2 generation after their complete concealment in the hybrid parents is evidence of this segregation. This is a close parallel of the separation of synaptic mates during the reduction division in gametogenesis.

3. The characters derived from hybrid parents are a recombination of characters represented in some way in both parents. In the formation of gametes and their union to form new individuals the possibility of similar rearrangement of parental chromosomes is apparent.

In other details of inheritance such as linkage, crossing over and sex-linkage the parallel is found to hold true. Behaviour of unit characters, however intricate, is found to be closely in accordance with the behaviour of the chromosomes as established by cytological studies.

Every known fact concerning the structure and function of the chromosomes singles them out as the seat of the hereditary properties of the organism. They are not independent of the cytoplasm, but are rather a controlling center. As the artist needs a medium of expression, so must they use the cytoplasm as a medium for the production of structures in the individual. There is nothing to show that other parts of the cell exert a shaping influence in themselves.

Summary. The chromatin contained in the nucleus of a cell is made up of many granules called chromomeres which are aggregated during cell reproduction to form chromosomes. The behaviour of these bodies follows a definite plan in ordinary cell reproduction and in the sexual reproduction of the individual. In sexual reproduction germ cells are formed which may differ in everything but chromosome content. This alone is the same in all germ cells of a species, with the exception of sex chromosomes. Since the heritage is derived equally from male and female parents, it is logical to interpret the chromosomes as its physical basis in the germ cells. Moreover the behaviour of the chromosomes in reproduction is so much like the behaviour of unit characters in the course of heredity as to establish the theory beyond reasonable doubt. The chromosomes are now commonly interpreted as the conveyors of determiners of hereditary characters, and there is no evidence to show that other parts of the cell act in a similar capacity.

REFERENCE

McClung, C. E., Cowdry's *General Cytology*, Section X, 1924.

CHAPTER XVII

GENES AND CHARACTERS

What Is Inherited? The resemblance of successive generations leads us to say that certain characters are inherited. For all practical purposes this is true and sufficiently accurate. We do receive from our parents the character in question to the extent that in ourselves the character develops from material received from the preceding generation and perfected during ontogeny. A more accurate analysis shows that the thing actually handed down from one generation to the next is some constituent of the germinal chromosomes which is capable of bringing about the development in the new generation of the same character which its progenitors had developed in the old. That some such entity exists for every unit character is evident from the facts already cited. The determiner in general is called a factor; the material entities within the chromosomes are called genes.

What Are Genes? During the stages of cell reproduction when the chromosomes are evident as distinct bodies they are more or less compact. At other periods the chromatin is obviously granular, made up of particles called chromomeres. Thus far the cell has refused to yield up other secrets of chromatin structure to the cytologist, and the gene remains a theoretical, not to say hypothetical, unit. Like the molecules, atoms, and electrons of modern physics and chemistry, genes lend themselves admirably to the explanation of phenomena of inheritance, and in the logical results obtained on this basis lies the justification for the use of the term.

Genes may be defined as chromatin units occupying definite parts of the chromosomes. Their function is to bring to expression in the developing organism the unit characters for which they stand, and to give rise to other similar genes which shall constitute the link with the next generation.

Somatoplasm and Germplasm. This distinction between genes and the characters resulting from their presence in the body emphasizes a distinction which has attained great importance in the literature of genetics and evolution. It has been shown that the course of development after fertilization results in the early separation of those cells which develop into the body proper and those which remain undifferentiated for the production of new germ cells. In *Ascaris*, for example, it has been found that one cell of the sixteen-cell stage develops into germ cells alone, while the remaining fifteen develop into the body proper. Within the germ cells the genes never gain expression; in the differentiated cells of the body they reach their fullest expression.

The body has been termed the *soma* or *somatoplasm*, and the germ cells are commonly called the *germplasm*. It is obvious that in all cases of sexual reproduction the former originates anew with every generation from the latter while the germplasm continues without evident change from generation to generation. The association of inherited genes with their respective characters is akin to this relation of soma and germplasm. Whether or not those transmitted to the next generation may have been influenced by the generation which has carried and nurtured them is one of the great unsettled problems of heredity. Mechanism for such influence has not been demonstrated, but neither has the lack of such mechanism.

Unit Characters and Their Factors. The behaviour of allelomorphic characters in hybrids indicates that each is represented by at least one factor. Evidently then a given unit character in a homozygous individual is also represented by two genes which segregate in the germ cells with the synaptic mates containing them.

Complication of Mendelian ratios has disclosed several other relationships which are valuable both from the practical and the purely scientific points of view. These include linkage and sex-linkage, crossing over, multiple allelomorphs and multiple genes.

Linkage. Different unit characters, when combined in hybrids, usually separate from each other in varying combinations in the F_2 generation. Thus the two characters of shape and color in Mendel's peas, with their allelomorphs, give a $9:3:3:1$ dihybrid ratio. Occasionally however two such characters remain normally associated and give only the monohybrid $3:1$ ratio. This phenomenon was briefly mentioned in Chapter XV. It is called linkage.

Linkage is easily explained by the chromosome theory of heredity, for organisms may have many dozens or hundreds of unit

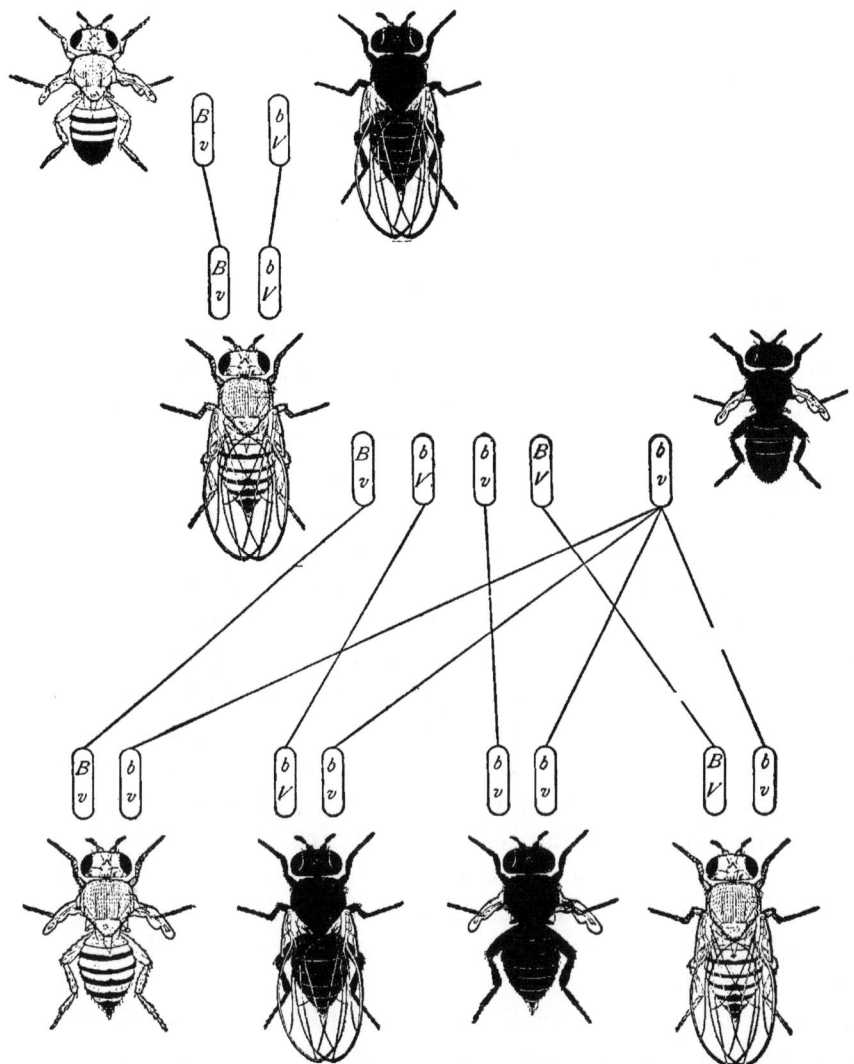

FIG. 175.—Back-cross of F_1 female (out of black by vestigial) with black vestigial male. B indicates normal color, b black, V normal wings and v vestigial. An illustration of crossing over. (From Morgan et al, *Mechanism of Mendelian Heredity*, with the permission of Henry Holt and Company.)

characters but only a few chromosomes. Each chromosome in such cases must necessarily bear the genes of many different characters, and since the chromosome and not the gene is the unit vehicle of transmission, all genes included in a given chromosome must necessarily retain their association with each other during reproduction. Modification of this relationship cannot occur without irregular behaviour of the chromosomes.

In the pomace fly, *Drosophila melanogaster*, many cases of linkage have been observed. Morgan and his associates record among these the case of black body color and vestigial wings with details of the behaviour of these characters when combined in hybrids with their normal allelomorphs, gray body color and normal wings.

Members of the F_1 generation of such a cross are all gray and long-winged. When the F_1 males are crossed with recessive, i.e., black-vestigial, females, two types of offspring are produced. One half are black-vestigial and one half gray-long. This proportion is exactly the same as would be expected if only a single character were involved, and indicates that the genes for these characters are located in the same chromosome. Linkage implies no functional relationship between these genes.

The reciprocal cross of an F_1 gray-long female with a black-vestigial male, however, gives wholly different results. In this case the offspring are of four kinds, black-vestigial, gray-long, black-long, and gray-vestigial. The ratio in which these kinds appear in experiments is not, however, even remotely similar to the Mendelian dihybrid ratio. Black-vestigial and gray-long appear in equal numbers, making up a total of 83 per cent of the generation; the remaining 17 per cent are also equally divided between the characters black-long and gray-vestigial (Fig. 175). This case involves an interchange of normally associated genes, and implies also an interchange of substance between the chromosomes involved.

Crossing over is the term applied to this phenomenon and to the accompanying chromosomal behaviour. The fact that interchange of the characters occurs in a majority of cases points to it as the abnormal type of inheritance, which would also be true of interchange of substance between synaptic mates. The equality of distribution of the reciprocal combinations is also exactly what would result from an interchange of chromosomal substance.

During synapsis the related chromosomes have been repeatedly observed tightly coiled about each other. McClung states that this coiling is readily visible at one stage of synapsis in the Orthoptera. It is easy to see that the separation of these chromosomes during the reduction division might be accompanied by rupture at some point of crossing and the formation of new chromosomes, each made up of parts of the two original synaptic mates (Fig. 176). If the genes for linked characters are located on the opposite sides of such a break, as indicated by the letters AB and ab, a reassociation would result giving Ab and aB linkage in all cells where crossing over has occurred.

According to Morgan and his associates crossing over occurs between many other linked characters in *Drosophila*, although in varying percentages. It has not been observed in the male, but

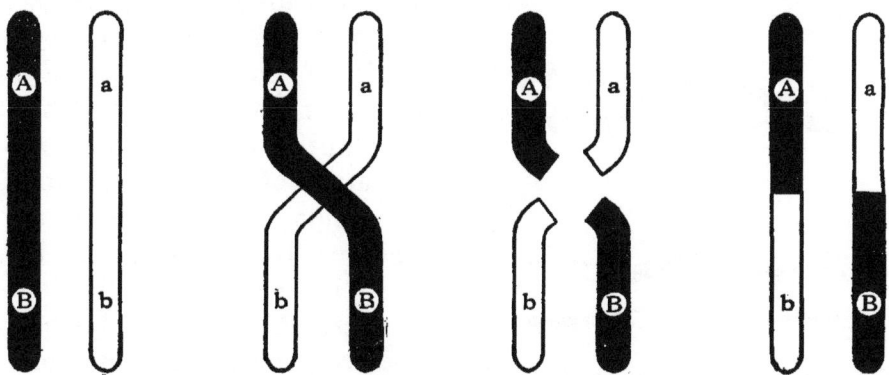

FIG. 176.—Diagram to illustrate the mechanism of crossing over.

occurs ordinarily in a constant percentage of germ cells in the female, and is independent of the way in which the characters are at first combined.

Multiple Crossing Over. Since the chromosomes during synapsis are twisted spirally about each other it is possible for a rupture to occur at one point of crossing as easily as at another, and perhaps at more than one point during the same cell division. Such an interchange would result in new chromosomes made up of alternate pieces of the two original synaptic mates. In the case of a double crossover (Fig. 177) involving three genes, two of the characters would remain linked as before, but the middle one would be interchanged, so that different linkage groups would be produced. Multiple crossing over would necessarily have a different effect on the percentage of interchange between linked characters, and in *Drosophila* it has been found adequate to explain several departures from the expected Mendelian ratios.

310   EVOLUTION AND GENETICS

Linkage in Other Organisms. Scientists have recorded linkage in other organisms. Among animals in which it is known to occur may be mentioned poultry, pigeons, rats, mice, rabbits, silkworms and other insects; among plants, sweet peas, snapdragons, primroses, corn, tomatoes, etc. The chromosome explanation would lead to the belief that it is exceedingly common, but the accumulation of evidence is a slow process. Moreover, *Drosophila* with only four kinds of chromosomes is an especially favorable species for the illustration of this phenomenon, since one quarter of its characters may be associated with a single chromosome. Man, with twenty-four kinds of chromosomes, may have only one twenty-fourth of his characters in one linked group.

Localization of Genes. An important result of the theory of crossing over, based on studies in *Drosophila*, is the idea of localization of the genes. Morgan and his co-authors have published chromosome maps of *Drosophila* indicating the theoretical location of genes for dozens of characters on all four of its chromosomes (Fig. 178). These maps are based on a very logical analysis of percentages of crossing over between the various characters.

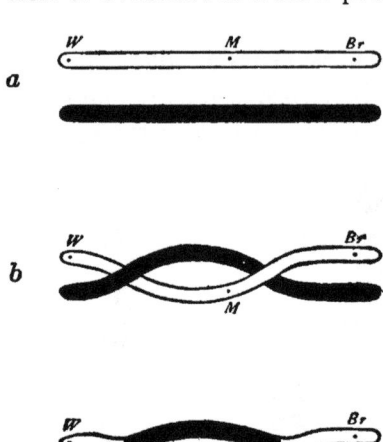

FIG. 177.—Diagram to illustrate double crossing over. In *a* the positions of three genes, *W*, *M* and *Br*, are indicated. When the chromosomes twist about each other between these genes as in *b* and break to form a new association of parts, as in *c*, *W* and *Br* remain in the same chromosome while *M* is shifted to the other. (From Morgan et al., *Mechanism of Mendelian Heredity*, with the permission of Henry Holt and Company.)

It is obvious that two filaments cannot twist about each other in a space less than their diameter, although they can be twisted several times within a relatively short length. Consequently if genes are located close together on a chromosome, there is less reason to expect crossing over between them than if they are remote from each other. It is not illogical to suppose that the percentage of cases in which crossing over actually occurs is in proportion to the distance between the genes.

Crossing over may occur between any characters of a linkage

group, hence a study of various combinations should indicate the spatial relations between their genes in the chromosome. If,

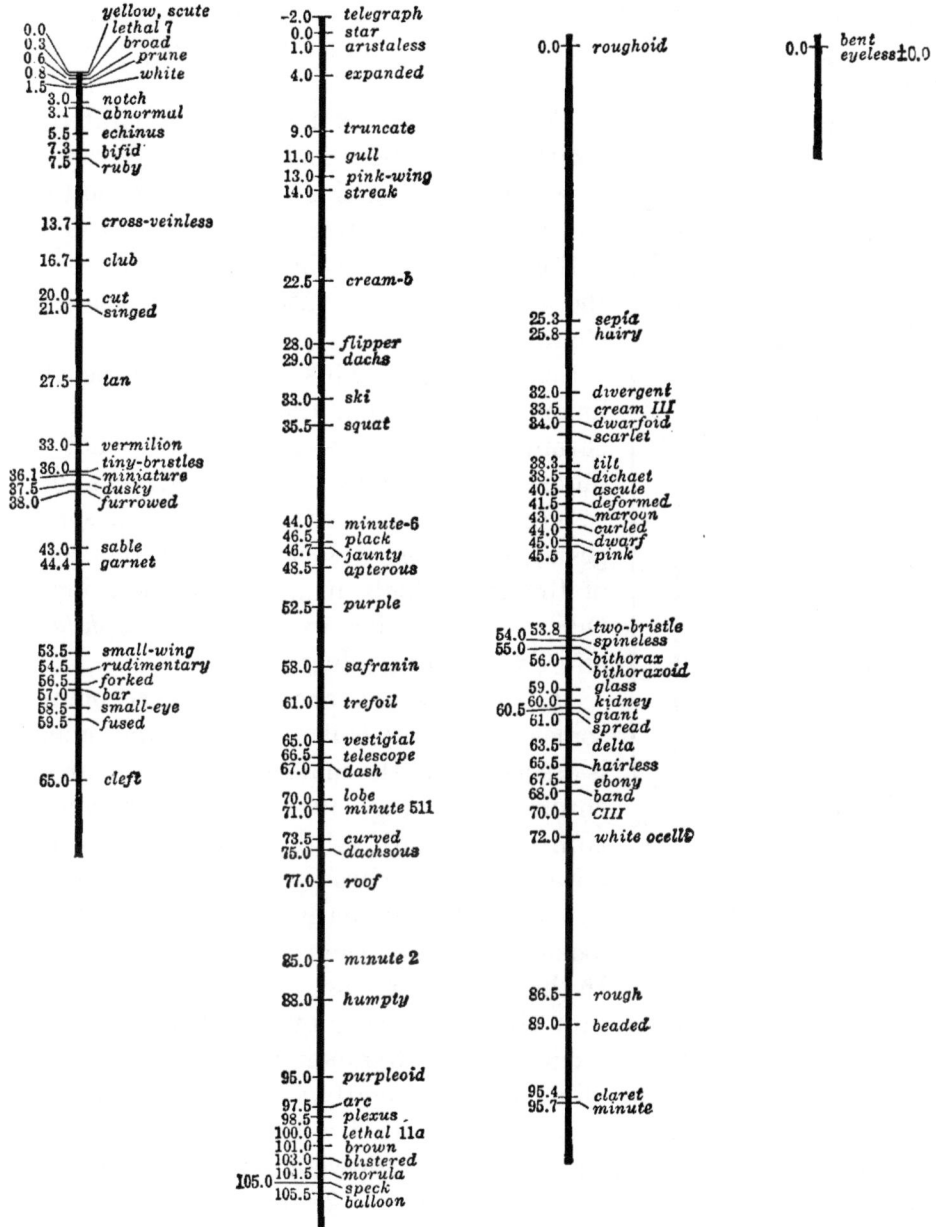

FIG. 178.—Chromosome maps of *Drosophila*. (From Walter, after Sharp.)

for example, crossing over occurs between B and C in 5 per cent of cases, and between A and C in 25 per cent, we may conclude that C is five times as far from A as from B. Whether A and B are on the same side of C or on opposite sides is not disclosed by these data but if an additional test gives 20 per cent of crossing over between A and B we may conclude that they are on the same side (Fig. 179).

Some genes of *Drosophila* are shown to be widely separated, yet cross over much less than is to be expected in such cases when handled together. Double or multiple crossing over accounts for these cases without invalidating the data derived from the study of more intimately associated characters.

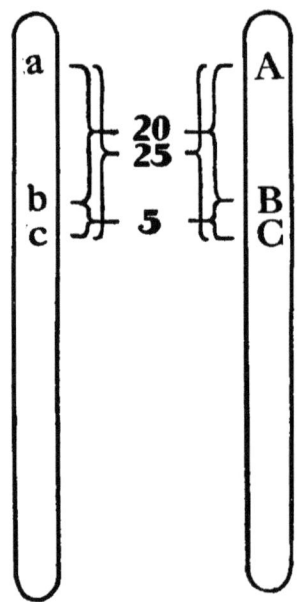

Fig. 179.—Diagram to illustrate the relationship between percentage of crossing over and the location of genes in the chromosomes.

Sex-Linkage. This interesting type of linkage results from the retention by sex chromosomes of genes related with wholly different characters. The inheritance of sex-linked characters is, as the term suggests, intimately associated with the sex of the individuals in a given generation. Our now familiar example, *Drosophila*, affords numerous illustrations.

In this species one of the chromosomes, commonly called the x chromosome, is associated in the male with a dissimilar y chromosome (Fig. 168). The y chromosome apparently lacks the functional powers of the x chromosome, although the two are synaptic mates, and is sometimes looked upon as a degenerate x chromosome. In the female there are two x chromosomes.

The gene for the mutant character white eyes in *Drosophila* is located in the x chromosome. This character is recessive to the normal red eyes of the wild fly, consequently a white-eyed female is a homozygous recessive. When mated with a red-eyed male all females of the F_1 generation are red-eyed, but all males are white-eyed. The results in this and the F_2 generation are graphically expressed in Figure 180. F_1 males produce two types of gametes, one containing the x chromosome bearing the gene for recessive white eyes, the other bearing the y chromosome without genes. The female produces two types of eggs, each with the x chromosome but one bearing the gene for red eyes and the other for white. Any egg which receives the x chromosome of the male will produce a female, therefore the F_2 generation contains

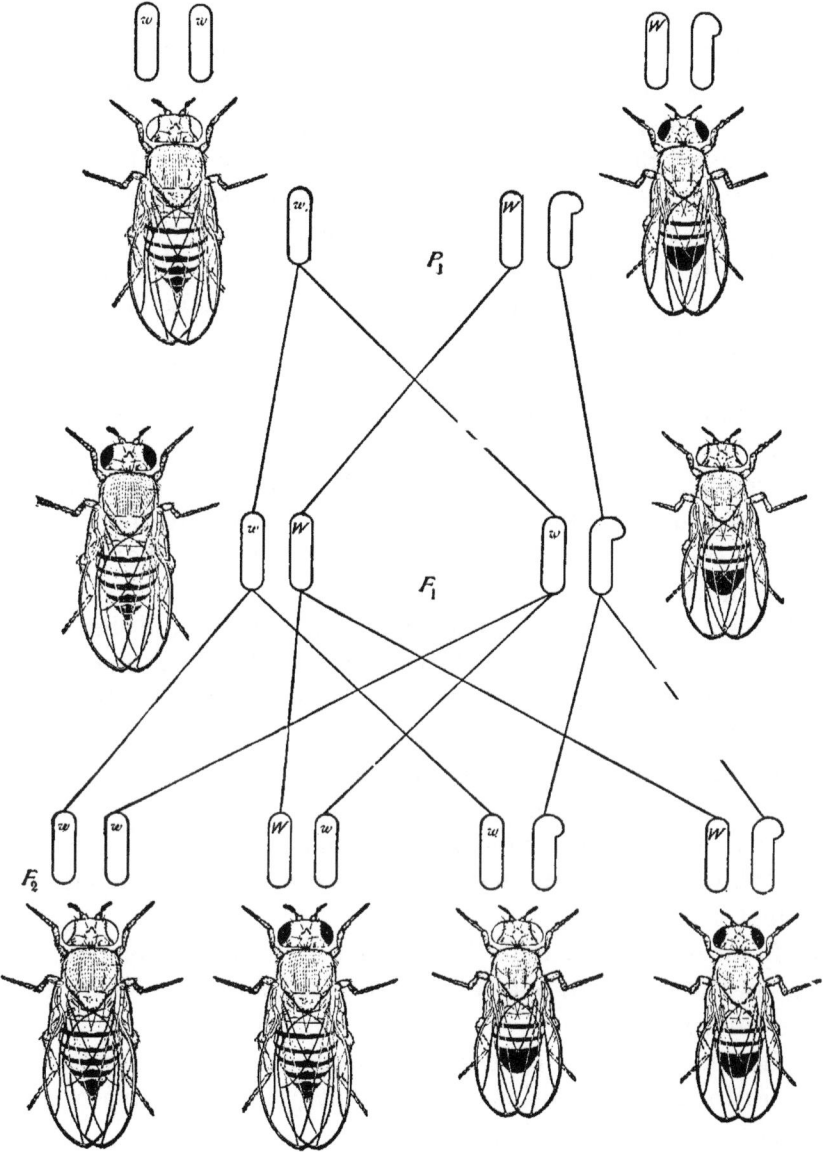

Fig. 180.—Diagram of the cross between a red-eyed female and white-eyed male of *Drosophila melanogaster*. W indicates red and w white. (From Morgan et al., *Mechanism of Mendelian Heredity*, with the permission of Henry Holt and Company.)

one half red-eyed heterozygous females and one half white-eyed recessives. Those eggs which receive the y chromosome from the male develop into males, which are one half white-eyed and one half red-eyed. In this case the condition of the male is determined entirely by genes received from its mother.

The reciprocal cross of a homozygous red-eyed female with a white-eyed male produces only red-eyed individuals. The females are heterozygous, and so produce two types of eggs as in the preceding example, while the males produce germ cells bearing the x chromosome with the gene for red eyes and the y chromosome with no genes, respectively. The F_2 generation therefore contains only red-eyed females, one half homozygous and one half heterozygous, while the males are one half red-eyed and one half white-eyed (Fig. 181).

Sex-linkage is known in other organisms, including man, but not all characters associated with sex are necessarily due directly to sex-linked genes. Secondary sexual characters may result from hormones produced by the gonads, and may appear in the opposite sex under abnormal conditions, although there is no reason to suppose that the chromosomes are modified by the unusual stimulus.

Multiple Allelomorphs. Most known allelomorphic characters appear in pairs which are independent of all other characters. Exceptions have been discovered in *Drosophila* of which the following data quoted from *The Mechanism of Mendelian Heredity* by Morgan, Sturtevant, Muller and Bridges are an illustration:

"1. If a white-eyed male of *Drosophila* is mated to a red-eyed female, the F_2 ratio of three reds to one white is explained by Mendel's law, on the basis that the factor for red is the allelomorph of the factor for white.

"2. If an eosin-eyed male is mated to a red-eyed female, the F_2 ratio of three reds to one eosin is also explained if eosin and red are allelomorphs.

"3. If the same white-eyed male is bred to an eosin-eyed female, the F_2 ratio of three eosins to one white is again explained by making eosin and white allelomorphs."

The assumption that these three characters are allelomorphic to each other carries with it as a necessary corollary the assumption that their genes occupy the same position in the chromosome. Data on crossing over between these and other characters bear

GENES AND CHARACTERS 315

out this conclusion. Under such conditions it is apparent that only two of the characters can be represented in any one individual.

The characters that have given such a wealth of material to the geneticist in *Drosophila* are mutations. There is no reason

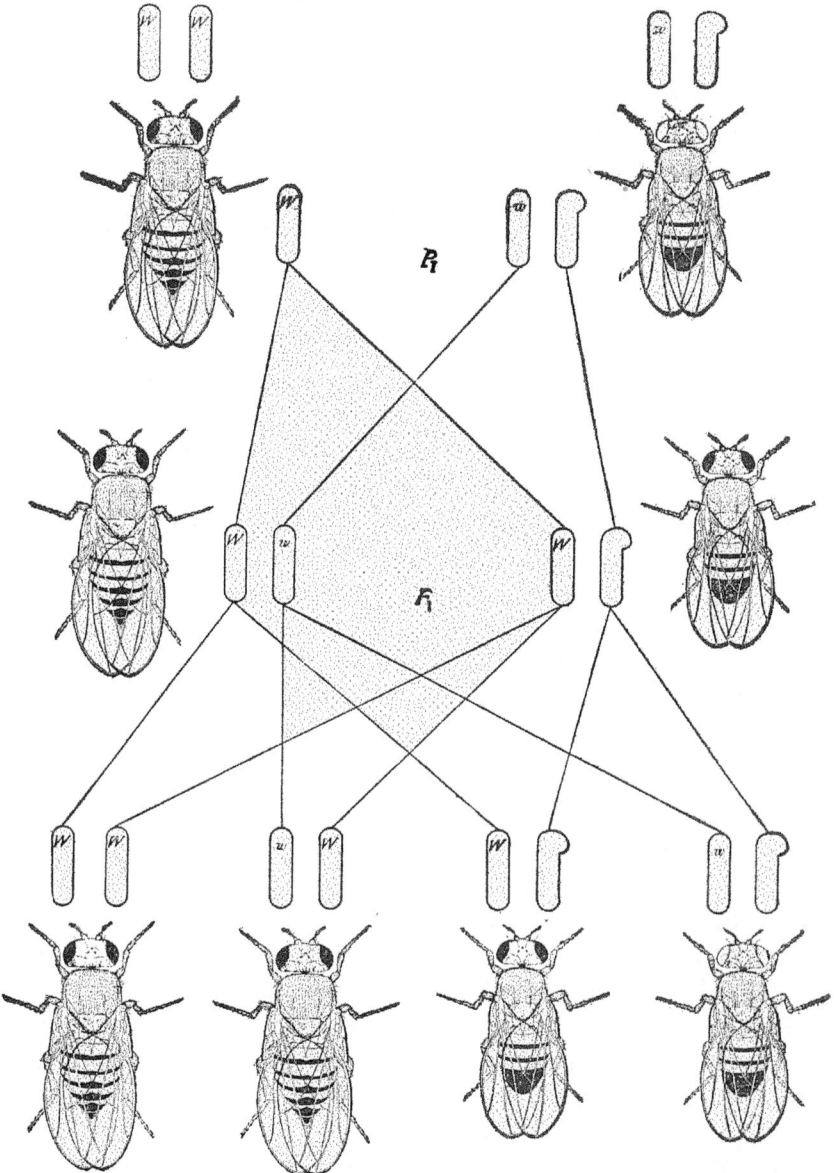

FIG. 181.—Diagram of the reciprocal cross of Fig. 180, i. e., white-eyed female and red-eyed male of *D. melanogaster*. W indicates red and w white. (From Morgan et al., *Mechanism of Mendelian Heredity*, with the permission of Henry Holt and Company.)

to suppose that a character which has given rise to one mutation is thereby prevented from giving rise to others. Multiple mutations of a character, since they are allelomorphic to the original character, would logically be allelomorphic to each other when brought together in one individual. This explanation is in accordance with all available data in the chromosome theory of descent. It is adequate without further amplification, which may be found if desired in the work cited above.

Multiple Factors. Although the laws of heredity thus far considered have to do with characters determined by single factors, there are many different cases on record which must be explained on the assumption that more than one factor is concerned in the production of a single character. Extensive data have led to the discovery of four kinds of multiple factors, known respectively as duplicate, cumulative, complementary, and supplementary.

Duplicate Factors. Shull discovered that the determination of seed-capsule shape in shepherd's purse is accomplished by the action of two pairs of factors. The seed capsules of this common plant are usually triangular in outline but may be fusiform. When plants of the two varieties are crossed, the F_1 generation has triangular capsules, and only one individual in sixteen in the F_2 generation reverts to the recessive spindle-shaped condition. This is reminiscent of the dihybrid ratio, but an examination of Figure 182 shows that the presence of a single dominant determiner results in the appearance of the dominant character, and only the homozygous recessive reveals its genotypic character. The behaviour of factors in this case is similar to that of all multiple factors, but the appearance of the resulting characters differs in this and the three following cases.

More than two pairs of duplicate factors may govern a character; their distinctive quality is the similarity of their effect, no matter how many of the dominant determiners are present.

Cumulative Factors. Those factors which bring a character to expression in the soma in proportion to the number of dominant determiners present are called cumulative factors. The case of Nilsson-Ehle's wheat is an old and excellent example.

Nilsson-Ehle found that a race of wheat with red grains and a race with white grains, when crossed produced an F_1 hybrid with intermediate pale red grains. In the F_2 generation very few white grains appeared, so that the color was obviously not due to a single pair of determiners. In one experiment, according to Babcock and Clausen, seven families with a total of 440 plants produced white seeds on only one plant.

Omitting further details, it has been found that color in this case was determined by three different factors. The genotypic formula for the parents may be indicated as AABBCC and aabbcc,

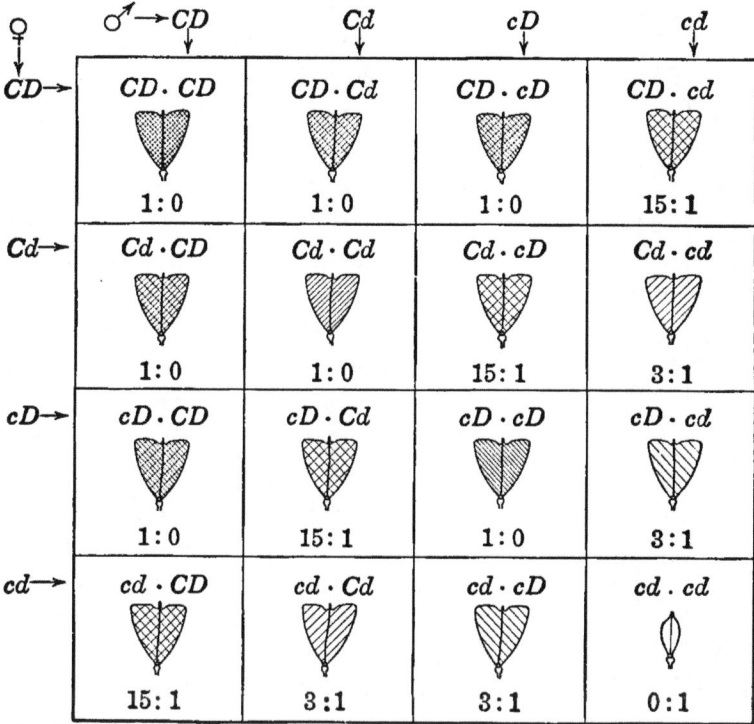

Fig. 182.—Punnett square with diagrammatic figures to illustrate the inheritance of seed capsule shape in *Bursa* (*B. bursa-pastoris* x *B. heegeri*). The two duplicate factors are represented by oblique ruling in opposite directions, and the number of factors present by the spacing of the lines. The ratios indicate the expectation in the F_3 generation from self-fertilized individuals. (After Shull from Babcock and Clausen's *Genetics in Relation to Agriculture* by permission McGraw-Hill Book Company, Inc.)

and that of the F_1 generation as AaBbCc. In the F_2 generation the recombination of these factors occurs in sixty-four ways according to the law of trihybrids, but only one out of the sixty-four has the homozygous recessive organization which produces white grains. All heterozygous individuals, such as AABbCc, AaBbcc, aaBbcc, etc., produce red grains, varying in depth of color according to the number of dominant factors present. Out of these red grains only one of the sixty-four can be of the depth of color of the original dominant parent. The construction of a Punnett square for this case will indicate that there are only six degrees of red present, in addition to the one white individual, and that most of the sixty-four chance combinations fall within the three intermediate groups of the seven (Fig. 183).

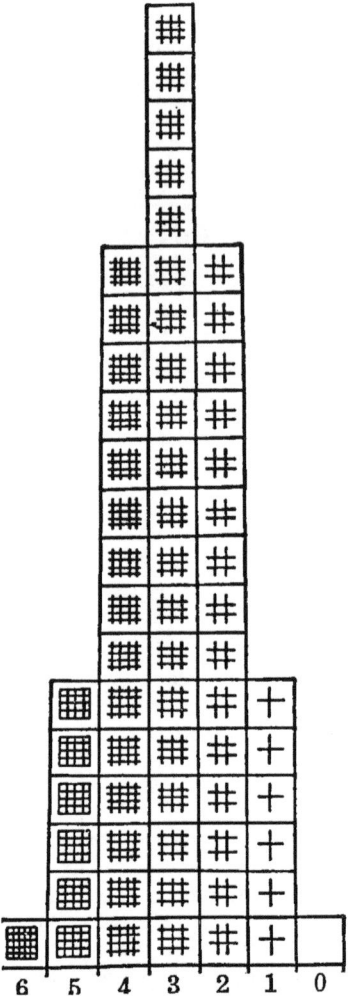

Fig. 183.—The distribution of the sixty-four possibilities in the F_2 generation when three similar determiners act to produce a given character. The numbers indicate the number of determiners present in the individuals represented by the column above. (From Walter.)

Ear length in rabbits was studied by Castle with similar results. The cross between lop-eared and short-eared parents gives an F_1 generation with intermediate ears, while the F_2 generation produced no individuals showing the ear length of either of the grandparents. It has been pointed out by various writers that if four factors are involved in the determination of ear length, only one individual out of two hundred and fifty-six in the F_2 generation can be expected to resemble either grandparent, while with six factors such resemblance can occur only once in 4096 times! As Castle concludes: "it would be remarkable if under such conditions the extreme size were ever recovered from an ordinary cross."

Inheritance of this type has been called blending inheritance, but it is to be distinguished carefully from the type of blending referred to in Chapter XV. The pink four-o'clocks cited as an example in that case are due to the fact that neither the red nor the white color factor dominates the other in heterozygous individuals, and the result is a complete expression of the Mendelian 1:2:1 monohybrid ratio. When cumulative factors are involved, however, the intermediate individuals are not all the same, but fall into graded classes whose number depends upon the number of factors. The F_2 ratio is always a modification of some Mendelian ratio above the monohybrid.

Complementary Factors. Two factors that have no effect when they occur separately but produce a definite character when they are brought together in the same individual are said to be complementary. An example often cited is that of Bateson's white sweet peas and their red offspring.

The two factors in this case may be interpreted as a color factor C and a red factor R. The white parents are CCrr and ccRR, individuals which produce gametes bearing the genes Cr and cR respectively, so that the F_1 hybrid is heterozygous for both factors. Its CcRr organization combines the two complementary factors and results in the production of red flowers. The F_2 generation of this hybrid corresponds genotypically to a dihybrid, but since the presence of both factors is necessary for the production of color, all combinations of C and R with recessives of the opposite type included in the second and third groups of the 9:3:3:1 ratio are white, and a 9:7 ratio results (Fig. 184).

	CR	Cr	cR	cr
CR	CR CR red	Cr CR red	cR CR red	cr CR red
Cr	CR Cr red	Cr Cr white	cR Cr red	cr Cr white
cR	CR cR red	Cr cR red	cR cR white	cr cR white
cr	CR cr red	Cr cr white	cR cr white	cr cr white

Fig. 184.—Diagram showing the behaviour of complementary factors. The genotypic ratio is the same as in Fig. 161 of a normal dihybrid but the phenotypic ratio is 9:7.

Similar factors have been discovered in corn, rabbits and other organisms.

Supplementary Factors. This last group consists of those which condition the expression of others without being essential to the production of the character in question. In the sweet peas just described, for example, color is produced by the factors C and R, but a factor B occurs in some sweet peas which modifies the color by adding blue.

Rabbits have a complex series of color and pattern genes which illustrate both supplementary and complementary factors. In them the factor C is necessary for the production of color. A cc individual is always an albino, although it contains factors for yellow, black and brown pigment. The albino may also carry a factor A, or agouti, which determines the distribution of pigments in individual hairs; this is characteristic of the cotton-tail rabbit and is known as the wild gray type. A black rabbit crossed with an albino bearing the agouti factor produces offspring of the wild gray type. Other factors influence the depth of color and its distribution on the body (Fig. 185).

Lethal Factors. Factors have appeared in a few cases whose presence in the proper combination results in death. Among animals these factors are usually evident through reduction of the usual number of progeny but among plants the congenital absence of chlorophyll is a lethal factor which does not interfere with the development of the seedling until food stored in the seed has been exhausted. Several kinds of plants, including corn, are known to produce these individuals. The character is in all cases, whether plant or animal, a recessive which can be perpetuated through heterozygous individuals.

Several lethal factors have been discovered in *Drosophila*. The effects of a sex-linked lethal character in these flies is shown in Figure 186. Since the male bears only one x chromosome it cannot be a carrier, but dies if recessive for the lethal character. When a female carrier, the only individual capable of perpetuating such a character, is mated, two-thirds of her offspring are females instead of one-half as under normal conditions, and of these females one-half are normal and one-half carriers. In the figure L indicates the normal condition and l the lethal recessive.

The Effect of X-rays on Genes. In connection with the study of heredity in *Drosophila* Professor H. J. Muller has recently shown that the treatment of flies with X-rays has a direct effect on the behaviour of hereditary characters. His experiments prove that the rays affect germ cells at any stage, even including the spermatozoa contained in the seminal receptacle of the female. Mutations appeared in the flies developed from gametes so treated

Constant Genes				Alternative Genes				Gametic Formula	Phenotypic Character when Crossed with the Same Kind of Gametic Combination
1	2	3	4	5	6	7	8		
Br	B	Y	C	E	I	U	A	AUIEC YBBr	Gray
							a	aUIEC YBBr	Black
						u	A	AüIEC YBBr	Gray spotted
							a	auIEC YBBr	Black spotted
					i	U	A	AUiEC YBBr	Blue-gray
							a	aUiEC YBBr	Blue (Maltese)
						u	A	AuiEC YBBr	Blue-gray spotted
							a	auiEC YBBr	Blue spotted
				e	I	U	A	AUIeC YBBr	Yellow (with white belly and tail)
							a	aUIeC YBBr	Sooty yellow (with yellow belly and tail)
						u	A	AuIeC YBBr	Yellow spotted
							a	auIeC YBBr	Sooty yellow spotted
					i	U	A	AUieC YBBr	Cream
							a	aUieC YBBr	Pale sooty yellow
						u	A	AuieC YBBr	Cream spotted
							a	auieC YBBr	Pale sooty yellow spotted

FIG. 185.—The factor hypothesis applied to colors of rabbits. (From Walter.)

Explanation: Br = a gene acting on C to produce *brown* pigmentation.
B = a gene acting on C to produce *black* pigmentation.
Y = a gene acting on C to produce *yellow* pigmentation. The three genes, Y, B, Br, are present in every rabbit gamete and up to date have not been separable as independent unit characters, although they have been separated out in guinea-pigs and mice. There are no brown rabbits, because black always goes linked with brown, covering the brown factor. Yellow rabbits result, as explained below, through the action of factor e.
C = a common *color* gene necessary for the production of any pigment. It was discovered in 1903 by Cuénot.
c = the absence of C which results in albinos, regardless of whatever pigment gene may be present. By changing C to c, sixteen kinds of albinos would be added to this catalogue, an addition of one phenotype and sixteen genotypes, all looking alike but breeding differently.
E = a gene governing the *extension* of black and brown pigment, *but not of yellow*.
e = the absence of extension or *restriction* of black and brown pigment to the eyes and the skin of the extremities only, while yellow remains extended and visible. Demonstrated by Castle in 1909.
I = an *intensity* gene which determines the degree of pigmentation. It can be transmitted independently of C through an albino. Discovered by Bateson and Durham in 1906.
i = the absence of intensity or *dilution*. Dilute black = blue. Dilute yellow = cream. Dilute gray = blue-gray.
U = a gene for *uniformity* of pigmentation or "self-color" discovered by Cuénot in 1904.
u = the absence of uniformity which results in *spotting with white*.
A = a pattern gene for *agouti*, or wild gray color, which causes the brown and black pigments to be excluded from certain portions of each hair, resulting in the gray coat. When present in the rabbit it is also associated with white or lighter color on the under surfaces of the tail and belly. It was demonstrated by Castle in 1907.
a = the absence of the agouti or pattern gene.

at a rate up to 150 times the normal. Among the mutations produced from treated flies there were many lethal characters and recessives, but such familiar things as multiple allelomorphism were also produced.

The experiments are of great interest in evolution because they show definitely that genes may respond directly to conditions of

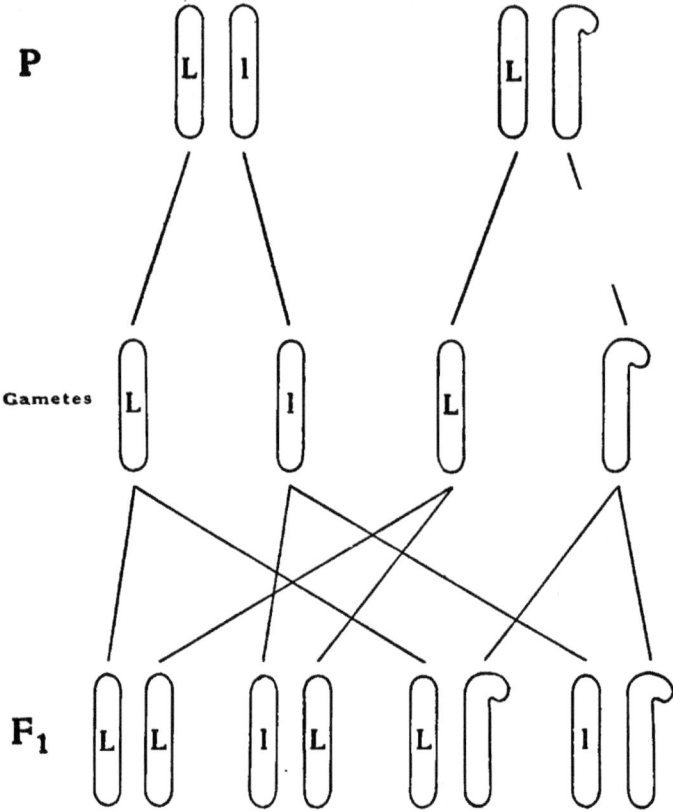

Fig. 186.—Diagram to show the inheritance of a lethal factor in *Drosophila*. l indicates the lethal factor and L its normal allelomorph. The male at the right in the F_1 generation fails to develop so that the ratio of sexes in such a generation is two females to one male.

the external environment. The appearance of reverse mutations, moreover, establishes the reversibility of slight evolutionary changes.

Genes in the Development of the Individual. Beyond these facts of inheritance, showing close association of characters in the individual and genes in its chromosomes, we may well inquire *how* the genes act during ontogeny to condition the differentiation of somatic cells. Lillie summarizes modern knowledge of this subject in a recent article and concludes that "we have no present working hypothesis effective in this most fundamental aspect of the life history."

According to modern views the complex of genes which is present in the zygote is carried into all cells of the body. The differentiation of the cytoplasm which gives rise to the multitude of specialized adult cells is therefore not accompanied by nuclear differentiation of equal degree. In spite of this fact there is evidence that the genes are active at various periods of life in connection with cytoplasmic differentiation for adaptation to definite environmental conditions. The ability to become tanned, for example, is inherited although it does not become operative until the proper environmental stimulus is received.

Goldschmidt has attempted to explain this interaction by two theories, one of which assumes that the genes are so conditioned that they enter into activity at a definite rate and time for each, while the other suggests that the genes act only with specific cytoplasmic materials. As Lillie points out, the latter assumes cytoplasmic differentiation as a condition for activity of the genes, yet differentiation is what it attempts to explain. The former theory also appears weak in the light of modern embryology.

Lindsey has suggested that the action of genes is not limited to the cells containing them, but that all genes of a kind in the body, through the coördination which is evident in more obvious physiological processes, act together to bring a character to expression. Some facts are apparently contradictory to this theory but not necessarily inimical to it.

A detailed discussion of these matters has no place in such a work as this, especially since no theory is generally accepted and none is more than a partial explanation. The facts given are sufficient to illustrate our ignorance of the connection between genes and the characters which they produce. So far we must content ourselves with the knowledge that both exist and that the appearance of characters is in accordance with definite laws which are harmonious with the behaviour of the chromosomes and the genes.

Summary. Characters of organisms are determined by definite things in the body, called factors. These factors are based upon hypothetical bodies called genes, located in the chromosomes, and occur in pairs except in cells which have only the haploid number of chromosomes. Here a single gene may represent a character. Characters are said to be linked when their genes occur in the same chromosome and therefore usually remain together. Linkage relations may be modified by interchange of material between two chromosomes of a kind during the reduction division; this phenomenon is called crossing over, and gives evidence of definite localization of the genes. When genes are located in the sex chromosomes the characters which they control are said to be sex-linked. In addition to characters determined by a single pair of factors many are known which involve two or more pairs. These multiple factors behave in the normal Mendelian manner, but the phenotypical ratios differ from those produced by the same number of independent factors. The difference depends upon the number of factors and the nature of their interaction. The way in which genes act to produce characters in the individual has been the subject of some speculation but is still unknown.

REFERENCES

CASTLE, W. E., *Heredity of Coat Characters in Guinea-Pigs and Rabbits*, Pub. Carnegie Inst., No. 23, 1905.

CASTLE, W. E. and WRIGHT, S., *Studies of Inheritance in Guinea-Pigs and Rats*, Pub. Carnegie Inst., No. 241, 1916.

MORGAN, T. H. and BRIDGES, C. B., *Sex-Linked Inheritance in Drosophila*, Pub. Carnegie Inst., No. 237, 1916.

BABCOCK, E. B. and CLAUSEN, R. E., *Genetics in Relation to Agriculture*, 2nd edition., 1927.

MORGAN, T. H., STURTEVANT, A. H., MULLER, H. J. and BRIDGES, C. B., *The Mechanism of Mendelian Heredity*, revised edition, 1922.

MORGAN, T. H., Cowdry's *General Cytology*, Section XI, 1924.

CHAPTER XVIII

THE DETERMINATION OF SEX

The association of chromosomes with sex has already been mentioned in Chapter XVI. For many years their behaviour has been regarded by many scientists as an adequate explanation of sex determination, although anomalies of various kinds have required ingenious explanations to harmonize them with the facts of chromosome behaviour.

At present the state of experimental work in this field enables us to say that the chromosomes are definitely associated with sex determination *under ordinary conditions* and that sex is therefore inherited in many cases as a Mendelian character; we now know, however, that this is only a part of the story and that other factors, possibly many others, are concerned with sex determination. These other factors have been discovered in some cases and placed under control with such accuracy that the sex of an individual, not merely superficial but also fundamental characters, has been determined independently of its chromosomal complex.

What Is Sex? Among the fundamental principles of biology reproduction is recognized as a primary characteristic of living matter. In the simplest unit, the cell, reproduction is a common function and in complex organisms every individual plays some direct part in the production of new individuals unless it is specialized for other purposes in a colony of different kinds of individuals.

In its simplest form, the process is one of subdivision into two new individuals as nearly as possible like each other and like their common parent. This process of subdivision is carried to a point where only a small portion of the parent buds off to form a new individual; the parent does not lose its identity in the process, but may continue its normal vital processes even while it gives rise to a new generation. It is only a step from this process of budding to the development of specific reproductive cells or *spores*, each of which after separating from the parent is potentially a new organism. Between this and sexual reproduction there is

326 EVOLUTION AND GENETICS

a great gap, for the formation and union of gametes demand a differentiation of reproductive cells which the simpler processes of asexual reproduction do not require.

In the simplest forms of sexual reproduction all individuals are apparently the same and the gametes which they produce show

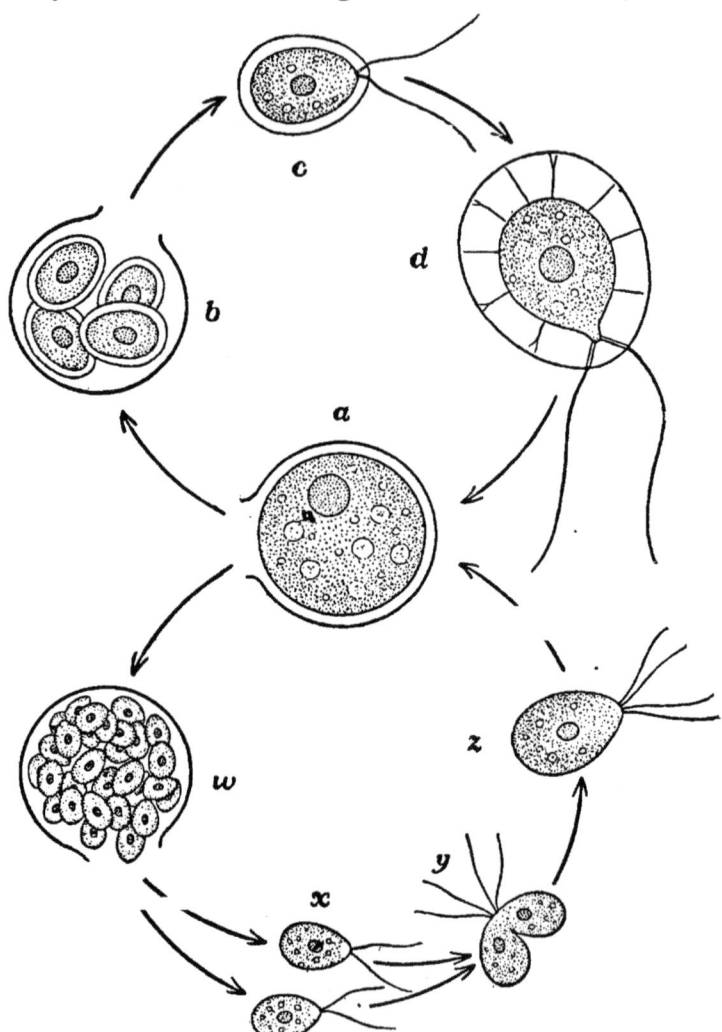

Fig. 187.—Life history of *Sphaerella lacustris*. *a, b, c, d*, asexual cycle; *a, w, x, y, z*, sexual cycle. *a*, dormant cell; the protective cyst has ruptured to permit the escape of the protoplast; *b*, division of the protoplast to form four spores (*c*), each of which grows and takes on the adult form (*d*). This process may be repeated, but ultimately the cells become dormant again. Under some conditions the protoplast divides as in *w*, forming 32 or 64 small cells. These are gametes. They escape (*x*) and fuse in pairs (*y*) to form a zygote (*z*) which develops a cyst as in *a*. (From Woodruff.)

few or no differences. The gametes of *Sphaerella*, for example, are both motile cells of the same form, but they are not potential individuals for two must unite to bring about the development of a complete organism (Fig. 187). In higher organisms the female gamete becomes more and more specialized as a passive cell in which food for the developing embryo is concentrated, while the male gamete is a highly specialized motile cell, unhampered by stored food and therefore able with maximum efficiency to seek out the egg cell. The difference may be extreme, yet as we have seen in Chapter XVI, the cells are potentially equal in their contributions to the new generation.

Among the more primitive organisms and among some higher species, differentiation of the individuals producing the two kinds of gametes is conspicuously lacking, even to the extent of hermaphroditism. In this condition the same individual produces both male and female germ cells. In the higher animals, however, and in some of the higher plants, complete separation is accomplished into male and female sexes each of which produces only one kind of gamete. The sexes are also often conspicuously differentiated in superficial character. These superficial differences may be without evident connection with the essential sexual processes.

It is evident that the one fundamental difference between the sexes is the ability to produce the different kinds of germ cells, yet when we consider that sex cells are, in their essential constituents, the same except in connection with sex determination, the difference seems entirely an adaptation to specific conditions of reproduction. The demonstrated occurrence of sex reversal in the domestic fowl and frogs further complicates the problem. Since some birds and frogs have been both father and mother during their lifetime, sex seems biologically much more a matter of convenience than of necessity.

The one great contribution of sexual reproduction for which there is no known substitute is the reassortment of characters with each generation through the combination of those present in two individuals to produce a third. To conclude with this statement is to confess our entire ignorance of the development of sex. We can only point out the existing gradations in living organisms and the one biological phenomenon which seems to be an adequate reason for sexual reproduction. It remains for us to consider the peculiarities of the germ cells and of the individuals producing them, and the conditions which have been found to play a part in the behaviour of these things.

Individuality of the Gametes. In most cases of sexual reproduction the function of each gamete is wholly complementary to that of the other. The cytology of these cells suggests that if the nucleus is, as supposed, the essential structure for the development of hereditary characters, the presence of a complete haploid set of chromosomes would enable either kind of gamete to develop into a complete organism. The fact that the cytoplasm is necessary for the differentiation of the somatic structures implies a limitation which actually does prevent the development of male gametes alone, but inheritance under certain experimental conditions and the inheritance of sex in special life cycles show that each gamete is fundamentally and potentially an individual.

The fertilization of enucleate eggs of sea urchins by the sperms of other species and their subsequent development of the characters of the male parent shows that the male germ cell is equal to the female except in its reduced and specialized cytoplasm. This must be regarded as incidental to the conditions of reproduction, and not as a fundamental difference in reproductive power. It is, of course, very effective in limiting the capacity of highly developed male gametes.

Parthenogenesis. Development of the unfertilized ovum, or parthenogenesis, occurs normally in many insects and other Arthropoda, in some worms and possibly in other phyla. Artificial parthenogenesis has been brought about by the application of various stimuli such as unusual chemical contacts and mechanical manipulation. The production of males would naturally result from parthenogenesis when this is the heterogametic sex, but in some cases females are produced.

Parthenogenesis in the Honey-Bee. Parthenogenesis here is very well known because of the economic importance of the species. It has been proved that all drones, or male honey-bees, are produced from unfertilized eggs, while the queens and workers, both females, are developed from fertilized eggs. A fertilized egg can develop into either queen or worker according to the food received by the growing larva and queen breeders have produced intermediate individuals. Some of these are said to have mated and deposited fertile eggs, although scarcely distinguishable from the workers, which never mate.

THE DETERMINATION OF SEX 329

Gametogenesis occurs normally in the queen bee, consequently her eggs contain the haploid number (16) of chromosomes. It follows that the drones developed from these eggs have only the haploid number although the queens and workers have the normal diploid complex of thirty-two chromosomes. The production of normal male gametes demands some compensating phenomenon, which is found in the abortive reduction division. At this stage all of the chromosomes pass into one spermatocyte so that all spermatozoa receive the normal haploid number, and at fertilization establish the normal diploid complex of the females.

The Life Cycle of Aphids. This life cycle demands another type of compensation in behaviour of the chromosomes. Sexual individuals appear at intervals during the cycle, but they are separated by long successions of parthenogenetic females. The stem mothers hatch in the spring from fertilized eggs which have passed the winter; no males appear at this season. The stem mother heads the succession of parthenogenetic females, and in the fall males and females appear which produce the winter eggs. The diagram (Fig. 188) shows the behaviour of the chromosomes accompanying these steps. The parthenogenetic eggs from which the males and females develop are produced in one case by a normal mitotic maturation division which results in one ovum and one polar body. The egg in this case contains the same chromosomal complex as the cells of the parent. In the other case one entire x chromosome is extruded in the polar body, and consequently the egg receives the odd number of chromosomes characteristic of the male. In order that females alone may develop from the winter eggs, maturation in the male is accompanied by degeneration of the secondary spermatocytes which contain no x chromosome. All spermatozoa therefore contain an x chromosome and the diploid number is restored in every case by fertilization.

Elimination of the Male. It is evident that the parthenogenetic production of females, which occurs in some insects and rotifers, might in extreme cases result in the limitation of reproduction to this method. In such an Amazonian species males would be superfluous, and this is a possible explanation for the failure of biologists to discover the males of certain rotifers. The evidence available is not, however, conclusive.

Artificial Parthenogenesis. The occurrence of natural parthenogenesis in sexual organisms suggests that the stimulus of fertilization need not be an extreme one. Loeb's researches in this field showed that a slight increase in the potassium or sodium chloride concentration of sea water was enough to bring about parthenogenetic development of the eggs of sea urchins and

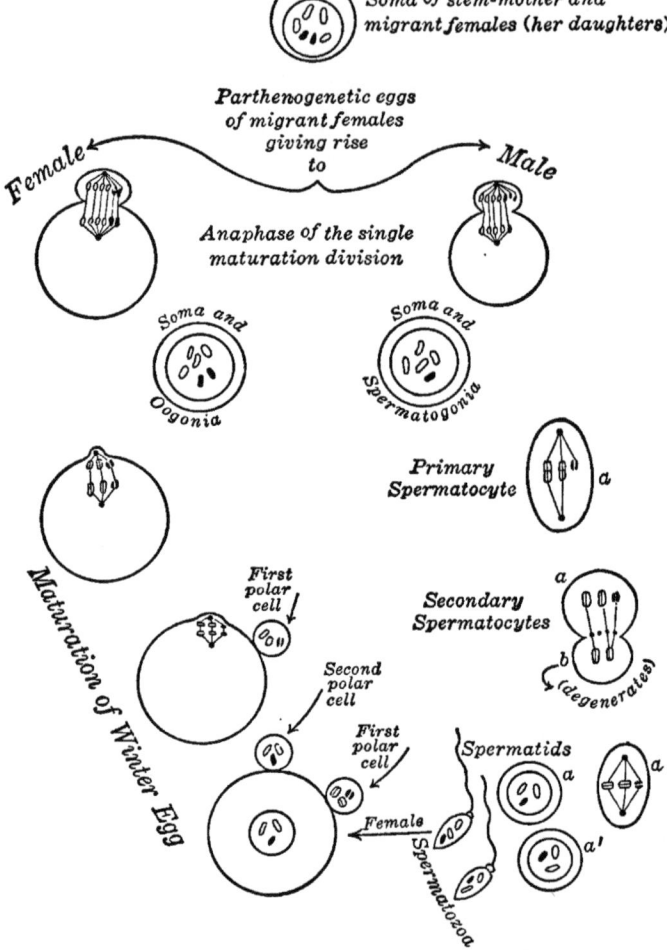

Fig. 188.—The chromosome cycle in the parthenogenetic insects, aphids and phylloxerans. (From Walter.)

annelid worms. He further records that the moderate agitation caused by transferring the eggs of some starfishes from one dish to another was sufficient to cause cleavage without fertilization. Such cases could be multiplied from the findings of other investigators, but there is little variety in them. Development proceeds to various degrees, but in Loeb's researches the chemical stimulus which brought about abnormal parthenogenesis was usually so severe as to cause early death.

The Purpose of Sexes. The foregoing data show that the different germ cells are fundamentally similar in so far as the fact of reproduction is concerned, and different only in accord with the parts which they play in the details of its accomplishment. The sexes which produce them cannot, therefore, have different capacities with respect to the fundamental fact of reproduction, but only with respect to the degree of specialization of the process in the species to which they belong.

This being the case, it is apparent that the differentiation of the sexes should be commensurate with their rôles in normal reproduction. The similarity of sexes in the lower organisms and the occurrence of hermaphroditic animals are therefore not surprising. Nor is the extreme differentiation of sexes in the higher animals and the concomitant necessity for sexual reproduction in a great majority of species any more than a normal development in harmony with the high degree of differentiation characteristic of these organisms.

Differentiation of Sexes. When a phenomenon is constant in occurrence and apparently necessary to the existence of the species some provision for its perpetuation is to be expected, for it is quite in harmony with the fundamental principles of biology. The chromosome theory of sex determination therefore deals with facts which seem entirely normal. It is not entirely clear how the presence of two chromosomes in the cells of an individual can cause it to belong to one sex while the presence of either alone would cause its development into the opposite sex, but the association of chromosomes with sex is well established.

The fact that the allosomes in one sex are the same as those in the other indicates one important consideration. While we are forced to the conclusion that allelomorphic characters are based upon qualitatively different genes, there is nothing to indicate that the sexes are in any way allelomorphic. The chromosomal difference between them is apparently quantitative. Under ordinary conditions it has a very definite effect upon the differentiation of the sexes, but we must recognize that the same active constituents are present in the allosomes of the two sexes. This interpretation is favored by the occurrence of hermaphrodites and abnormal individuals intermediate in sex.

Hermaphrodites are individuals which bear functional gonads of both sexes, and produce both male and female gametes. It does not follow that they are capable of self-fertilization, although this is sometimes a possibility and sometimes apparently the normal method of reproduction. Hermaphroditism is common among the lower phyla. Sponges and coelenterates, flatworms and annelids include many striking examples.

Since the individual can have only one chromosome complex, the occurrence of these bisexual forms is absolute proof that chromosomal difference is not an essential of sex. The same processes that produce the gametes in different individuals produce them here in one, and the same result of recombination of parental characters is attained as when the sexes are separate.

Abnormalities of many kinds occur in nature or can be produced by artificial means which further substantiate this view. Wherever they occur in nature they are apparently due to accidents of development or to abnormal behaviour of the germ cells. Experimentally they have been produced by operative methods and by variation of the environment.

Gynandromorphs, sometimes incorrectly called hermaphrodites, are occasionally produced among insects of various orders and among other animals. They are individuals which display the characters of both male and female in different parts of the body, and are incapable of carrying on the normal reproductive functions of either sex.

Gynandromorphs are usually bilateral, one half being male and the other female. In species with distinctly different sexes this results in very striking contrasts as shown in Figure 189. This distribution of sex characters is accompanied by a similar asymmetry of chromosome distribution, the female half having the female complex and the male half the male complex. How such an abnormality occurs it is difficult to determine but one logical explanation is available.

According to this interpretation the first cleavage of the ovum may be hastened by some force so that two groups of chromosomes are present at the consummation of fertilization. The union of the male pronucleus with one of the female pronuclei would establish the diploid complex in one half of the individual while the other half would have only the maternal haploid complex. The diploid complex might readily contain two allosomes and so be of the homogametic sex while the other side would necessarily be of the opposite sex, since it could not have more than one allosome.

Mosaic gynandromorphs in which the characters of the two sexes are distributed at random through the body are baffling. The one valid conclusion to be derived from them is found in McClung's statement that "like other bodily functions . . . sex

Fig. 189.—A gynandromorph of *Papilio turnus* Linn. The left half of the insect is of the normal yellow male sex while the right half is of the black female form *glaucus*. (Through the courtesy of the Philadelphia Academy of Sciences.)

is a matter of cell activity, and must ultimately be explained in terms of cell performance."

The known facts concerning gynandromorphs have led to the conclusion that in insects at least, the secondary sexual characters are determined directly by genes. Any allosome, however, may contribute definitely to the determination of either sex. The chromosomal difference between the sexes is quantitative, as has already been pointed out, unless we conclude that the y chromosome is an active determiner of sex and such a conclusion is not in harmony with the available data. The y chromosome is not constant in occurrence; inheritance of sex is the same whether the y chromosome is present or absent, and finally sex-linked characters are transmitted by the x chromosomes. In view of these facts it is impossible to accept the conclusion that genes of the allosomes alone are directly responsible for the development of the secondary sexual characters even in the insects.

Although no adequate explanation of the entire phenomenon of gynandromorphism is available, we cannot fail to see that the chromosomes are not alone in their effect upon the organism as determiners of sex.

The secondary sexual characters of birds and mammals are known to be dependent for their development upon hormones secreted by various ductless glands. The gonads, although they are more evidently cytogenic glands, also have the function of secreting hormones which influence the development of the body, and the other endocrine glands, such as the pituitary, may be related so closely to sex as even to affect the development of the genitalia. Obviously any effect produced by a hormone is likely to be expressed in all possible parts of the individual, since the hormones are distributed by the blood stream. The activity of these substances would seem to be inimical to the development of gynandromorphs.

Insufficiency of hormones results in complete or partial failure of the individual to develop the characters dependent upon them. Since the characters of the sexes are in many cases antithetical expressions of the same structure, the effect of insufficiency is often an approximation of the secondary sexual characters of the opposite sex.

An extreme degree of insufficiency is caused by removal of the gonads. Since this is a common practice in the breeding of animals for food its effects are well known. In all domestic animals it results in the development of males which resemble females instead of normal males. Removal of the ovaries is almost without effect. Removal of the testes of most fowls likewise results in the development of some female characteristics, but early removal of the ovaries results in the almost complete development of male characters in the female.

Males of Sebright bantams have tail feathers similar to those of most hens, but when the testes are removed they develop the long curved tail feathers characteristic of most cocks. In this case a type of cell usually found only in ovaries is present in the testes. It apparently produces a hormone which influences the development of the tail feathers and the removal of the testes or ovaries in which it occurs frees the individual from this influence.

When we consider that the males of mammals are heterogametic while in the birds this condition is true of the opposite sex, the contrasting effects of castration are seen to be quite in harmony with the relationship of chromosomes and sex.

Freemartins afford another striking illustration of the effect of hormones. They are infertile female cattle, twin-born with males. Lillie has concluded in his studies of twinning that the condition is due to the mingling of blood in the two foetal circulatory systems through their close approximation in the maternal uterus. The hormones produced in either body would be carried through both by the fusion of the systems. Those produced in the male appear to inhibit the normal development of the female, with resulting sterility, although they are not capable of bringing about complete reversal of sex.

The indifferent stage of development through which all individuals pass, and the fact that the primary sexual organs of the two sexes show definite homologies still further indicate that both male and female have a common beginning. The abnormal cases cited show that residual capacity for the expression of one sex may remain even after the complete differentiation of its opposite. In these cases it is evident also that the differentiation of a heterogametic sex is an advance over that of a homogametic sex.

Sex Reversal. Man has not yet succeeded in determining the sex of animals under his control, but the facts so far presented suggest that such control might be possible, and records of sex reversal in fishes, amphibia and fowls show that the chromosomes are not an insurmountable obstacle. One of the most striking instances of sex reversal on record is that published by Crew in the *Journal of Heredity*. This case records that an adult hen from whose eggs normal chicks had hatched began to show the secondary sexual characters of the male and later became the father of chicks (Fig. 190). As in other cases of apparent sex reversal in fowls and pigeons this change was due to the destruction of the ovary by tuberculosis. Crew states that fresh growths of sex-cords from the peritoneum enter the ovary in the fowl, and that the absence of growing oöcytes permits the development of these cords into male structures. For a detailed consideration of sex reversal Crew's paper should be consulted.

Schaffner's Plants. The conclusive results obtained by Schaffner in his experiments on the determination of sex in plants must be regarded as among the greatest contributions since the discovery of the sex chromosomes. These experiments have been conducted

Fig. 190.—A Buff Orpington hen that became a cock. Two stages early and late, in the reversal of sex. This hen laid eggs from which normal chicks were hatched, and later became the father of normal chicks. (From Shull's *Heredity*. McGraw-Hill Book Company, Inc., by permission.)

chiefly on hemp (*Cannabis sativa* L.), but corn and various other plants have also been used. Plants have been reared under various conditions of environment and the results show that it is possible to determine the sex of the individual at will by changing the food supply, or to bring about reversal in a plant which has previously developed an inflorescence of one or the other sex.

One of the most striking of Schaffner's experiments was the production of Siamese twins of opposite sex in the jack-in-the-pulpit (*Arisaema triphyllum* L.). The result was obtained by very careful control of the food supply of the two individuals. Since the plants developed from a forked corm the control could not be accomplished through quantitative treatment but only through artificial reduction of the leaf surface of one plant and such other factors as might influence its capacity for photosynthesis and food storage. In spite of the fact that *Arisaema* twins are normally of the same sex or similarly monoecious, the result

Fig. 191.—Siamese twins of Jack-in-the-pulpit (*Arisaema triphyllum*). Male, left; female, right. The smaller size of the male plant is the only character readily seen in the figure. (Courtesy of Professor John H. Schaffner.)

of the experiment was the production of distinctly male and female individuals, still joined by the same forked corm (Fig. 191).

Conclusions. As we consider the entire succession of organic forms it is evident that the more highly organized a species may be, the less directly dependent it is upon the conditions of environment. The lower organisms and even such complex animals as the insects are directly influenced by temperature but the birds and mammals have surmounted this limitation. Because they are homothermous they can withstand a great range of external temperature without conspicuous modification of their activities. In man this process culminates, for through his intelligence he exerts control over all phases of his environment.

To a certain extent this independence of the organism is due to the fact that the complex body itself establishes an environment for its several parts through which the exigencies of life are tempered. This internal environment has become a definite part of the heritage, since it is in reality the result of correlation of heritable structures. Since the chromosomes are the carriers of heredity it seems wholly logical that all inherited characters should be in some way connected with them, and the internal environment may be regarded as a product of chromosomal activity.

No characters, however, are developed unless the proper conditions of environment favor the normal expression of the chromosomes. An animal without air cannot live; a nucleus without cytoplasm is equally incomplete. Since these conditions are resident in the body in the case of characters which appear automatically in every generation it is not a simple matter to determine what they are, but it is logical to believe that when a condition can be discovered, or when an external condition on which it depends can be determined, the behaviour of the genes with which it is associated can be modified. There is no reason to suppose that sex is any different in this respect from other characters.

It is therefore possible to conclude that sex is determined by the chromosome complex reacting to definite environmental conditions which are in most cases determined by the organism itself. The modification of these conditions, whether within or outside of the organism, results in the modification of sexual characters.

Modification of a sex is, however, subject to definite limitations. In most animals sexual organs are developed only once during the life of an individual, instead of annually or more frequently as in many of the flowering plants and some of the lower animals. The degree of differentiation of the primitive indeterminate structures in the higher animals also is extreme. Although the sexes of the birds retain a fairly primitive condition of the reproductive organs which offers no structural obstacle to reversal, the highly differentiated organs of mammals can hardly be expected to submit to such an extreme change. It is probable that if the mammalian embryo could be subjected to the proper stimulus while its sexual organs were still in the indeterminate stage its sex might be controlled. The freemartin most nearly attains this condition, but since male hormones appear to be the determining factor it is evident that sexual differentiation must precede the formation of these substances; the female must be definitely female before it is subjected to the action of the hormones of its male twin.

The entire subject is one of little practical importance. One sex is sometimes more useful than the other among both plants and animals and man has always shown a desire to predict or control the sex of his own offspring, but there is so little possibility of simple and effective methods of controlling sex that at present it has only scientific value. In this it is of primary importance as a demonstration of chromosomal functions.

Summary. Sex is not essential to reproduction as a biological phenomenon, although in most of the higher organisms reproduction is accompanied by sex phenomena. Both types of germ cells and the sexes that produce them are equivalent, and their differentiation must therefore be explained by the specialization of the reproductive process. The most important result of sexual reproduction is the recombination of the unit characters making up the species. These conclusions are borne out by a large number of unusual cases, both natural and artificial, including hermaphroditism, gynandromorphism, the effects of endocrine imbalances and sex reversal. All available evidence combines to show that sex is in many cases determined by the chromosomes but that these require definite conditions for their normal expression. Whether these conditions are also determined by the body or come from without, their modification may be expected to influence the sex of the individual, but determinate sexual structures may limit the extent of modification.

REFERENCES

KORNHAUSER, S. I., *Denison U. Bulletin, Jn. Sci. Lab.* XX, 1–21, 1922.

WALTER, H. E., *Genetics*, revised edition., 1923.

SCHAFFNER, J. H., "Siamese Twins of Arisaema Triphyllum of Opposite Sex Experimentally Induced," *Ohio Jn. Sci.* xxvi, 276–280, 1926.

—— "Sex and Sex-Determination in the Light of Observations and Experiments on Dioecious Plants," *Am. Nat.* LXI, 319–332, 1927.

SHULL, A. F., *Heredity*, 1926.

ROBB, R. C., "Y-chromosome Inheritance," *Am. Nat.* LXI, 568–571, 1927.

CREW, F. A. E., "Abnormal Sexuality in Animals," *Quarterly Review of Biology*, I, 315–359, 1926; II, 249–266, 427–441, 1927.

CHAPTER XIX

THE PRACTICAL VALUE OF GENETICS

The human race was successful to a remarkable extent in shaping other organisms to its needs and desires before the advent of science. Many of our domestic animals and plants were developed without the slightest knowledge of the underlying principles of heredity and breeders even now continue to apply the practical principles of their craft without such knowledge. As in so many other fields of human activity, theory in this field is not essential to practical success.

The science of genetics has shown us not only how the results of the past have been obtained but also why many desired results have never been secured. By clearly establishing the fundamental principles of inheritance it has shown what we can hope to do in the future, how it can best be accomplished and in some cases how human efforts in this field are hedged about by apparently insurmountable barriers. Some things known to practical breeders and scientists alike can only be determined by experiment, but in general the best results can be obtained only through the combination of scientific knowledge with sound practical methods. As in all fields, limited knowledge of any kind cannot hope to compete with broad knowledge of all phases of the subject.

The methods of plant and animal breeding are dependent upon the two general methods, hybridization and selection. In a work of this kind it is impossible to mention the details of method involved; discussion must be limited to these fundamental processes, some of their results and their common limitations.

Hybridization. This process, as the behaviour of Mendelian unit characters shows, makes it possible to secure combinations of characters found in nature only separately in different species or strains. The combination desired may easily be secured if the difference between the parents is slight, or it may involve a multitude of difficulties if the parents belong to different species. As a general rule any combination is *possible*, even though it may not be *practicable* to secure it.

A

B

Fig. 192.—Results of crossing two inbred strains of corn. *A*, at the left are two rows, each containing an inbred strain. The tall corn at the right is the result of crossing them. *B*, the basket at the right represents the average production of the two inbred strains after three generations of inbreeding—sixty-one bushels per acre. The basket at the left shows the first generation yield from the hybrid—one hundred and one bushels per acre. (From Walter, after East and Hayes.)

The diagrams of Mendelian hybrids (Fig. 159, 161, 163) show that any combination, no matter how many unit characters are involved, appears in a limited number of individuals in each generation after the first filial in the homozygous state. When once secured, individuals of this kind bearing a new combination of characters are the potential heads of a new race.

If, for example, one of our American roasting-ear enthusiasts were cast on an uninhabited island with only yellow field corn and white sweet corn in his possession he could secure at least an imitation of his beloved yellow bantam sweet corn in the second generation. By cross-fertilizing the two strains he would secure a yellow-starchy hybrid of the zygotic formula YySs. When these seeds were planted and the new generation carefully inbred, four kinds of grains would appear, viz., yellow starchy, yellow sweet, white starchy and white sweet. Among the yellow sweet grains would be two thirds Yyss and one third YYss. The latter could produce nothing but yellow sweet corn.

The sources of valuable characters are those already mentioned. The range of fortuitous variation supplies some, as in the case of corn which varies in starch, sugar and oil content. Mutations useful to man have also appeared. Babcock and Clausen record the following mutants of cultivated plants:

Early maturing varieties of the Florida Velvet Bean.

Tobacco, including one mutant of Connecticut Cuban which showed an increase of 90 per cent in yield.

Sugar beets with increased sugar content.

Various useful mutations of cotton, hemp, rye and sunflower.

Plant Hybrids. Such hybrids are valuable for new combinations and for increased vigor. Hybrids of different strains of corn are commonly reported to show a marked increase in yield over either of their parents, sometimes amounting to 250 per cent (Fig. 192). Crosses have been made of many different varieties of corn, and in most cases this increase in yield has been observed. Such increase of vigor in hybrids, whether plants or animals, has been called *heterosis* by Shull. The following table gives the results of some of the crosses of corn.

YIELDS OF MAIZE CROSSES COMPARED WITH PARENTAL YIELDS
(Modified from Babcock and Clausen, after Collins)

Name of Hybrid	Yield of Female Parent, Pounds	Yield of Male Parent, Pounds	Average Yield of Parents, Pounds	Yield of Hybrid, Pounds	Percentage of Increase of Hybrid Over Average of Parents, Per Cent
Maryland dent by Hopi	1.19	0.74	0.965	1.25	29
Tuscarora by Cinquantino	0.53	0.24	0.385	0.75	95
Kansas dent by Chinese	0.99	0.39	0.690	1.09	58
Chinese by Chihuahua	0.39	0.69	0.540	0.95	76
Hopi by Chinese	0.74	0.39	0.565	1.28	126
Chinese by Xupha	0.39	0.63	0.510	0.54	6
Brownsville by Chinese	0.77	0.39	0.580	1.16	100
Brownsville by Guatemala Red	0.77	0.31	0.540	0.49	−9
Huamamantla by Hairy Mexican	0.40	0.18	0.290	0.31	7
Hairy Mexican by Chinese	0.18	0.39	0.285	0.61	114

Many other plants have been improved in the same way. Tomatoes, cucumbers and strawberries not only give vigorous plants in the F_1 generation but produce heavier yields than the most prolific parents. Hybrid varieties of some of these plants are commonly used, but difficulties arise if the plant must be raised from seed because of the segregation of different combinations of characters in the F_2 generation. This has limited the commercial utilization of heterosis, although plant breeders have shown that in many cases the additional difficulty of securing seed is more than compensated by the productivity of the hybrid. The simplest method of producing seed is to maintain the two parent strains and cross them whenever seed is required. Since the fertility of most seeds lasts several years it is not necessary to hybridize every year.

No example of hybridization for character combinations is more striking than that of the Concord grape. "Ephraim Wales Bull produced the Concord grape as a result of eleven years of patient work in crossing the native species, *Vitis labrusca*, with European varieties, raising the seedlings and testing selections. 'From over 22,000 seedlings there are 21 which I consider valuable,' he writes. Although the hybrid nature of the Concord and other derivatives of *Vitis labrusca* has been questioned, the evidence from extensive tests of selfed seedlings of this and several other standard American varieties as reported by Hedrick and

Anthony seem to indicate that they are really hybrids between American species if not between *V. labrusca* and *V. vinifera*. Whatever the origin of the Concord may have been, its sterling value is evidenced by its history. Introduced in 1853, 'ten years later the Concord grape was spread over the entire northern part of the United States and is now widely used in the temperate regions of most parts of the earth.' Ephraim Bull's service to his fellow men seems to have been all but forgotten while he was still living, since 'he died neglected, in poverty, broken in spirit.' Vast as would be the value of his contribution if it could be computed, even more valuable was the inspiration he gave, 'which has helped to make plant breeding one of the great forces in cheaply feeding the world.'" (Babcock and Clausen.)

Nor has food supply alone been the object of plant hybridization. An inestimable number of beautiful varieties of flowers have been given to us through this medium, and even now fanciers of peonies, irises, roses and many other plants find new offerings available every year from the gardens of plant breeders who experiment tirelessly with hybrids of promising varieties. Within the last few years the beautiful yellow hybrid tea rose, Souvenir de Claudius Fernet, has been acclaimed by lovers of flowers throughout the world. Even more recently there has been added to the already magnificent array of tulips a new class, late-flowering hybrids, produced by crossing the Darwin and Cottage varieties. The poetaz narcissus was produced by crossing poeticus ornatus and polyanthus varieties. It combines the large clusters of the latter with the hardiness of the former and has an exquisite odor of its own (Fig. 193). New varieties of *Iris germanica* are also constantly appearing.

Animal Hybrids. The problems of the animal breeder are very different from those of the plant breeder, but in general the same fundamental methods are open to him. While selection plays a very large part in the development of improved strains of animals, we are not without familiar examples of animal hybrids whose value to the human race is permanently established. Mules, for example, are produced by crossing the male ass with the female horse and can be produced in no other way. The mule breeding industry attained a value of $500,000,000 in the United States in 1915. The mule is a more vigorous animal than either of the parent species, and is more resistant to adverse

Fig. 193.—Varieties of Narcissus. *A*, polyanthus; *B*, poeticus; *C*, poetaz. Various forms of poetaz narcissi have been produced as hybrids of different varieties of polyanthus and poeticus. (Through the courtesy of the A. B. Morse Company, horticultural printers, St. Joseph, Mich.)

environmental conditions; it combines morphological characters of both parents.

Few other interspecific animal hybrids are of more than potential value. Cattle have been crossed with the American bison, the zebu, and other species with excellent results, but the hybrids are not in common use (Fig. 194). Of these crosses Babcock and Clausen say: "By long-continued selection it would be possible to transfer many of the excellent qualities of the bison such as superior coat, greater hardiness, resistance to tick and insect

FIG. 194.—Quinto Porto, five-eighths bison, three eighths polled Hereford. (With the permission of the *Journal of Heredity*.)

infestation, and superior beef qualities to domestic cattle.' The hybrid between the zebu and our common cattle is sufficiently resistant to tick-borne disease to be very valuable in the southwestern states according to Lush. Many other hybrids have been recorded between domestic animals, such as the sheep and goat and various species of fowls, but they are chiefly of scientific interest.

An interesting effect of hybridization occurs in bees. The Italian bee has long been recognized as superior to the black or German bee and has become the most popular and widely kept variety in the United States. It is resistant to one of the two serious bee diseases and is not seriously affected by the bee moth, while the black bee is susceptible to both diseases and when colonies become weak they may succumb to the inroads of the bee moth. Hybrids between the two have no proved superiority and retain the characteristic nervousness of the black bee which makes them difficult to handle as compared with Italians. In addition they are the most ill-tempered of all three. Italians vary greatly in temper but are in general mild, while the hybrids sting readily under any conditions. This cross is usually looked upon as intervarietal, although the systematic rank of the various kinds of honey-bees is by no means clear.

Heterosis is sometimes as marked in animals as in plants. It must not be thought, however, that because crossing may increase vigor, inbreeding must reduce it, for many strains of domestic animals have been intensively inbred for many generations without reduction of vigor or fertility. In some cases, in fact, these qualities have been improved through inbreeding accompanied by careful selection. In general, crossing results in heterosis while inbreeding may be practiced with varied results, hence the conclusion has been reached that vigor and fertility depend upon unit characters which are much more likely to be isolated in homozygous combinations through inbreeding than through the mixing of various strains.

Limitations of Hybridization. The crossing of different species is attended by many difficulties which limit its value. Some are insurmountable while others can be met by special methods of procedure. Hybridization within a species, whether between varieties or with respect to single characters, is a much simpler process.

Interspecific infertility is the most serious of these limitations, for when normal union of the germ cells cannot take place no hybrid can be produced. It is such a common phenomenon that some biologists have recognized in it a criterion for the limitation of species. While such an extreme interpretation is not favored by the available evidence, the fact remains that many species cannot be crossed. The facility with which hybridization may be accomplished is often, if not always, in proportion to the degree of relationship between the species involved.

The reasons for infertility are various. In some cases the sperm cell is unable to penetrate the surface of the ovum of a different species. In others the spermatozoön not only enters the ovum but also initiates development, although it makes no material contribution to the new individual. In no case can we conclude that chromosomal insufficiency is responsible, for the haploid complex of either germ cell may contain a complete set of determiners for the production of a new individual. Hybrids have been produced many times between species with different numbers of chromosomes. The general reason for infertility between species may therefore be stated as some lack of harmony in the accessory phenomena of reproduction.

Since varieties and strains of the same species are similar in all fundamental structures and processes this difficulty cannot hinder the production of hybrids between such groups.

Infertility of Hybrids. Although difference in the chromosome complexes of the parents need not affect the production of the hybrid, it may well be expected to have serious results in the production of the F_2 generation. Since the process of maturation of the germ cells hinges upon synapsis and the resulting reduction of the chromosomes, it is easy to see that any asymmetry which affects the consummation of this delicately adjusted series of events may prevent the formation of normal germ cells. The number of chromosomes found in ova of the horse is said to be nineteen, and the number in the spermatozoa of the ass thirty-two or thirty-three. The mule therefore has an asymmetrical chromosome complex. Male mules are not known to produce functional germ cells. Cases are on record of fertile female mules, and they may occasionally occur, although they are open to doubt. The rarity of even doubtful cases is in itself suggestive.

When the species are closely related and have similar chromosomes there is no reason to expect infertility in their hybrids, but the actual occurrence of infertility even in such hybrids forces us to the conclusion that physiological differences in the chromosomes may exist even when visible morphological differences are lacking. The production by some species of fertile female hybrids and infertile males is another puzzling complication of the problem. This is true of the hybrids of domestic cattle with the bison.

For practical purposes it is evident that the infertility of hybrids is not insurmountable. It is, indeed, no more serious than the inconstancy of desirable heterozygous strains. Either demands the maintenance of pure parent stocks and the production of new hybrids solely by repetition of the cross unless the breeder, in the case of heterozygotes, is willing to breed from hybrids and discard the fifty per cent of homozygous individuals in every generation.

Asexual Propagation. Repeated hybridization is naturally more complicated and more expensive than the normal course of reproduction. It is avoided by commercial plant breeders through asexual propagation of their hybrid stocks. Hybrid fruits are propagated by grafting scions of the desirable stock onto hardy root systems, sometimes of entirely different species. The beautiful varieties of French lilacs are grafted onto roots of the common lilac or privet. Chrysanthemums are easily raised from cuttings, roses from cuttings or by grafting, peonies and other flowers by division of the roots and crown of the plant, and bulbs through their natural asexual increase. Plants which can be produced only from seed are obviously subject to the same limitations as animals.

Hybridization for the production of new combinations of characters is limited only by the difficulty of isolating homozygous strains. If the desired type is complex this difficulty is great and it is necessary to resort to asexual propagation if possible. Many desirable hybrids are simple, however, so this does not limit the uses of the process to organisms which can be produced asexually.

Perhaps no useful hybrid is a better illustration of complexity and the value of asexual propagation than Burbank's Alhambra plum. The ancestry of this variety is incorporated in the following diagram by Babcock and Clausen:

$$
\text{Alhambra} \begin{cases} d \ldots \ldots \begin{cases} \text{Nigra} \\ \text{Americana} \end{cases} \\ e \ldots \ldots \begin{cases} c \ldots \ldots \begin{cases} \text{Triflora} \\ \text{Simoni} \end{cases} \\ b \ldots \ldots \begin{cases} \text{French Prune} \\ a \ldots \ldots \ldots \begin{cases} \text{Pissardi} \\ \text{Kelsey} \end{cases} \end{cases} \end{cases}
$$

What an impossible task it would be to fix the desirable combination of characters in any other way!

THE PRACTICAL VALUE OF GENETICS 351

Selection. The process of selection has been practiced for many centuries for the production of improved strains of cultivated plants and domestic animals. It is a logical consequence of the fact that "like produces like" to a marked degree. Cows which produce a large quantity of rich milk and bulls of the same strain are much more likely to produce good dairy cattle than those which possess other qualities. Sheep with fine and heavy fleeces are obviously more valuable for the production of wool than those whose fleeces are coarse and light. By the early recognition of these facts man has produced beef and dairy cattle, draft and race horses, dogs of many breeds and a multitude of other distinct varieties of relatively few natural species (Fig. 195).

Methods of Selection. Before the discovery of scientific principles of inheritance selection was necessarily based upon observed characters, and hence may be called phenotypic selection. It is usually known as mass selection because the best individuals from a given group of organisms are selected as the parents of future generations and reproduction of the poorer individuals is prevented, but no further attention is given to the details of parentage.

Closer attention to individual parentage brought about the refinement of method known as line selection, which is closely allied to the most modern and scientific method, genotypic selection.

Mass Selection. The English scientist, Hallet, associated selection with environmental effects by giving plants the best possible environment and selecting those which did best under these favorable conditions. Rimpau, on the other hand, subjected his grains to unfavorable or merely average conditions, and selected those which showed the ability to do well in spite of adverse surroundings. Either method results in the improvement of the organism, but the latter in particular is valuable for it discloses something of the inherent possibilities of the individual.

Selection as it has unavoidably been limited in the honey-bee is a fine example of the effectiveness of mass selection. Since the functional sexes are merely reproductive and the individuals which are directly of use to man do not reproduce, the breeding stock in this case can be judged only by its progeny. The mating of bees occurs in flight so that only the female parentage of a colony can be definitely known and selected. Only within the

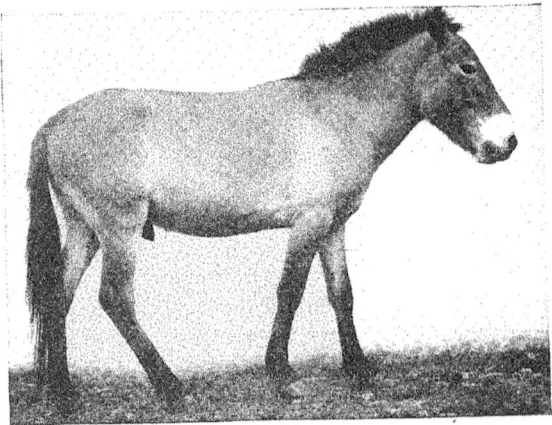

A. Prjevalsky Horse, a wild Asiatic species. (Through the courtesy of the New York Zoölogical Society.)

B. Lou Dillon, a trotter.

C. Benedict, a Clydesdale stallion. (B and C from Plumb's *Types and Breeds of Farm Animals*, with the permission of Ginn and Company.)

Fig. 195.—The results of selection in the horse.

last few years has a method been devised for the artificial control of mating, and this has not yet been widely used.

Selection has been practiced for various characters in the honeybee, such as light color, industry based on the amount of honey stored, color of wax produced, temper, and swarming propensity. By rearing queens only from mothers whose colonies best expressed the desired characters and by restricting the production of drones (males) in other colonies as much as possible, many distinct strains of Italian bees have been produced. Some are called three-banded leather-colored Italians, others golden Italians because they very nearly lack the black abdominal bands and are pale in color. Some sting readily while others are mild tempered and sting only when conditions for handling bees are very unfavorable. A most desirable result of selection is the reduction or elimination of swarming instinct, which is the bee-keeper's greatest source of annoyance. Some strains swarm readily and often, while others will go through a season under conditions entirely favorable to swarming without attempting it.

Line Selection. The fundamental principles of Mendelian inheritance disclose the necessity of knowing the genotypic organization of an individual for accurate control of succeeding generations. Even before Mendel worked out his laws the value of his discoveries in relation to selection was anticipated by the work of Vilmorin, near the middle of the nineteenth century. He selected single plants whose offspring were isolated for comparison. The principle was later applied by various plant breeders, and in the hands of Hjalmar Nilsson at the experiment station of the Swedish Seed Association at Svalöf it has produced many valuable strains of wheat, peas, potatoes and other plants. The method is also known as pedigree breeding.

The obvious value of line selection is that the isolation of offspring of single individuals is much more likely to produce a uniform variety. Even superficially identical individuals, as we have seen, may be genotypically different and therefore capable of producing different offspring.

Genotypic Selection. While any method that takes into account the character of the succeeding generations in relation to their known ancestry is to some extent genotypic, in the strict sense, this term should apply to the type of selection which is used in connection with known facts of Mendelian inheritance.

It must be used in connection with hybridization for the isolation of desirable characters and character combinations in the homozygous state, and if the heterozygous individuals are desired it may be used for the elimination of their homozygous offspring.

The Pure Line. Darwin's theory of natural selection influenced thought in such a way that for many years selection, through natural or artificial means, was supposed to result in actual modification of the line selected. Galton's law of filial regression also suggested very strongly the possibility of shifting the general character of a group of organisms by always selecting extreme individuals as the parents of the next generation. The Danish botanist, Johannsen, tested the accuracy of this view and in doing so discovered the existence of pure lines.

Johannsen used for his experiments a cultivated bean (*Phaseolus vulgaris nana*). The weights of seeds planted were recorded and the entire lot of seeds produced by every plant was carefully harvested and weighed. In general the largest beans produced the largest offspring, but Johannsen was impressed by two important facts. In the first place, the seeds produced by a single plant sometimes fluctuated about a mean quite different from that

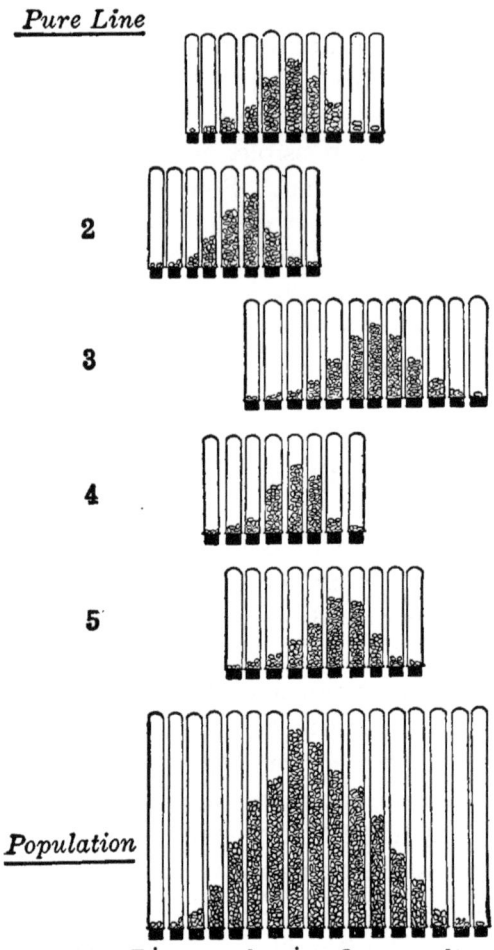

FIG. 196.—Diagram showing five *pure lines* and a *population* formed by their union. The beans of each pure line are represented as assorted into inverted test tubes, making a curve of fluctuating variability. Test tubes containing beans of the same weight are placed in the same vertical row. (From Walter, after Johannsen.)

suggested by the size of the parent seed, and in the second, parent seeds of the same weight in some cases produced beans with very different ranges of variation. In spite of the fact that the results were reasonably harmonious with Galton's law of regression Johannsen concluded that he was dealing with mixed stock and conducted further experiments to explain his discoveries.

In these experiments plants produced from beans of known weight were self-fertilized and their offspring for several generations were treated in the same way until it was certain that the lines were homozygous. The result indicated that the original stock was made up of nineteen different kinds of beans, each kind varying between certain extremes of weight which might overlap with others, but every one of the nineteen fluctuating about a different mean. These nineteen groups were called pure lines. Johannsen defined a pure line as the progeny of a self-fertilized homozygous individual. An aggregation of pure lines such as that with which he first dealt was called a population. Figure 196 illustrates graphically the difference between five of Johannsen's pure lines and the population formed by mixing them.

It is evident that intensive line selection in any plant may bring about the isolation of pure lines (Fig. 197). The difficulty of maintaining such lines under ordinary conditions is obvious, however, and it is doubtful that true pure lines often occur in nature. They must certainly be restricted to those plants in which elaborate adaptations for self-fertilization are present.

Equivalents of the Pure Line. The self-fertilization of a homozygous individual is the same in result as the union of gametes with the same complex of determiners from different individuals, consequently *homozygous crosses* produce equally homogeneous groups of offspring. It is also similar in effect to reproduction without fertilization, since here there is no possibility of different characters being brought in. Reproduction of the latter type includes *parthenogenesis* and *agamic reproduction,* such as fission and budding. The individuals descended from one ancestor through a series of asexual generations collectively constitute a *clone.*

Under natural conditions none of these pure-line equivalents are more likely to be maintained than the typical pure line, but they may exist for a considerable time in many species. At the end of a summer, for example, the offspring of each stem-mother aphid constitute a clone. Such animals as the Protozoa and *Hydra* also undergo a series of asexual divisions which give rise to clones before sexual reproduction intervenes to bring about a reassortment of characters.

Homozygous crosses are not only uncommon in nature but also difficult to obtain in the laboratory. With respect to single char-

Fig. 197.—Typical heads from seven pure lines of Defiance wheat. (From Babcock and Clausen's *Genetics in Relation to Agriculture*. McGraw-Hill Book Company, Inc., by permission.)

acters this difficulty is not encountered and lines homozygous for one or a few characters have been secured many times and in many species.

Selection in Pure Lines. When Johannsen had isolated his nineteen pure lines of beans he found that no matter what the size of the parent seed, those which it produced fluctuated about the mean for the pure line to which it belonged. Walter has graphically indicated these results in a diagram which is reproduced in Figure 198. In all kinds of pure lines the effects of selection have been likewise negative. Aphids, daphnids, *Droso-*

THE PRACTICAL VALUE OF GENETICS 357

phila, Paramecium and many plants have been the basis for the conclusion by many biologists that selection within the pure line is without effect.

The Pure Line as a Limit of Selection. The fact that selection has proved a valuable method of improving organisms according to most experimental evidence, rests upon the possibility of isolat-

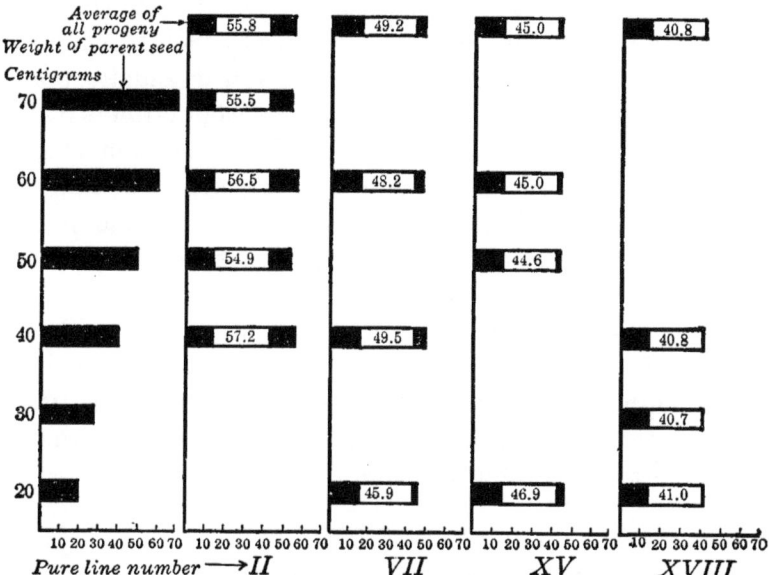

Fig. 198.—The result of selection in four pure lines of beans. The vertical columns, representing the average progeny from different sizes of parents all derived from the same pure lines, contain groups nearer alike than the horizontal columns, representing progeny from parents of the same size but of different pure lines. All of the numbers indicate weight in centigrams. (From Walter, after data from Johannsen.)

ing those pure lines which best express the desired character. It has been found that environmental conditions may bring about differences in the development of individuals within the pure line, hence this factor may also enter into improvement by selection, and finally supplementary Mendelian factors may act upon characters in a pure line to produce an entirely different result. The pure line is therefore only a partial limitation of selection and even without the inconstant modifications mentioned there is a lack of absolute evidence that it is permanently fixed in nature. Mutations occur to change pure lines, and there is a growing feeling that the environment may be important in the modification of inherent qualities.

Summary. Hybridization and selection have played an important part in the establishment of varieties of plants and animals of use to man. Hybridization is useful for two purposes, viz., increase in vigor and productiveness, called heterosis, and the combination of useful characters. Its effectiveness in animals is limited by the segregation which occurs during reproduction. This necessitates repeated hybridization whenever the heterozygous condition is the one desired. In plants even this limitation is offset by the fact that asexual propagation is usually possible. Hybridization is also limited where sexual reproduction is unavoidable by the frequent infertility of hybrids. Selection is accomplished by several methods, all directed toward the isolation of the most favorable individuals as the parents of the following generation. It is practiced in connection with hybridization for the isolation of desired combinations of characters, and within established species for the isolation of the best strains. Selection is apparently limited by the pure lines of which a species is composed, although the isolation of these lines may constitute an effective degree of modification. Even pure lines are known to be susceptible to some modification.

REFERENCES

BABCOCK, E. B. and CLAUSEN, R. E., *Genetics in Relation to Agriculture*, 2nd edition., 1927.

An extensive bibliography is published in Babcock and Clausen's work.

CHAPTER XX

HEREDITY IN MAN

The study of heredity in the human race is hindered in a number of ways. Because of the span of generations, results approximating those secured in the laboratory in the study of other organisms are impossible. It is rare to hear of five generations or even four alive at the same time in a human family, and most of us never know more than three, so it is necessary to fall back upon records and these are at present pitifully incomplete. Genealogies furnish some valuable data, the records of public institutions are also a dependable source of information, and within the present century such institutions as the Eugenics Record Office at Cold Spring Harbor, Long Island, have begun the work of making accurate scientific records in this field.

Even the accurate observation and recording of natural phenomena cannot, however, give results like those obtained in the study of laboratory animals. It is and will probably always be impossible to control the reproduction of human beings except in extreme cases which demand the action of organized society for its own protection. Fortunately such methods, undesirable from the normal human viewpoint, are not essential to an understanding of human inheritance. We are animals and there is every reason to believe that the laws of heredity in other organisms are equally applicable to ourselves. The corroboration of this relationship by the available data is adequate.

One unfortunate feature of our knowledge of human heredity is that extreme cases and particularly abnormalities are most likely to make an individual the object of scrutiny. Data are more abundant concerning the inheritance of supernumerary digits and mental defects than on the behaviour of valuable qualities, although the latter are by no means lacking. Fortunately accurate data, whatever the characters recorded, are a valuable basis for the application of general laws as worked out in other organisms, and at least a partial indication of the trend of heredity in general.

What Is Inherited? The application of conscious thought to the problems of existence is a complicating factor which it is exceedingly difficult to avoid in a scientific consideration of human behaviour. Every-day interpretations of factors in human life must be translated into the accuracy of scientific observation. We hear of the inheritance of drunkenness, disease and special ability, as well as of structural characters, whereas these things are phases of behaviour, and behaviour can only be the expression in the individual of its hereditary properties. There is necessarily some basis in the individual for anything which it does, and that basis is at least indirectly associated with some inherited character. It is not enough to say that a great musician inherits musical talent; he inherits an exceptional sense of hearing, great manual dexterity and the wonderful nervous coördination which any skilful performance demands. The use to which he puts these things is response. The skilled mechanic who builds an instrument of precision has an equally fine inheritance, but his ability attracts less attention.

Man, like other organisms, has a heritage and an environment to which the heritage responds. Responses are the things that interest us chiefly. They are so conspicuous that they usually obscure the heritage from which they arise but careful consideration will show that a structural heritage is present for every function. Mental activity, although it is exceedingly complex, is no less definitely based on structure than other functions. It is none too well understood in detail but of its anatomical source we need have no doubts. Whatever the inherited structure, in so far as it finds the proper environment for its expression, it will manifest itself in the same way in successive generations.

The distinction between heritage and response, since they are so likely to correspond in successive generations, is not essential from the popular point of view. In effect, a talent may be inherited. From the point of view of the scientist, however, no talent is the simple thing into which ordinary language resolves it. In no case can exceptional ability be looked upon as a unit character, nor can conditions of mental deficiency always be so simply handled. These things are the result of many conditions present in the body. They may be based upon unit characters, but are due to complex immediate causes. The things actually handed down from generation to generation in man as in other organisms are chromosomal determiners which are capable of a certain response if the right conditions prevail.

Unit Characters. The complexity of human responses is so baffling that it is often impossible to discover all of their underlying causes. In simpler conditions the behaviour of heritable characters can often be traced through several generations, so that the occurrence of unit characters in man is well established. Eye color, hair color, pigmentation of the skin, curliness of hair, polydactyly and symphalangism are among the well known structural characters in this category.

The chromosomes of man are also similar to those of other animals and furnish a basis for heredity of the same type. Various investigators have studied the cytology of human cells with the result that the chromosome number is commonly accepted as forty-eight. Of these, forty-six are autosomes and two allosomes. There are an x and a y chromosome in the male and two x chromosomes in the female. The gametes therefore contain either $23+x$ or $23+y$. This is only a moderately large number but considering only one determiner to a chromosome it affords the possibility of 4^{24} or more than two hundred thousand billion recombinations. The diversity of human beings is not surprising!

Eye Color. The color of the eyes depends upon the presence of two pigments, brown and blue. Brown pigment varies greatly in quantity, so that brown-flecked blue eyes are common, but when distributed through the entire area of the iris it masks the blue because it lies in front. It is dominant over lack of brown, which is, under ordinary conditions, equivalent to dominance over blue. The two are not allelomorphic, but the allelomorph of brown permits the blue pigment to show and only in albinos can the absence of blue be seen. Brown-eyed parents may be heterozygous and are therefore able to produce blue-eyed children.

Hair Color. Hair color is also due to two pigments. Its behaviour is not thoroughly understood because of the occurrence of various modifying conditions, but darker colors are dominant over light hair and inheritance is in general similar to that of eye color. Black-haired parents may produce blond children but blond parents cannot produce brunettes.[1]

---

[1] A recent article by Hausman (*Am. Nat.* LXI, 545–554, 1927) contains many interesting facts on the pigmentation of human hair.

Pigmentation of Skin. Davenport's studies of negro-white crosses are the chief source of information on this subject. He reached the conclusion that pigmentation depends on two pairs of cumulative factors which may be designated as AA and BB (Fig. 199). Since the white race is not totally devoid of pigment

	A B	A b	a B	a b
A B	A B / A B — 70	A b / A B — 55	a B / A B — 53	a b / A B — 38
A b	A B / A b — 55	A b / A b — 40	a B / A b — 38	a b / A b — 23
a B	A B / a B — 53	A b / a B — 38	a B / a B — 36	a b / a B — 21
a b	A B / a b — 38	A b / a b — 23	a B / a b — 21	a b / a b — 6

Fig. 199.—Punnett square showing the expected shades of color in the possible offspring of two mulattoes. $A=18$, $B=17$, $a=2$, and $b=1$ per cent of black pigment. (From Walter, after data from Davenport and Danielson.)

it follows that the factors aa and bb do not stand for albinism but only for slight pigmentation. The percentage value ascribed to each determiner by Davenport is indicated in the diagram. A pure African black would have the formula AABB for color, while a white would be represented by aabb. The hybrid mulatto has the formula AaBb. Figure 199 represents the F_2 generation derived from mulatto parents. In this diagram it is evident that there are three kinds of mulattoes with the formulae AAbb, AaBb and aaBB, each differing slightly from the others in pigmentation, as well as intergrades between these and the dominant and recessive combinations. The names for these intermediates are quadroon for individuals with only one dominant determiner and mangro or sambo for those with three. It is also evident from the diagram that mulatto parents have one chance in sixteen of producing a black child and one of producing a white. Individuals of the latter class are called by a number of names, including

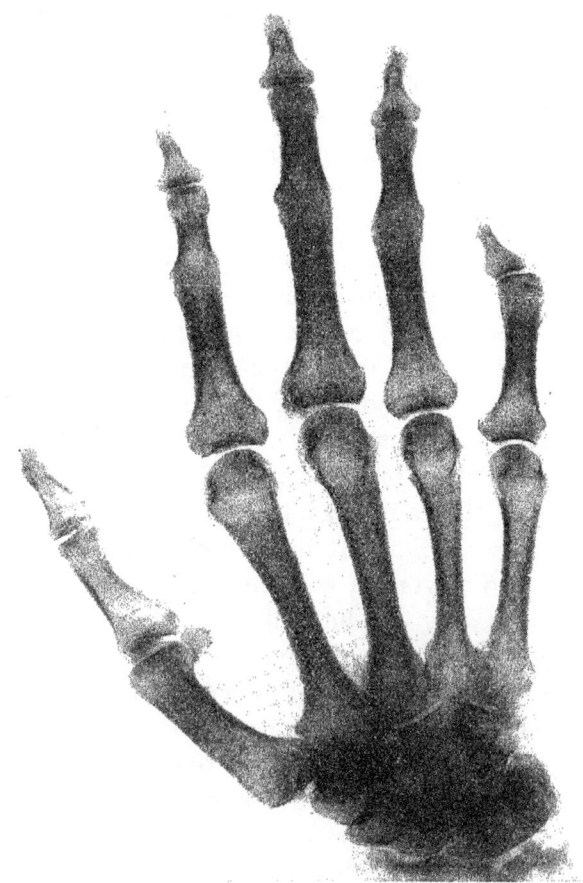

FIG. 200.—Radiograph showing symphalangism in man. The two proximal phalanges in each of the four fingers are fused. (Through the courtesy of Professor R. A. Hefner.)

pass-for-white and octoroon. Their negroid ancestry is usually plainly evident in other characters than color.

Polydactyly, Brachydactyly, and Symphalangism. These conditions are modifications of the fingers and toes which are dominant over the normal condition. The first is multiplication of the usual number of five digits. An extra thumb or great toe is frequently the added member. Brachydactyly is extreme shortening of the fingers, which sometimes lack one of the three phalanges and sometimes have an extremely short terminal phalanx. Symphalangism is fusion of the phalanges of toes or fingers so that one of the usual joints is stiff.

Hefner has traced the last condition through six generations of a family in which it behaved as a Mendelian dominant and appeared in both males and females (Fig. 200). In his report on this case he cites another remarkable record: "John Talbot, first Earl of Shrewsbury, was supposed to have had fingers with stiff joints. He was killed in battle near Bordeaux in 1453, by a blow on the head, received after his thigh had been broken. He was buried in Shrewsbury Cathedral. Recent alterations made it necessary to disturb his grave, when tradition was confirmed and his bones identified by the fused finger-joints, the cleft skull, and the broken thigh-bone. By a strange coincidence this work was under the direction of one of Talbot's direct descendants in the fourteenth generation, the joints of whose fingers were fused like those of his remote ancestor. . . ."

Sex-Linkage. Color blindness, a sex-linked character, is inherited in man in the same way that other sex-linked characters are inherited in *Drosophila*. It is recessive to normal vision. A color-blind man and a woman with normal vision cannot produce color-blind children but one-half of their daughters are carriers and can produce color-blind sons even if mated with normal men. Their daughters would be one-half normal and one-half carriers. Color-blind females can be produced only when both parents supply factors for this condition, since one x chromosome comes from each parent. The following diagrams show how the character is transmitted in the four possible crosses (Fig. 201).

These diagrams make evident a number of interesting phenomena. A shows that a color-blind parent may have children with normal vision. Even a color-blind female, as shown in B, may have some children with normal vision, and if she had only daughters all would be apparently normal although able to transmit the defect. Diagram D shows how parents with normal vision may produce color-blind sons. A combination of such cases as these shows how a sex-linked defect may be transmitted generation after generation through a female line, to crop out in an occasional male. Because of the small size of human families there is so little chance for a complete expression of Mendelian ratios that this may easily occur, while a high rate of reproduction would be almost certain to bring out all possibilities within one or

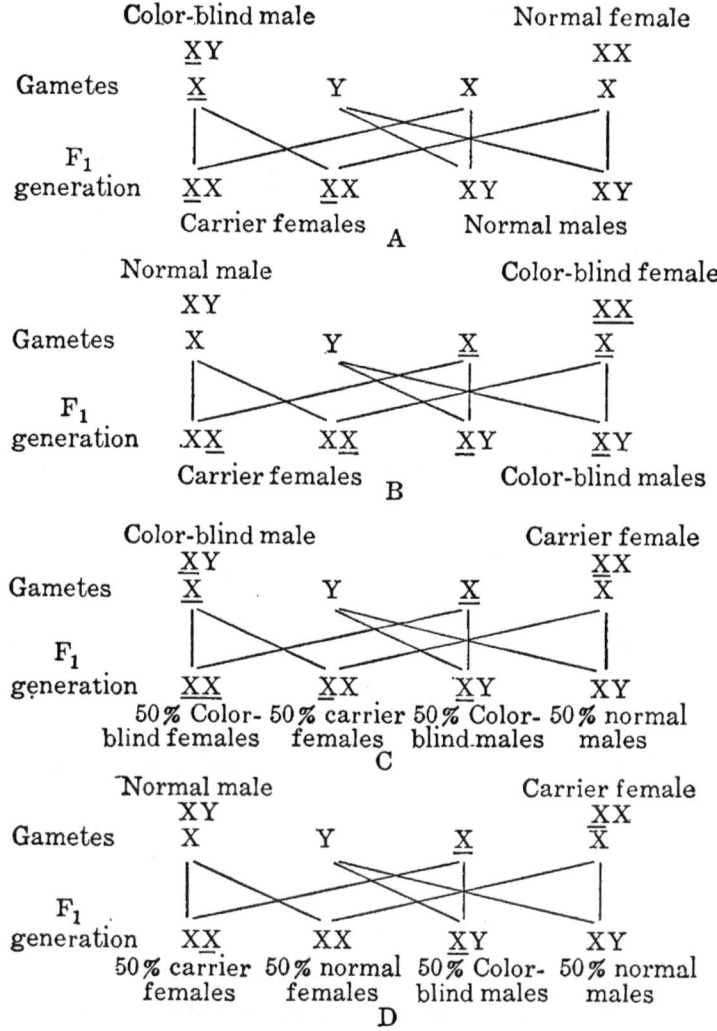

FIG. 201. Diagram showing the four possible combinations of color blindness and the way in which the condition is inherited from these crosses. $\underline{X}$ represents the x chromosome bearing the determiner for color blindness and X that bearing the determiner for normal vision which is dominant over $\underline{X}$.

a few generations. The small size of human families also makes possible the elimination of the defect from a line of descent in some cases.

Inheritance of Defects. The lesson taught by these cases of human inheritance is rather obvious. Many defects are heritable which are not necessarily serious but we must recognize the occurrence of other heritable defects which are not only a serious handicap to the individual but a menace to society. Like the simple morphological defects these may appear or be concealed in an individual; if the carrier finds a normal mate, the latent defect need never appear in his descendants, but if, as so often occurs, similar individuals mate, there is little hope that they will produce normal offspring. Their appearance depends, of course, on the method of inheritance. If dominant a defect must appear more often than if recessive. The following table indicates the expectation for all possible crosses.

THE MENDELIAN EXPECTATION FOR DEFECTS
(After Walter)

		IF THE DEFECT IS POSITIVE (DOMINANT)	IF THE DEFECT IS NEGATIVE (RECESSIVE)
When both parents show the defect	1	DD DD = all DD	dd dd = all dd
	2	DD Dd = $\frac{1}{2}$ DD $\frac{1}{2}$ Dd	
	3	Dd Dd = $\frac{1}{4}$ DD $\frac{1}{2}$ Dd $\frac{1}{4}$ dd	
When one parent only shows the defect	4	DD dd = all Dd	dd DD = all Dd
	5	Dd dd = $\frac{1}{2}$ Dd $\frac{1}{2}$ dd	dd Dd = $\frac{1}{2}$ Dd $\frac{1}{2}$ dd
When neither parent shows the defect	6		DD DD = all DD
	7	dd dd = all dd	Dd DD = $\frac{1}{2}$ DD $\frac{1}{2}$ Dd
	8		Dd Dd = $\frac{1}{4}$ DD $\frac{1}{2}$ Dd $\frac{1}{4}$ dd

Individuals with determiners for dominant defects cannot be unaware of their defectiveness. There is a possibility that defective parentage may produce normal offspring in some cases, but it is obviously much more certain if only one parent shows the defect. When the defect is recessive it is impossible to know of it's presence in a carrier except through his ancestry or his offspring. If the defect is evident the individual is certainly a homozygous recessive and unless mated with a normal individual some of his offspring are certain to be defective.

Human Pedigrees. The heritability of serious defects, such as insanity, epilepsy, cretinism, pauperism and the socially less

HEREDITY IN MAN 367

serious but none the less unfortunate defects of deafness, tendency to disease and similar conditions is illustrated by a large number of recorded pedigrees. Kellicott reproduces several from Whetham's *Treasury of Human Inheritance*, including the following record of inherited deaf-mutism (Fig. 202). This case is adequately illustrated in the diagram, but attention should be given especially to the frequent marriage of defectives in this line and to the fact that even normal unions produced defective offspring.

In contrast to the unavoidable state of deaf-mutism other pedigrees show the constant recurrence of tuberculosis. This disease must be acquired by every individual. Possibly nobody

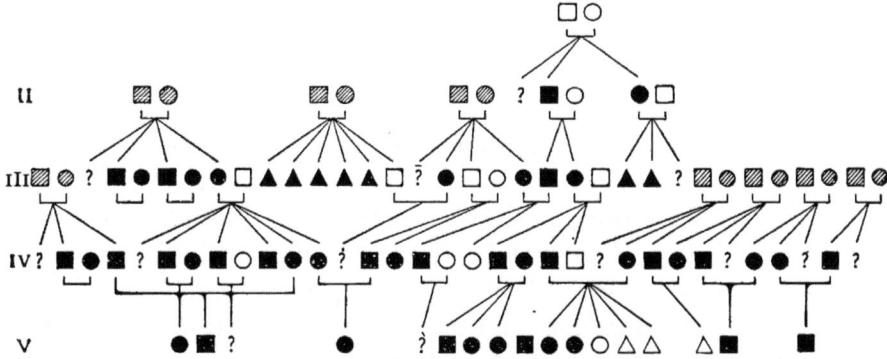

Fig. 202.—A family history showing the inheritance of deaf-mutism. Males are indicated by squares, females by circles, and individuals of unknown sex by triangles; deafness is shown by black, normal hearing by white, and uncertain data by shaded areas. (Modified from Kellicott, after Whetham.)

goes through life without infection but some persons lack the inherent qualities to resist the attack of the bacillus and so develop tuberculosis. The recognition of such an inherited weakness of resistance, or as we usually say, tendency to the disease, should be a warning which would lead to the proper steps for avoiding its serious results. Choice of location, occupation, and recreation might well offset the inherent defect.

Inbreeding. The preceding cases are very good illustrations of the effects of inbreeding in the human race. Defective lines, whether the defect is dominant or recessive, are improved by intermarriage with normal lines. Since recessive defects are not evident in heterozygous individuals it is impossible to predict what the results of marriage of normal individuals may be within lines known to have such defects. These marriages are very likely to result in the reappearance of the defect, while marriage of the heterozygous members of a line to members of normal lines would be certain to keep any but a sex-linked character submerged and might even eliminate it entirely.

The same must be said of desirable and indifferent qualities. Whether evident or not a character is much more likely to be preserved through the mating of closely related individuals than through the crossing of different strains.

It is fortunate that modern social customs do not favor cousin marriages, although they are tolerated as the closest permissible inbreeding. Such an attitude tends to prevent the expression of inherent defects. Even though good and bad alike are perpetuated through inbreeding it is doubtful that the benefits derived from the marriage of closely related persons are sufficient to offset the risks.

A moderate degree of inbreeding cannot be avoided, but the persons involved are usually so remotely related that the effect is practically the same as that of marriages within a social or intellectual class. Too great contrast between individuals is inimical to happy married life, and happiness must remain a fundamental consideration in this important relationship. Like will continue to seek like and to beget like. The latter process is the one certain result, whether good or bad, of inbreeding.

The Jukes. Several families have become famous in connection with the study of human heredity, among them the Jukes. The history of this family was first reported by Dugdale in 1875 and more recently by Estabrook in 1916. Dugdale's interest was first aroused by the frequent recurrence of the same name (the name Jukes is fictitious) in prison records. His original investigations covered 709 individuals of whom "180 were paupers or had received poor relief to the extent of 800 years, 60 were habitual thieves, 50 prostitutes, 7 murderers, and the total cost to the state was estimated at $1,308,000.00" (Holmes).

When Estabrook monographed the family in 1916 he was able to report on 2,094 individuals, of whom not more than one-half were living. The general quality of the family was the same as in its earlier years. Criminal records, intemperance, pauperism, and prostitution abound in the story of these people. Feeble-mindedness is very common, especially among the criminal members of the family, and combinations of feeble-mindedness, illegitimacy, pauperism, and criminality are pitifully frequent. It is obvious that they have been reared under the poorest environmental conditions, but no less obvious that their heritage is very deficient; it is doubtful that they would be able to respond adequately to the finest of surroundings.

In writing of this and similar families Holmes sums up their significance in heredity in the following words: "People with good stuff in them very often rise out of their vicious environment, while others under the best of conditions seem to take instinctively to evil pursuits. We should bear in mind in studying degenerate families and their unfavorable surroundings, that bad environment tends to be created by a bad heredity. Given stocks with an inheritance of low mentality, feeble inhibitions, and more or less mental disorder, in a few generations such stocks would gradually sink into the ranks of dependent or outcast humanity, and would soon develop traditions of vice and immorality which would make it especially hard for an individual to rise in the social scale. When we consider a single individual born amid such unfavorable surroundings, we might be prone to attribute his shortcomings to his poor opportunities. We might be able to point to many cases in which members of degenerate strains have become worthy citizens when given better chances for obtaining success. Such cases, in fact, are not infrequent. But this fact would in no wise controvert the assertion that heredity is primarily responsible for the condition of these degenerate families. Under the conditions that prevail in our civilized society, there is a general tendency for families of good inheritance to rise into higher ranks, whatever misfortunes may have been responsible for their inferior position in the social scale. Families of bad inheritance, although they may be endowed with wealth and social standing, tend after a time to sink into the lower social strata."

Illustrious Families. After such a depressing picture as the Jukes it is a pleasure to turn to some families of the opposite type. Galton was one of the earliest writers to consider the inheritance of ability, and in his work on *Hereditary Genius* he shows that there is a striking tendency for the reappearance, generation after generation, of high ability in the same line of descent. Superior ability in almost every line of human endeavor has been shown to follow this rule. A particularly appropriate example for such a work as this is the family of Charles Darwin. His grandfather,

Erasmus Darwin, has already been mentioned as one of the great contributors to the early history of evolution. Two sons of Erasmus, one the father of Charles, were distinguished men in their chosen fields. Charles himself needs no mention; the fact that his name is almost synonymous with organic evolution in the popular mind is enough evidence of his greatness. Charles Darwin's wife, Emma Wedgwood, was his cousin. Her grandfather was the founder of the famous Wedgwood pottery works. The four sons born to this union were prominent in as many activities.

Winship's data on the family of Jonathan Edwards, an eminent minister, are a similar evidence of inherited ability. Of 1394 descendants identified in 1900 there are listed 295 college graduates, 13 presidents of leading colleges and many in similar offices of less importance, 60 physicians, over 100 clergymen and religious workers, 75 officers in the army and navy, 60 writers, over 100 lawyers, 30 judges, 80 public officials including a vice-president of the United States and three senators, and many officials in business enterprises of various kinds.

The Kallikak Family. The record of this family is even more convincing evidence of the potency of heredity in determining the value of human beings, for it contains contrasting lines of descent from a single ancestor, Martin Kallikak. (This name is also fictitious.) Kallikak, although of good family, became the father of a feeble-minded son by a feeble-minded woman. The descendants of this son have been traced, and out of several hundred none have been above average ability, most have been below average, and more than a quarter were feeble-minded. Later Kallikak married a girl from a good family and the known issue of this union, numbering almost the same as his other descendants, have been almost without exception respectable citizens of normal ability (Fig. 203).

The Basis of Mental Traits. Many students of human heredity have attempted to analyze the inheritance of mental qualities without gratifying success. The behaviour of such characters cannot be explained on the basis of simple Mendelian laws, although it is impossible to avoid the conviction that a Mendelian foundation is present in some degree of complexity. Even though authorities disagree on the subject it seems that complexity is the keynote to human ability and mental traits. So many different kinds of ability combine to make a skilful surgeon, for example, or an expert engineer or mechanic or musician, that in the absence of exact knowledge we can only admit our ignorance of these things. Of this fact we can be certain, that there is a heritable

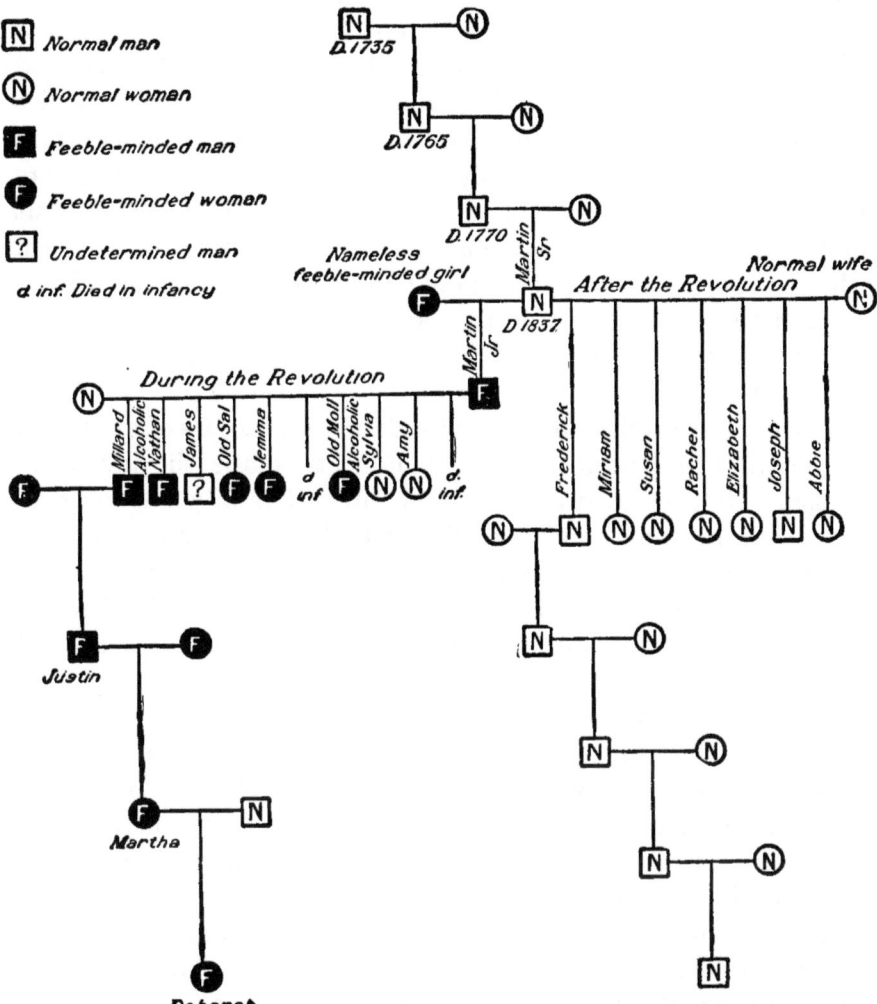

Fig. 203.—The Kallikak family. Of the 480 descendants of this branch of the family 143 (21%) were feeble-minded; only 46 (9%) were normal; of the rest 189 (68%) are still undetermined. 24 were confirmed alcoholics; 3 were criminals; 3 were epileptics; 82 died in infancy; 41 were degenerate. Of the 496 descendants of this branch of the family, none were feeble-minded and all were good citizens. Among them were educators, physicians, lawyers, judges, traders, land-owners—men and women prominent in every phase of social life. Only 2 were alcoholics. (From Goddard.)

basis for the mental qualities involved as well as for the physical structures.

Environment and Human Life. Writers have disagreed on the part played by environment in the development of human characters. We must realize that intelligence has given us a degree of control over our environment which removes us almost completely from the direct influence of natural conditions, and such control cannot fail to have an effect. Through the facilities of modern agriculture, transportation, and storage we maintain an almost constant food supply whose seasonal fluctuations are chiefly among the luxuries of diet. Engineering methods give us summer temperatures in our homes during the winter, and warm clothing protects us in proportion to our needs. Our supply of light and water is also carefully regulated. It would be difficult to provide more uniform and favorable surroundings than those of civilized man.

Environment is only one of the fundamental factors of existence, however, and we cannot expect it alone to shape the life of an organism. If a child grows up in a musical family any musical ability that he may possess is likely to be expressed, but if he lacks ability there is no possibility that he can be made musical. He may be forced through a course of training in music and become a mediocre performer, but of such material true musicians are not made.

With the breadth of opportunity available in modern society it is difficult to avoid the conviction that as man has shaped his own environment, so will the individual shape his. The least that the individual can do is to seek the most favorable environment within his reach, and reach in America is a matter of inherited capacity. The environment of early life may facilitate or retard his progress but eventually the life of the individual is bound to be an expression of his heritage. If he has not made himself a good environment, he has shown himself deficient in a fundamental quality of the human race.

Summary. While human inheritance cannot be studied as readily as that of other organisms, enough is known to show that it is based upon the same principles. There is a material basis similar to that of other animals in the chromosome complex. Anatomical unit characters of several kinds have been traced repeatedly through several generations, and in some cases more complex forms of Mendelian heredity have been demonstrated. Human responses are of more interest than structures. These are, however, based upon the structure and functions of various organs, and for that reason have a heritable basis whose transmission from generation to generation is essentially the transmission of the response. Mental defects and other undesirable qualities, as well as unusual ability, are so transmitted. Many records are available to prove these facts. While environment may have some effect upon the progress of individuals, the power of man to control his own environment is evidence that ultimate development is largely the expression of the heritage.

REFERENCES

KELLICOTT, W. E., *The Social Direction of Human Evolution*, 1911.
GODDARD, H. H., *The Kallikak Family*, 1912.
DAVENPORT, C. B., "Heredity of Skin Color in Negro-White Crosses," *Pub. Carnegie Inst.*, No. 188, 1913.
DAVENPORT, C. B., "The Feebly Inhibited," *Pub. Carnegie Inst.*, No. 236, 1915.
ESTABROOK, A. H., "The Jukes in 1915," *Pub. Carnegie Inst.*, No. 240, 1916.
CASTLE, W. E., *Genetics and Eugenics*, 1921.
HOLMES, S. J., *The Trend of the Race*, 1921.
SHULL, A. F., *Heredity*, 1926.
GUYER, M. F., *Being Well-Born*, 1927.

CHAPTER XXI

EUGENICS

It is a safe premise that man's chief interest is himself. We are no longer at the mercy of the elements. Wild beasts have ceased to be a daily menace. Even physical conflict among ourselves is insignificant in comparison with the eternal struggle of wild creatures. The instinct of self-preservation is still strong within us but it has taken on a new significance which finds expression in better homes, higher standards of living, and education.

Perhaps it is only natural that we should regard ourselves as the chosen species and feel that all else exists for our benefit. Certainly this attitude prevails, whether natural or not, and we go blithely through the years, making and modifying to suit our needs. We have a right to a modest degree of conceit over our attainments, for even the brief span of the twentieth century is crowded with progress. But through it all thoughtful minds have noted the very human tendency to choose pleasant rather than useful activities, save under the pressure of necessity, and a proneness to avoid fundamental facts of our very existence.

Civilizations have come and gone. Their contributions to our own are of no mean value and their examples, good and bad, are before us. We can see their errors and their greatness. We know more than ever before of the foundations for both. We have discovered how to secure and perpetuate desired qualities in other organisms; can we do the same for ourselves? An earnest attempt is being made to answer this question in the science of eugenics. Its material is complex and in many cases elusive, but there is every reason to suppose that it will some day be an important factor in human welfare. It is literally the science of good birth.

The Problems of Eugenics. *What Is Desirable?* A fundamental requisite of any attempt to improve the human race is accurate knowledge of what constitutes good birth. The complexity of our social organization requires very careful judgment of this matter because the responses on which ordinary judgment is based must be analyzed in terms of hereditary qualities by the eugenist. It requires very little insight to note that the expert mechanic has a number of qualities in common with the skilful surgeon. The keenness of senses, manual dexterity, and nice coördination demanded of each in the performance of his duty are not at all different fundamentally. The *knowledge* which each man uses and the *training* which has given him the ability to express these inherited powers are very different. When such basic similarity can exist between such remote walks of life, who can determine what qualities, what activities, are the most valuable in our social structure from among the many heritable fundamentals?

The Ultimate Goal. This is a matter which cannot fail to concern us in a broad application of the principles of genetics to our own future welfare. If we are to control our own destiny our attempts should be directed toward the highest realization of our powers and the maximum efficiency of our organization. Such a program is at present too idealistic to be regarded as practicable, but it is the only possible goal for eugenics. If we are to succeed where other civilizations have failed, we must do so through the elimination of their fatal errors from our own racial lives and through the highest possible development of the qualities that build civilizations.

Present Possibilities. In this quest for betterment of the human species we are fortunately not limited to idealistic pursuits. They are fascinating, it is true, and have claimed all too large a place in the popular literature of eugenics, but scientific sponsors of the movement are not blind to their dangers. A few things are clearly valuable steps toward betterment of the human heritage and from these a sound foundation for eugenic progress can be built even with the limited knowledge now available.

Since man's evolution has become almost entirely a matter of intellectual development, mental capacity is of primary importance and the maintenance of a high level of intelligence is essential to continued progress. Whatever may be the field of activity, the individual with the greatest mental ability is certain to excel in the performance of his duties, other factors being equal. The details of such a generalization must be extremely variable;

Goddard points out that the intelligence of a moron, for example, fits him admirably for special education for the performance of menial tasks in which he can feel a sense of accomplishment, while a more intelligent person would find the same tasks drudgery. We cannot avoid the fact, however, that intelligence is a valuable asset. A man must have the physical capacity for his occupation, but within limits he may even overcome physical inferiority by the exercise of intelligence.

The physical basis for human activities finds its greatest expression in health, since we have now so largely replaced the need of great physical strength with machinery. Normal bodies to begin with and normality of functions throughout life are not only a blessing to the individual but a real asset to society. Their lack is not inimical to great accomplishment, as has been shown over and over in the lives of great men, but it cannot fail to be a handicap even to those who successfully overcome it.

Abnormalities, both of body and of mind, are fundamentally abhorrent to the eugenist. Abnormalities of body are usually nothing more than handicaps and so can hardly be considered as justifiable material at present for eugenic control; they must be relieved and corrected as far as possible but humanitarian considerations forbid any further control. There are some very serious hereditary structural defects whose perpetuation is a matter of grave concern but these, fortunately, are not common.

Abnormalities of mind, however, are much more likely to be insurmountable. The intelligence of a moron may not make him a public charge. Indeed, if we follow Goddard's teaching we must regard morons as an asset when properly trained. Idiocy and congenital insanity are on an entirely different plane; they are certain to throw a considerable burden on the public and the public should therefore have something to say about the proper control of their lives. In 1910 there were 187,791 insane in hospitals in the United States alone. Here is a case in which eugenists can logically and humanely urge restriction. In addition to these extremes there are many socially undesirable traits including dipsomania, paranoia, immorality, etc., which are not necessarily a matter of public concern in every individual displaying them.

The amelioration of existing conditions is rather a matter of sociology than of biology, but in many cases conditions are such that their perpetuation from generation to generation seems unnecessary and undesirable. Here eugenics is attempting to awaken humanity to the need for the sensible application to ourselves of the principles which have been used so successfully to shape the development of domestic animals and plants.

Limitations of Eugenics. The situation encountered in this attempt is unique. No other animal consciously attempts the control of its own future. Man has acquired the ability to do so but the same qualities that have made it possible have, paradoxically, made it difficult if not impossible. Control of environment is responsible for all of our material acquisitions; abundant and constant food supply, physical comfort in spite of natural conditions, and luxuries of all kinds are due to the fact that we need not take things as they come but may shape them as we desire. Every step in this direction has carried us farther and farther from the primitive conditions of human existence. Social as well as individual efforts tend to emphasize the individual and yet the subordination of the individual alone can further the best interests of humanity.

Strangely it is not the limitation of defectives that is the chief source of difficulty; this is rather the most hopeful field for eugenic progress in the near future. It is the preservation and improvement of desirable classes that seems to be an insurmountable difficulty because of the divine rights of the individual! Democracy, the highest affirmation of equality of rights, puts the matter squarely up to everyone. Will *you* do the best thing for humanity where you yourself are concerned? If not we can do little to guarantee continued progress unless conditions change enormously.

The Differential Birth Rate. Since eugenics is concerned with the quality of the heritage a knowledge of the birth rate, especially of different classes, is important. Moreover it is the finest possible illustration of the facts just presented.

Decrease in Birth Rate. It is significant that the birth rate of entire populations undergoes a gradual decrease with the advance of civilization. European countries in general show this tendency, France to such a degree that her inadequate birth rate is a matter of common knowledge. In the United States official statistics on the birth rate are lacking but the following table is indirect evidence of a similar decline.

DECREASING PROPORTION OF CHILDREN IN THE UNITED STATES
(From Holmes, after Willcox)

Date	Number of Children under 5 per 1,000 Women 16-44 Years of Age
1800	976
1810	976
1820	928
1830	877
1840	835
1850	699
1860	714
1870	649
1880	635
1890	554
1900	541
1910	508

This is not, in itself, a serious matter. Natural conditions tend to maintenance of species rather than to their increase, and there is nothing in human welfare that demands any more than the maintenance of our present population. The birth rate could continue to decrease in the United States for many years without bringing about a decrease in population.

Reproduction of the Unfit. A most unfortunate factor in the decreasing birth rate is that the least desirable members of society do not share in the common decrease. Many eugenists have called attention to the fecundity of mental defectives, which can probably be explained by the fact that they lead a more nearly animal existence than persons of normal intelligence. Families of five to eighteen children have been reported among such people and even unmarried individuals contribute to their undesirable stock through illegitimate unions. When associated with the frequent transmission of their mental defectiveness to their offspring the fertility of these people becomes a serious problem. Whetham says: "Most of these children inherit the mental condition of their parents, and where both parents are known to be feeble-minded, there is no record of their having given birth to a normal child. In one workhouse there were sixteen feeble-minded women who had produced between them one hundred and sixteen children with a large proportion of mental defect. Out of one such family of fourteen, only four could be trained to do remunerative work."

The Birth Rate of the Mentally Normal. Even among the classes of desirable citizens a marked disparity may be noted.

The studies of various investigators show a higher birth rate among unskilled than among skilled laborers. Members of the latter class usually marry later in life, although this is not necessarily the direct cause of smaller families. Professional men usually have smaller families than any of the classes mentioned and consequently a lower birth rate, but the intellectual classes are the worst offenders in this respect.

Data derived from several sources on the percentage of marriages and the number of children per family among the graduates of several colleges and universities disclose the following facts:

1. Graduates of colleges marry later than persons who have not attended college.

2. A low percentage of college graduates marry, viz., about 75 per cent of the men and less than 60 per cent of the women.

3. The number of children in the families of college graduates is low. "The average for Wellesley graduates between 1875 and 1899 was .83 of a child." Cattell's studies of scientific men show that they have a little better record with almost 90 per cent married and 1.88 children each in the completed families.

Since an average of more than three children per family is necessary for maintenance without increase, it is obvious that the intellectual classes are not maintaining themselves. When we consider that college graduates include chiefly men who enter the professions or become capable business leaders, the conclusion becomes even more significant.

The Causes of a Differential Birth Rate. The gradual lowering of the birth rate in the ascending scale of ability and intelligence and the fact that it has been found impossible to correlate these things definitely with the fertility of individuals suggests that it is rather the economic status and the foresight of parents that determines the rate of reproduction. Parents who have little or no sense of responsibility for their children produce the most. The higher the standard of living, the less children are produced. Many other factors enter into the problem but this alone seems to be constantly applicable. The cost of raising children properly is high. Education, especially higher education, is expensive even though children of college age may be ready to assume a part of the burden of their own maintenance. It is only natural that these things should force parents with high ambitions and a keen sense of responsibility for their children to limit the size of their families in proportion to their means. All too often families are limited in spite of their means, and in this we must agree with Shull's statement: "If, in some cases, selfishness leads to a desire to avoid children, and if selfishness is inherited, as it presumably is to a large extent, such people are probably doing the race a service by permitting their lines to be extinguished."

It seems very probable that most persons do not give such matters serious attention, and that sheer indifference is a contributing factor in their failure to reproduce. There are so many attractive activities in the modern world to occupy the individual's time that mere conflict of interests may result in the limitation of family without any sense of race consciousness or economic pressure entering the matter. Such a tendency must probably be construed as a degree of selfishness; if so it is a very common fault. The same indifference undoubtedly contributes to the fecundity of the lower classes, who merely obey the powerful instinct of reproduction without thought of the social and economic consequences.

Immigration and the Birth Rate. Immigration is a national problem for more reasons than one which cannot be considered here, but it has had an important effect in the maintenance of desirable elements in the population of the United States which bears definitely on the problems of eugenics.

Such men as Carnegie and Steinmetz are fine evidences of our indebtedness to foreign countries even within recent years for men of exceptional ability. Our native American population must acknowledge its foreign ancestry within a few generations of ancestors, but even since the establishment of a fairly definite American stock we have drawn constantly upon the European nations. Opportunity has been the keynote of immigration; in America lay opportunity which the Old World could not furnish and many valuable men have taken advantage of it.

It is difficult to say what might have been the accomplishment of such men if they had stayed in their own countries. One can hardly imagine them contented with anything less than a leading rôle in their chosen fields, yet the class limitations of Europe are undeniably more stringent than those of America and can scarcely have afforded ample opportunity for the development of their genius. Given the opportunity of America, where achievement is limited only by ability, they have risen to prominent places in all walks of life.

Our leaders have increased in number with the increase of population. Since they have not been maintaining themselves, to what extent has immigration been responsible for the renewal of these classes? The number of foreign names added to the personnel of American institutions of learning since the world war suggests that it may even now be an important source of desirable citizens.

Immigration since the war has, however, shown an undesirable trend toward peoples of proved inferiority. The precautions taken in the admission of aliens are such as to prevent the entry of undesirable individuals, but we must expect a distinct reduction in the valuable types that have previously entered this country and a modification of the total effects.

It would be pessimistic to suggest that this change must have an early effect on the general level of intelligence, or that it might result in the extinction of those classes which are failing to reproduce themselves. It is possible that enough native-born citizens rise above the standing of their parents because of increased opportunity to offset that phase of the differential birth rate. We must recognize in immigration, however, an important factor which has previously exerted a valuable influence and now bids fair to change. Moreover it is impossible to point out with certainty anything which is taking its place.

What Shall Be Done? In meeting these problems the eugenist is confronted by the difficulty of securing a sympathetic audience. Prosperity tends to blind a people to their future problems, and when these problems are beyond the pale of their education, they are doubly oblivious. Relief measures, especially when they are of serious consequence to the individual, are not popular. It is all too easy to leave action to the other fellow.

Reduction of Defectives. In spite of this difficulty the eugenics movement has made a steady advance and offers several proposals which are well grounded and worthy of serious consideration. Among them the sterilization of mental defectives has a prominent place. The year 1927 has witnessed a reaction to this proposal which urges that not all mental defects are known to be heritable and that even such a taint does not prevent the production of valuable citizens by the defective line. Such a cautious attitude is desirable but probably unnecessary. The men in whose hands the administration of restrictive laws would be placed would necessarily be trained scientists in whom we could expect a maximum of ability and discrimination. The fact remains that too many records show fecundity, pauperism, and feeble-mindedness going hand in hand through the generations, and no adequate reason can be given for the perpetuation of such conditions.

Many states have already provided legally for the sterilization of defectives under prescribed circumstances, and many operations have been performed under these laws. It is a matter of record that some individuals have voluntarily applied for treatment.

Segregation of defective men and women is another measure which would have the same effect. It is favored by many eugenists, but with the exception of cases in which confinement in public institutions is necessary for other reasons it involves much greater expense. Neither measure would bring about rapid reduction of defectives even if rigidly administered, but either would be a step in the right direction.

Guyer presents the following opinions in favor of segregation: "It has been urged against vasectomy [sterilization] that it will work untold harm because it relieves of the responsibility of a probable parentage. This argument does not appeal to one as very weighty as far as the imbecile or other degenerate is concerned, because one of the very traits characteristic of such individuals is lack of any sense of responsibility. By this same token, however, we have a very good argument for sequestration as against sterilization, for the degenerate, even though sterilized, will not be restrained sexually and will be likely to disseminate venereal diseases or commit rape. Furthermore, there will be the temptation to sterilize and liberate certain types that would otherwise have been kept permanently in custody.

Education of the Fit. The world has long cared for its dependents, but if a valuable line ceases to perpetuate itself it is gone forever. For this reason the correction of the disparity in birth rate among desirable classes is greatly to be desired. Individuals in these classes are responsible and self-maintaining. Together they make up the backbone of social structure and their interdependence is so complete that it is impossible to say that one is more necessary than the other, however different their material rewards may be.

Nobody would think of urging forcible control of these classes, but education may ultimately succeed in awakening a sense of racial responsibility which will bring about the desired result. Attempts have been made to increase desirable families by subsidizing human reproduction and various laws have been suggested for their preferential treatment. None of these methods have had or can be expected to have an extensive effect, and since the classes with a low birth rate are not financially incompetent it is doubtful that the desired adjustment can be gained in this way. Moreover parents do not necessarily produce children as capable as themselves, so that application of such measures might well fail of its intended result.

Perhaps some of the educated people included in the classes with a low birth rate are really ignorant of their responsibility to society. Proper understanding of the prevailing conditions would have a desirable effect in such cases, but the impression that they do not constitute a majority is strong.

Education of the classes with a high birth rate, on the other hand, might readily bring about the reduction which would be a necessary step, in any case, in securing a satisfactory balance.

The Effects of Environment. Environmental conditions often have a great deal to do with individual accomplishment, and many agencies are directed toward the maintenance of proper conditions of life. Child labor movements, public health, physical training and other things are designed to offset the unnatural conditions which result from our rapid social metamorphosis before the body can adjust itself thoroughly to the change. They are valuable corollaries of eugenics, although we are not yet in a position to say that they actually contribute to the heritage. Certainly, since they are practicable, they should be emphasized as the nearest approach to the ideals of eugenics. They at least aim at the maximum realization of inherited possibilities.

Practical Aspects of Eugenics. Education of a considerable portion of any population in an unfamiliar movement is not easy. Human beings respond emotionally with conspicuous readiness, and to a subject like eugenics, demanding rigorous analysis of inalienable personal rights, the response is likely to be emotional opposition. Education must be gradual. Those who are able and willing to give careful consideration to such a movement should do so; they should also learn to distinguish between the careful decisions of well-informed scientists and the scientific quackery which is all too common in matters concerning the human race. No thoughtful eugenists would consider the wholesale regulation of marriages beyond protection against unfortunate consequences, yet the idea has gained a hold on the popular mind through various agencies that this is proposed by eugenics. Our knowledge of man is insufficient to accomplish the production of ideal types even if such a course were desirable. Many of the popular ideas of eugenics are myths which have had unjust and wholly unwarranted consequences.

The whole program of eugenics at present may be summed up as watchful waiting. Proposals of eugenic organizations recommend principally extensive research in all fields related to human heritage, education, both formal and popular, in heredity and eugenics, and a policy of delay in attempts to secure legislation in this field. Such a program is no more than the wise use of our intelligence in relation to ourselves.

Summary. Eugenics recognizes the importance of the heritage in man and proposes as an ideal goal the securing of an adequate heritage for every individual. The movement is young but it is even now possible to make positive proposals for the elimination of obviously unfit strains, such as mental defectives. The differential birth rate shows the need of eugenic measures in all classes of society, since the lowest classes are the most prolific and educated classes do not maintain themselves. In this field, however, nothing but sound education is at present possible. The program of eugenics therefore urges continued investigation, extensive education, and only such legislation as progress warrants.

REFERENCES

HOLMES, S. J., *The Trend of the Race*, 1921.
―――― *Studies in Evolution and Eugenics*, 1923.
―――― *A Bibliography of Eugenics*, 1925.
GUYER, M. F., *Being Well-Born*, 1927.

CHAPTER XXII

NATURAL SELECTION

Many theories of varying importance represent our attempts to explain the evident relationship of organisms and their apparent origin from a common source. Some of these theories were formulated without knowledge of some of the important biological principles which science now makes available for their evaluation. It is therefore important to bring to the study of evolutionary theory a sound understanding of the facts of organic relationship presented in the foregoing chapters.

The theories represent, in the main, two general tendencies which are foreshadowed by the two scientists, Lamarck and Darwin. The former placed emphasis on the environment as a factor in evolution, the latter upon inherent qualities. There is less of a gulf between their views than is indicated by this common treatment, but the fact remains that they have been regarded as the exponents of opposed schools of thought.

Origin of the Theory of Natural Selection. No theory has played a greater part, and none perhaps contains a greater measure of truth than the theory of natural selection which was formulated almost simultaneously by Charles Darwin and Alfred Russell Wallace and later developed by Darwin in his famous book, *The Origin of Species*. It was suggested to both men, strangely enough, by Malthus' ideas of overproduction in the human race. It is not to be supposed, however, that this alone was responsible, for the application of the principle to plants and animals and its elaboration to explain the origin of species with their infinite diversity required knowledge of a wide range of scientific facts. Both Darwin and Wallace were well informed in the field of natural science, and the range of material presented by Darwin in his books relating to evolution is truly remarkable.

Statement of the Theory. The theory of natural selection is based on the fact that more individuals are produced in every species than can survive, and on the occurrence of useful characters in the normal range of variation. It assumes that overproduction must result in a struggle for existence in which survival is determined by the inherent characteristics of individuals; those possessing useful variations will survive and those with harmful variations or merely without the useful characters will perish. The surviving individuals alone will perpetuate the species and so their characters will become characters of the species. Darwin added the belief that this process, through a succession of generations, would result in progressive development of a character. To this simple foundation may be added other principles which exert a similar selective action upon species, but in the beginning it was literally a theory of the "survival of the fittest."

Underlying Principles. *Variation.* The importance of variation was impressed upon Darwin's mind especially during the voyage of the *Beagle*. He noticed as he travelled not only that species varied in a given region but that as he passed from one limit of their range to another their general characteristics also varied according to geographical distribution. These facts showed him that species, far from being rigidly fixed entities, showed a tendency toward intergradation and developed the idea of changeability which was necessary to any thought of evolution.

In the classification of variations we have noted that some are a part of individual life, both as process and result. These have been called modifications and acquired characters, and are as aptly characterized by one term as by the other. Better still they may properly be looked upon as individual adaptations. Such characters appear in every individual in response to definite conditions of existence. Many examples are familiar to everyone, including such common things as tanning of the skin, calluses, muscular development, tolerance for poisons and the like. Mutilations have also commonly been included here, but for reasons to be considered later they cannot properly be considered as characters of the organism. The possibility of modifications affecting the heritage of the species has been one of the most bitterly contested points in evolutionary theory.

In contrast to this kind of variation are two which are not evident as processes within the individual but only as fully developed characters. To the extent that we are able to apply the idea of heritage and response to characters of the organism we must look upon them as a product of inherited powers responding to conditions within the body during development. They include adaptations of the species, but are sometimes non-adaptive in so far as external conditions are concerned. These two are ordinary fluctuating variations which appear independently of external conditions, and mutations.

Both fluctuating variations and mutations are available as materials for natural selection, since they are known to be heritable as they occur. The former differ from the latter in that they may arise, and probably do arise in most cases, from the recombination of unit characters made possible by sexual reproduction, but in that they are both the direct result of chromosomal determiners and inherited bodily qualities, both are definitely of the heritage.

Importance of Variations to the Individual. A further factor emphasized by Darwin in the *Origin of Species* is value to the individual. Since no indifferent character can be of vital importance to an individual, only useful or harmful characters can play an active part in the shaping of a species. Indifferent characters may, however, be incidentally or accidentally selected.

Among the indifferent characters of animals may be included slight variations in the color and pattern of insects. Some butterflies have brightly colored ocelli, or eye-like spots, on the hind wings which vary in number although not ordinarily to a sufficient extent to cause a distinct difference in appearance, since a row of spots is usually present. In the human race, however, variations in resistance to or tolerance of the typhoid bacillus is of primary importance. Some individuals suffer no ill effects from the presence of these bacilli, and become carriers, while to others the disease caused by their presence is fatal. A race of carriers would be, in effect, immune from the disease, while an ordinary population is very variable in susceptibility.

Segregation. The last condition necessary for natural selection is something to separate the individuals of a species into groups possessing different variations or to preserve only a limited portion of the individuals. The cause of segregation may be mere spatial isolation brought about by topographic or climatic change or migration, or it may be the relationships of organisms within a limited region. The two causes are closely linked with distribution and adaptive radiation respectively.

Spatial isolation may have varied effects upon the organisms that come under its influence. It was made the basis of a theory of evolution by Moritz Wagner, with some reason, but in considering Wagner's work Darwin looked upon isolation as of minor importance. He recognized, however, that it must necessarily be a contributing factor in natural selection in some cases. While it may perpetuate and even emphasize indifferent characters, it may also subject the organism to specific conditions of environment or organic association which some variations may meet more readily than others, and to this extent it is an effective stimulus to the selection of the useful characters.

Overproduction and Crowding. Organic relationships, based on the principle of overproduction and crowding suggested by Malthus' work, were chiefly emphasized by Darwin as the immediate cause of natural selection. Such relationships may, as we have already seen, be due either to the competition of individuals of the same species or those of different species, and may bring about very different responses.

The fact of overproduction cannot be doubted. If the eggs produced by single animals, or the seeds of single plants, were all to mature, in a few generations one species would crowd the earth to its own extinction. Darwin wrote: "Even slow-breeding man has doubled in twenty-five years, and at this rate, in less than a thousand years, there would literally not be standing room for his progeny. Linnaeus has calculated that if an annual plant produced only two seeds—and there is no plant so unproductive as this—and their seedlings next year produced two, and so on, then in twenty years there would be a million plants." But we have already considered cases of rapid increase of animals of economic importance when freed from natural checks. When we consider that oysters commonly produce 16,000,000 eggs, and some fishes even more, we realize the vastness of overproduction and the possible effects of interruption of the natural balance in such species. Lull states the striking facts that if all the progeny of one oyster survived and multiplied and so on until there were great-great-grandchildren, these would number 66,000,000,000,-000,000,000,000,000,000,000,000, and the heap of shells would be eight times the size of the earth!

Competition. Such increase can only result in competition of various kinds. Animals must have other organisms for food, and so much of the surplus is destroyed. Plants must have sunlight and moisture, and so they cannot be crowded beyond the available supply. Grass is a fine example of maximum crowding, but where the periwinkle vine grows, it crowds out even grass, and where the rainfall is slight, only sparse grasses can grow, since they must draw water from a greater area of soil in order to live.

The struggle between organisms may thus be passive competition for the same things, or open conflict. Between animals it may be either, between plants it is the former. Animals may fight for the right to a favorable range, and the conflict between carnivorous species and their prey can only be direct competition. In many cases, however, competition implies merely that different organisms seek the same thing. If the supply is ample, both succeed; if insufficient, then first come, first served, and the loser must die or change his mode of life. In all cases the relationship is essentially the same. The organisms concerned are actually pitted against each other, and whether the conflict is of tooth and claw, or merely a matter of the early bird's getting the worm, some individuals must lose out.

The Rôle of Useful Characters in Competition. It is in such competition that the factor of usefulness is preëminent. Every contact with the environment demands some power of response. If struggle is merely in the form of passive competition for a limited food supply, keenness of the senses involved in the search for food may decide the winner. Some slight variation in speed might well favor the faster animal in seeking the same prey. Broad leaves might enable a plant in crowded ground to secure necessary sunlight and at the same time deprive the small-leaved neighbor in its shade of an adequate supply. The fact that some plants vary to such an extent that they can meet the requirements of very different environments emphasizes the importance of such conditions. Some species of plants have finely divided leaves when immersed in the water, but when stranded by receding ponds, or when growing on muddy banks, they produce much less finely divided or even entire leaves (Fig. 204). They may live either as hydrophytes or mesophytes.

Harmful Characters. In contrast to useful characters there are some structures whose development may reach the point of inconvenience. In one extinct species, the Irish elk, the enormous antlers were probably a handicap. The broad wings of many butterflies make flight in a wind impossible, while others with narrow wings and strong flight muscles are hindered but little.

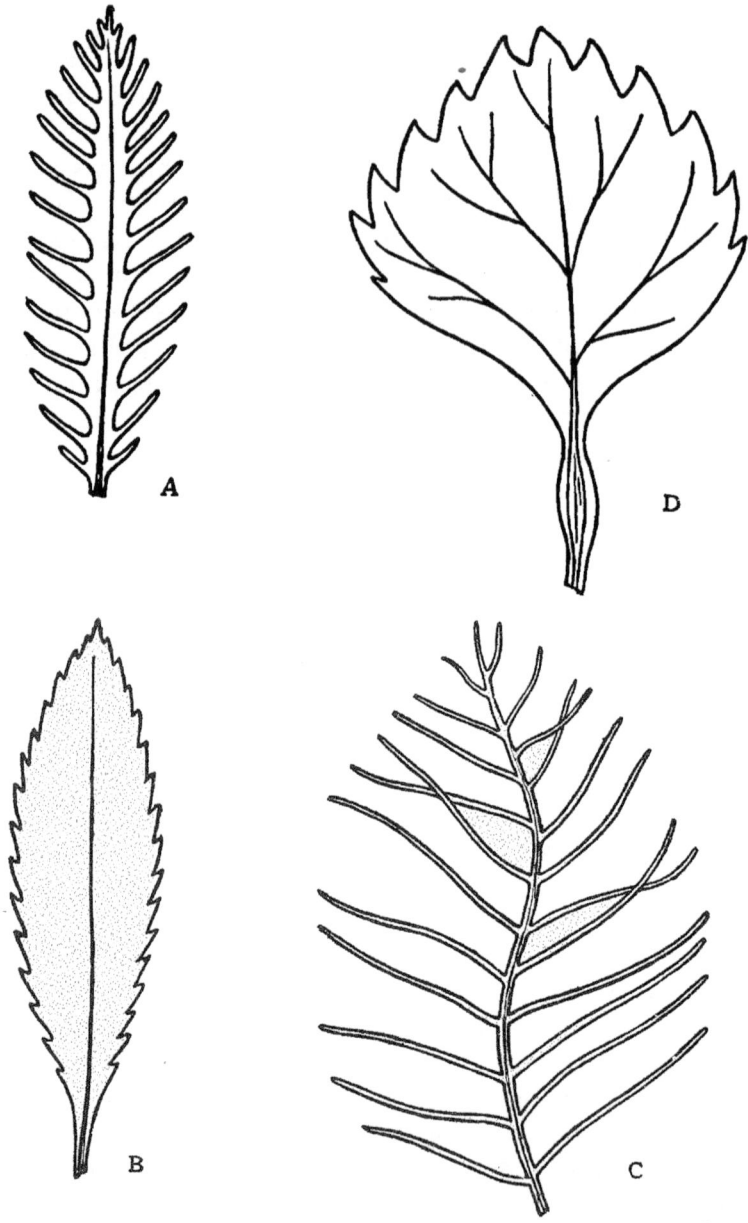

Fig. 204.—Transition from aquatic to aërial habitat as shown by leaf structure. A, Pr*oserpinaca palustris*, submerged leaf; B, same, aërial leaf; C, *Trapa natans*, submerged leaf; D, same, aërial leaf.

The blade-like canines of the sabre-tooth tiger were apparently so easily broken that their excessive development ultimately became a handicap (Fig. 205). In addition to such positive char-

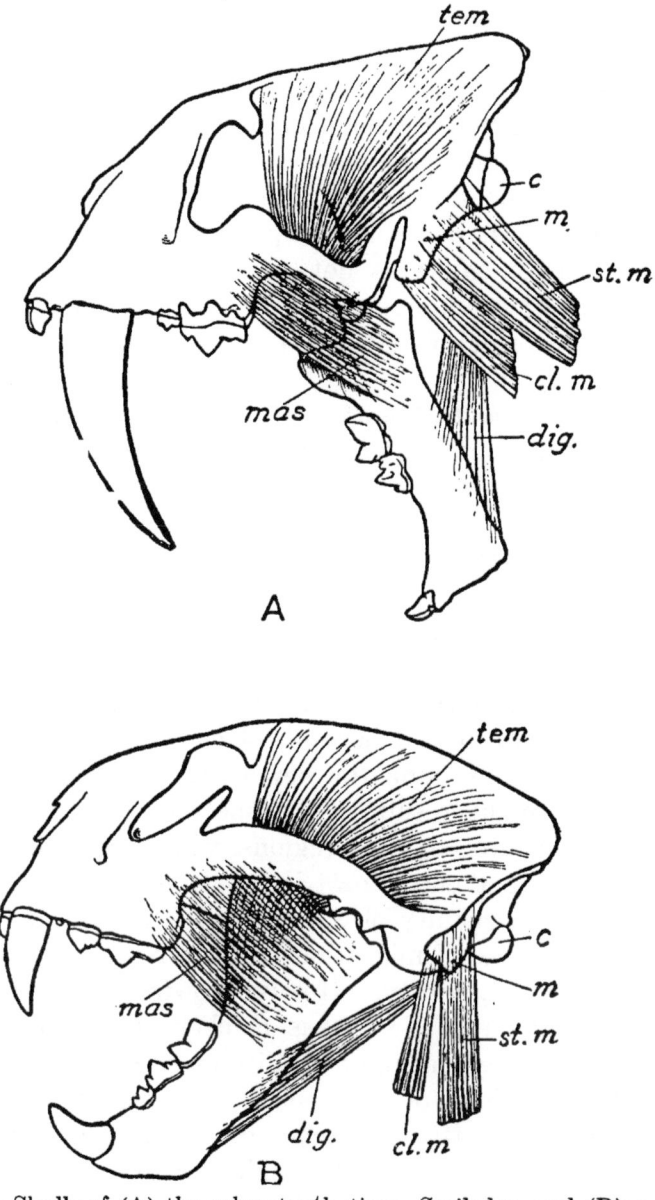

Fig. 205.—Skulls of (A) the sabre-tooth tiger, *Smilodon,* and (B) a cat, *Felis,* showing contrasting skull form and musculature. c, occipital condyle; cl. m., cleidomastoid; dig., digastric; m, mastoid process; mas, masseter; st. m., sternomastoid; tem., temporalis. (From Lull, after Matthew.)

acters, the extreme of variation opposite to any useful character is likely to be a hindrance, and is certain to operate in direct competition to determine the loser.

Accidental Destruction. Belief in the positive or negative value of variation does not mean that they are absolute determiners of individual success in all cases. Undoubtedly many well endowed organisms are destroyed by chance, both through contact with other organisms and through natural catastrophes. A volcanic eruption or a great flood is an extreme condition which cannot be met directly, but must be avoided if possible, and to many organisms the chance for safety is not given. In general, however, conditions of the environment are subject only to gradual fluctuation with which normal variations may be an adequate basis of adjustment.

Change of Habits and Migration. It is necessary to recognize the fact that overproduction and consequent crowding, in addition to determining the survivors in a given region, may result in individuals utilizing their fitness for other conditions of the immediate environment. If a species depends normally upon one kind of plant for food, as is often the case among the insects, crowding may result in the shifting to a new food plant of the individuals which are less successful in securing a share of the available supply. Shortage of food arising from any other cause might have a similar effect. This response would depend, of course, upon the ability of the animal to thrive upon the new diet. On the other hand, such conditions might result in the species spreading through gradual migration into adjacent regions, whereupon the factor of spatial isolation might play a part in its development. During a severe winter in the vicinity of Sioux City, Iowa, the writer once witnessed a minor case of migration induced by hunger, although it was, of course, temporary. Heavy snowfall, beginning early in the winter, had kept the prairie chickens from their normal supply of food. Under ordinary conditions it was unusual to see more than two or three of these birds in a season, but during that winter they had left their range, and on one day sixty-seven were seen within a distance of a few miles. Hunters also reported their abundance, but said that the birds shot were very thin.

Summary of Natural Selection. Natural selection makes use of all of the foregoing factors. The theory may be summarized as follows:

1. All organisms produce more young than can survive.
2. Struggle for existence results from overproduction.

3. All organisms vary, their variations including useful and harmful, as well as indifferent characters.

4. In the struggle for existence favorable variations are preserved and harmful eliminated, while indifferent variations may be incidentally influenced.

5. As a second result of crowding some individuals may be forced into other habitats involving the use of other characteristics, or

6. Some may be forced into adjacent regions, in which variation may be of immediate importance, either as a result of competition or by geological disturbances.

7. Any variation either isolated or emphasized by any of these processes will be preserved through heredity and made a character of the species or variety descended from the originally isolated progenitors.

8. Successive repetition of the process results in gradual change leading to an ultimately wide separation from the parent species.

Examples of Natural Selection. *Hypothetical.* As an example of the operation of this process, we may assume a case of a carnivorous species, of which ten thousand individuals live in a given area. The prey of these animals may be chiefly an herbivorous species, which, because of an epidemic disease or some other factor, is reduced in numbers to a point where it can suffice for the food supply of only one-half of the carnivorous species. Granting that the sole defense of the herbivorous species is speed, it is very probable that the five thousand carnivores best adapted for speed will be the survivors; they alone would be able to reproduce, and the inherited characters which gave them superior speed would become a quality of the subsequent generations. The herbivores likewise would be able to survive only if swift enough to escape, consequently they too would attain greater speed as a fixed quality of the species.

Wallace's Orchid. Wallace cites the case of a Madagascar orchid (*Angraecum sesquipedale*) with an extremely long and deep nectary. The flower is cross-fertilized only by long-tongued moths. In reaching for the nectar the base of the proboscis comes into contact with the anthers, and pollen is carried to the next flower. Wallace's explanation of the case follows: "Now let us start from the time when the nectary was only half its present length or about six inches, and was chiefly fertilized by a species of moth which appeared at the time of the plant's flowering, and whose proboscis was of the same length. Among the millions of flowers of the *Angraecum* produced every year, some would always be shorter than the average, some longer. The former, owing to the structure of the flower, would not get fertilized, because the moths could get all the nectar without forcing their trunks down to the very base. The latter would be well fertilized, and the longest would on the average be the best fertilized of all. By this process alone the average length of the nectary would annually increase, because, the short-nectaried flowers being sterile and the long ones having abundant offspring, exactly the same effect would be produced as if a gardener destroyed the short ones and sowed the seed of the long ones only; and this we know by experience would produce a regular increase of length, since it is this very process which has increased the size and changed the form of our cultivated fruits and flowers." Wallace carries the example on to greater lengths, but this alone is a sufficient illustration of the logic of the theory.

Sexual Selection. As a corollary to his theory of natural selection Darwin proposed the theory of sexual selection. This is based upon the same fundamental principles as his chief theory, but takes note of the choice exercised by the female in her selection of a mate as the factor which determines the variation to be preserved.

It is well known that animals at the mating season display their qualities to the best advantage and parade them before the opposite sex. Ordinarily the male alone finds it necessary to woo a mate by these means, but Beebe's observations of the tinamou, already mentioned, show the possibility of a complete reversal of these instincts. Even common birds are well known for their display of color and song during the mating season, and a few species, such as the ruby-throat humming bird and the yellow-breasted chat, go through wonderful evolutions in the air which show remarkable powers of flight. Obviously the female makes a choice from among her available suitors. Granting that there is a definite basis for this choice, and that the things attractive to one female are those normally attractive to others of the species, we may readily see that the variations in the selected males would be preserved in their offspring. The characters of rejected males would die out for lack of an equal opportunity to be perpetuated.

Objections to the Theory. Darwin recognized many objections to his theory, and a few others have been added in later times. The salient objections may be summed up as follows:

1. If species have descended from other species by fine gradations, what has become of the many transitional forms?
2. Can the production of relatively simple organs and such complex structures as the eye both be explained on this basis?
3. Can instincts be acquired and modified through natural selection?
4. How can the sterility of hybrids be accounted for?
5. How can the development of neuter castes be accounted for?
6. Can natural selection account for the occurrence of vestigial structures and overspecializations when they are of indifferent value to the organism?

Most of these objections have been adequately met, but a few indicate limitations of the theory.

Answers to Objections. *Transitional Forms.* Darwin's answer to the first objection was: "As natural selection acts solely by the preservation of profitable modifications, each new form will tend in a fully stocked country to take the place of, and finally to exterminate, its own less improved parent-form and other less favored forms with which it comes into competition. Thus extinction and natural selection go hand in hand. Hence, if we look at each species as descended from some unknown form, both the parent and all the transitional varieties will generally have been exterminated by the very process of the formation and perfection of the new form." The failure of these intermediates to persist in abundance in the geological record is readily explained by the incompleteness of that record; they do exist in small numbers, and transitional individuals occur among some living creatures.

Complex Organs. The production of complex organs was not an insurmountable difficulty to him, although he recognized the impossibility of explaining in detail just how the gradual elaboration of such an organ as the eye might have occurred. As a matter of fact, the eyes of modern insects range from a state of development which makes possible merely the reception of light stimuli and the determination of the direction whence they come, to the elaborate compound eyes of higher orders (Fig. 206, compare with Fig. 62 and 63). The very existence of stages of development bears out Darwin's answer to the objection. We may never be able to explain the exact steps in the development of these organs, but the belief that it has been gradual offers no greater difficulty than the belief that such elaborate structures have sprung suddenly into being.

Instincts. Instincts are difficult to explain, especially when they involve but a single action during the lifetime of the individual. Many larval insects produce cocoons as an exercise of just such instinct. Darwin answered this objection by pointing out that inherent mental traits vary just as do structural characters, and that useful variations might be selected as readily in the one category as in the other. The selection of a definite mental tendency would result in its becoming a part of the inheritance of the species. It would then be generally expressed by the individuals of the species, whether only once or often within a lifetime.

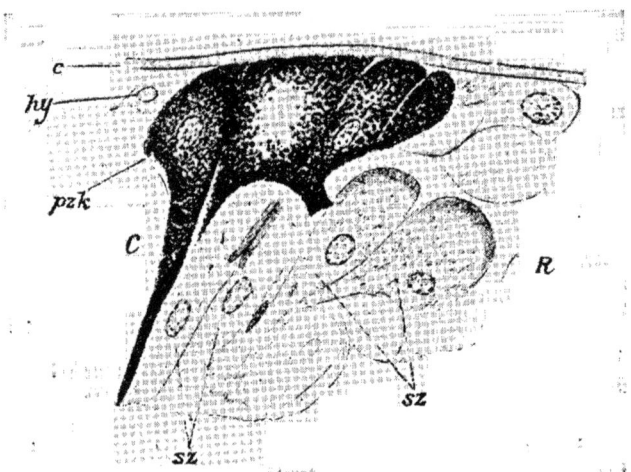

FIG. 206.—Median section through a primitive insect eye (*Orchesella rufescens* var. *pallida*). x 750. c, cuticula; *hy*, hypodermis; *pzk*, pigment cells; *sz*, visual cells; *C*, caudal; *R*, cephalic. (From Schröder's *Handbuch der Entomologie*, after Hesse, with the permission of the publishing house of Gustav Fischer, Jena.)

In support of this view Darwin mentioned the occurrence of peculiar instinctive actions in related species in widely separated regions; ". . . the Hornbills of Africa and India have the same extraordinary instinct of plastering up and imprisoning the female in a hole in a tree, with only a small hole left through which the males feed them and their young when hatched." The male wrens both of Europe and America build several nests, only one of which is occupied by the female. These facts point strongly to similar processes of development of the mental and physical traits of organisms.

Sterility of Hybrids. Such sterility and interspecific sterility are probably due to many causes. Now that the minute structure of the germ cells is known, we realize that the mechanism of reproduction differs in various species as well as gross morphology. Even in similar species this mechanism may differ, and the association and separation of chromosomes may take place, if it takes place at all, in such an irregular way that the normal process of reproduction is prevented. The mere fact that species have arisen from other species does not involve the maintenance of fertility between the divergent branches; if they differ in one way they may well differ in others.

Neuter Castes. In social insects, such as the ants and bees, neuter individuals do not ordinarily reproduce, but they are merely non-functional females and may in certain cases lay eggs. Worker bees do so when the colony becomes queenless and there is no way to rear a new queen. In such cases the eggs laid by the workers are not fertilized, since the insects never mate, and can develop only into drones. There is possibility here for the transmission occasionally of the specializations of the so-called neuter caste, but it is doubtful that the conditions would occur often in a natural state. Even discarding this possibility, however, we find in more primitive species of the same order an explanation of the development of worker structures and instincts. The functional female, or queen bumblebee, not only acts as the mother of the colony, but also carries on all of the duties of workers until she has reared a colony to relieve her of such work. It is highly probable therefore that the characters of workers appeared before limitation of reproduction to one or a few individuals. This would imply that no further change could occur in the workers, but we must recognize that the queens vary, and that there is possibility in this variation for the selection of characters to be expressed in the workers. On the basis of such selection carried on by bee-keepers, various strains have been established, as noted in Chapter XIX. The difference between this and non-social species is that the colony becomes the biological unit, instead of the pair. Colonies are preserved or destroyed through the qualities of the workers, which in turn are derived from the queen. The individual is wholly subordinated to the welfare of the group.

Vestigial Organs. Probably the most important objection to natural selection is its failure to explain the persistence of vestigial organs. These organs are obviously in a state of gradual reduction following the termination of their usefulness to the individual. They are usually indifferent, but too uniform in occurrence to be explained as indifferent structures incidentally retained. The human appendix, for example, might readily be influenced by natural selection to a point of indifference, since we see that it is now a harmful structure through its tendency to become infected. Beyond that point, however, its condition would be of no vital importance, would neither eliminate nor preserve the individual, and so could be changed only through the operation of other forces.

Overspecialization. Overspecialization may conceivably bring a structure to a stage of development which destroys its usefulness without rendering it actually harmful. The tusks of the Columbian elephant, now extinct, became very large, and in some specimens curved so extremely that their tips must have crossed (Fig. 207). Such organs could not be used for digging or for defense in the way that the tusks of existing elephants are used, yet there is nothing to show that they were sufficiently burdensome to this magnificent creature to account for its disappearance.

The theory of natural selection is therefore widely applicable, but is not a sufficient explanation of all evolution of species. Darwin, in his earlier works, looked upon it as the chief process of evolution, and some of his followers have given it even greater emphasis, but even Darwin came to realize that other views were not without a claim to serious consideration. The contributions of the twentieth century have added definite theories no less logical than Darwin's, if of less widespread application, and to these we may now turn our attention.

Darwin's theory may be looked upon as one of several partial explanations of evolutionary change. It is limited in that it does not account for the origin of characters, and only incidentally affects indifferent characters. It may readily be seen, however, that the useful and harmful characters of individuals play an important part in their lives, and that natural selection logically accounts for the preservation and elimination of such characters as components of the species. Within these limitations it may be applied extensively to problems of evolution, but beyond them it must give way to other theories. It is more a theory of adaptation than of species-formation.

Fig. 207.—Restoration of the Columbian elephant (*Elephas columbi*). (From Scott.)

Summary. The theory of natural selection is the most important of the great theories of evolutionary method. It is based on the fundamental principles of variation and influences tending to favor the development of a limited range of the variations of a given species. These influences may be either spatial isolation or the effects of overproduction and resultant crowding. Crowding brings organisms into competition in which useful variations aid in the preservation of the individuals possessing them and harmful variations hasten destruction. The result is the preservation of a limited number of individuals with some uniformity which is preserved in their offspring, and the range of variation in the species is thus shifted to that of the surviving individuals. Darwin interpreted this as a cumulative process which would ultimately bring about a greater development of the characters preserved. Various objections to the theory have been proposed and many of them have been satisfactorily answered, but they lead to the conclusion that natural selection is an effective process of adaptation rather than of species-formation.

REFERENCES

WALLACE, A. R., *Contributions to the Theory of Natural Selection*, 1870.
MIVART, ST. G., *On the Genesis of Species*, 1871.
DARWIN, C., *The Origin of Species*, 6th edition, 1880.
ROMANES, G. J., *Darwin and After Darwin*, 1892.
MORGAN, T. H., *Evolution and Adaptation*, 1903.
KELLOGG, V. L., *Darwinism Today*, 1907.
LULL, R. S., *Organic Evolution*, 1917.
NEWMAN, H. H., *Readings in Evolution, Genetics and Eugenics*, 1921.

CHAPTER XXIII

OTHER THEORIES OF GERMINAL SELECTION

In the preceding chapter we have seen that Darwin's theory of natural selection, while it appeals strongly to the logical mind as having a definite bearing upon the process of evolution, is by no means a complete explanation, as has sometimes been urged. In recognition of its limitations various scientists have proposed other theories to account for those processes of evolution which are not within the scope of natural selection. Some of these theories have returned to the Lamarckian concept of environmental influence; these will be treated in the next chapter. Others have assumed the heritability of the factors involved and consequently stand on the same basis as natural selection; these are our present concern, and they may be known as theories of germinal evolution since they look upon changes in the hereditary germ plasm as causes and not results of evolution.

The failure of natural selection to account for the origin of variations is shared in part by these other theories but this subject involves factors which can better be treated in a succeeding chapter.

The Limitations of Selection. Darwin's use of fluctuating variations as a basis for evolution has been a source of difficulty, for in artificial selection and the experiments of geneticists it has been shown that the possibility of change through selection of such characters is probably limited. Artificial selection was emphasized by Darwin as an example of the possibility of rapid change in organisms through selective processes. It is now recognized that the aggregations of individuals making up a species vary not only in actual range of characters, but also in the latitude of variation which they are able to impart to successive generations, and that the limits of artificial selection are the separation of the various pure lines. Short and tall races may be produced, for example, from a normal population, but the maximum and minimum cannot be extended indefinitely by selection alone.

This question is closely related to Galton's law of filial regression, and was clearly explained by Johannsen's work on pure lines, which was described in Chapter XIX.

The Mutation Theory of De Vries. Darwin recognized the occurrence of sudden departures from parental characters, or saltations, but made little use of them. They later became the basis for the mutation theory of De Vries, which disposes effectively of the question of specific change by disclosing the possibility of minute mutations serving as materials for natural selection. Such variations accomplish the actual shifting of specific characters which in ordinary fluctuating variations is limited to the isolation of pure lines.

De Vries, a Dutch botanist, conducted a long series of experiments with the evening primrose, *Oenothera lamarckiana*, as the basis for his mutation theory. He found that this species was highly variable in its natural habitat near Hilversum, in the vicinity of Amsterdam, where it was observed for several years. De Vries speaks of some of its variations as new species. "Some of them were observed directly on the field, either as stems or as rosettes. The latter could be transplanted into my garden for further observation, and the stems yielded seeds to be sown under like control. Others were too weak to live a sufficiently long time in the field. . . . These various methods have led to the discovery of over a dozen new types, never previously observed or described." (Fig. 208.)

To give an account of the differences between the species or mutants so produced, would be beyond the needs of this work. It is sufficient to note that the degree of difference between mutants and the forms from which they originated was great enough to lead De Vries to call them species; it was as great as differences between other related species of plants, and greater than exists between some divisions commonly called species. In addition a mutant once produced continued to produce offspring like itself—an essential factor which distinguishes true mutants from mere sports.

A significant feature of De Vries' results was the production of a species *scintillans*, which in turn gave rise to mutants. This species was produced, like others, from self-fertilized seeds of *Oenothera lamarckiana*, and among the mutants that it produced were many *lamarckiana*. some *oblonga*, and a few *lata*, in addition to a moderate percentage of true *scintillans*. Similar reverse mutation has since been secured by Muller in *Drosophila*.

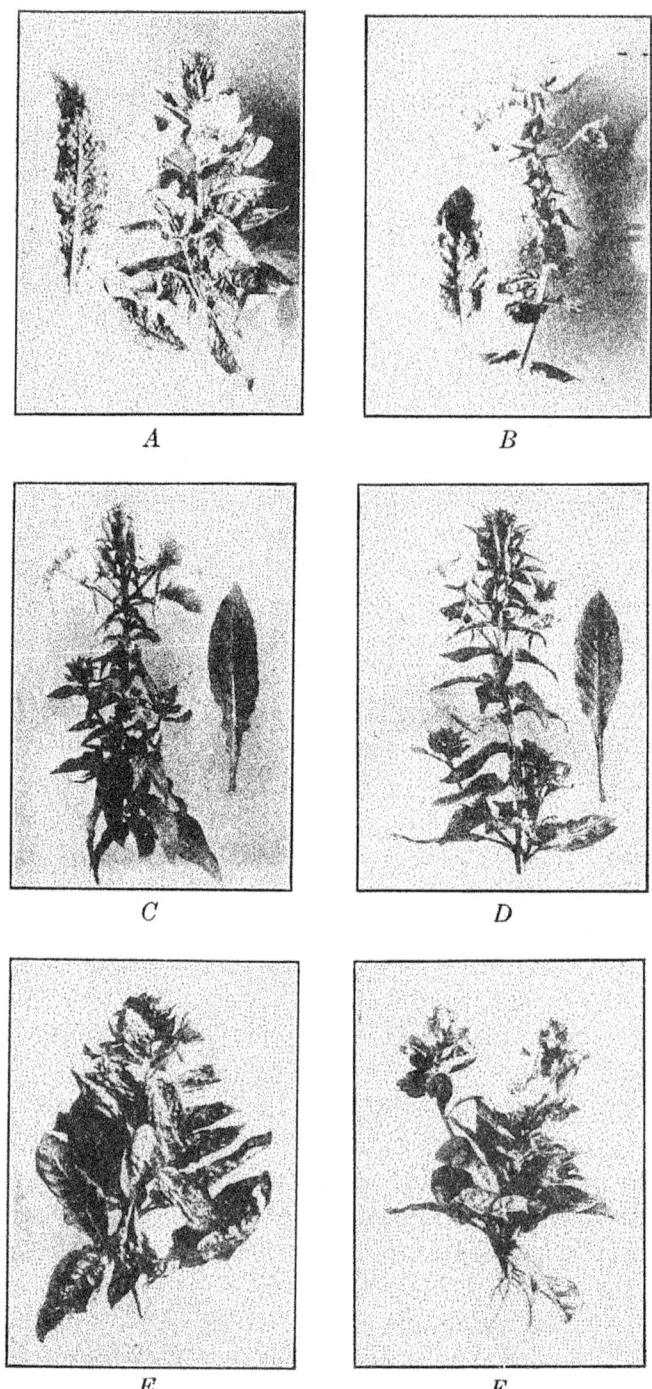

Fig. 208.—*Oenothera lamarckiana* and some of its mutants. A, *lamarckiana*; B, *gigas*; C, *oblonga*; D, *scintillans*; E, *lata*; F, *nanella*. (From Castle's *Genetics and Eugenics*, with the permission of the Harvard University Press.)

Objection to the mutation theory as an explanation of species-formation has been based chiefly on the repeated assertion that *lamarckiana* was probably a hybrid, and that the mutants were merely different strains separated from the hybrid. Such a view is plausible, for a homozygous recessive based on two determiners would have only one chance in sixteen of appearing, on three determiners only one chance in sixty-four, etc. However, a recessive once produced would be unable to produce the characters of its parents among its offspring, as was the case with *scintillans*.

Examples of Mutants. Fortunately, although species very productive of mutants are open to interpretation as hybrids, there are many examples of apparently stable mutants. Not the least of these are the classic examples of the Ancon sheep and polled Hereford cattle. Each is said to have originated from a single mutant. The sheep first appeared in the flock of Seth Wright, a Massachusetts farmer, in 1791, and the hornless Hereford mutant in a herd of horned Herefords at Atchison, Kansas, in 1889. The fact that the mutant character proved to be dominant over the normal condition in each case is another argument against the hybrid-species interpretation, since dominant genes, if present, are expressed. If separated as homozygous dominants from a hybrid race there is no somatic difference from the parent unless the hybrid character is a blending or mosaic.

In the vast amount of work done within recent years on the fruit flies of the genus *Drosophila* many mutations have been discovered which illustrate the possibility of very slight departure of a mutant from the characters of its parents. Such slight departures as have been recorded in these flies have not been interpreted as specific differences, in fact, in any case. Even a slight mutation, however, is inherited, and if of any use to the individual or individuals in which it appears, may well be a deciding factor in the preservation of its possessors.

Mutations in Natural Selection. It is the operation of natural selection upon such definite even though slight departures from the parental form, that seems most likely to account for the formation of species according to Darwin's ideas, rather than its operation upon fluctuating variations. If the short-legged Ancon sheep, for example, had occurred in a natural state and not under domestication, and if it had depended upon speed for safety, it would undoubtedly have been more quickly eliminated than any of its longer-legged relatives.

Mutations apparently do not depend on natural selection, however, for their potency in the formation of species. We must recognize that whether we are willing to call them species or merely varieties, some of the forms that have arisen this way are entitled to specific rank, provided only that they can satisfy other criteria for the limitation of species than mere morphological differences.

The Cause of Mutations. The possible cause of mutations has been the subject of much discussion. An early view compared the species to the individual in its development from infancy to old age. In its period of maturity the species was supposed to produce other species as the individual produces its offspring. Such an analogy is attractive but hardly a soundly scientific explanation. There is more reason to believe the conclusion of modern geneticists that abnormal behaviour of the chromosomes or modification of the genes by unexplained processes may be responsible. Even this opinion, while it leaves us with a definite basis for future investigation, also leaves us with a feeling that we have not explained the origin of mutations; rather we have merely paved the way for further study of the problem.

Given living organisms and their several kinds of variations we now know something about the processes of change which go on within them, but we have not yet penetrated beyond that starting point. Our progress is not discouraging for there is still much to be discovered by accurate scientific study of these matters, and additions are being made to our knowledge every year.

Weismann's Theories. Two other corollary or substitute theories of evolution were proposed by Weismann, one of the greatest followers of Darwin. One, the theory of germinal selection, is a frank attempt to meet the failure of natural selection to explain the modification of organs beyond the range of importance to the individual. The other, panmixia, was first developed for the same purpose and later discarded in favor of germinal selection. Weismann recognized that his theory of panmixia explained the modification of indifferent organs to a limited degree even though he found it necessary to discard it as an explanation of evolution.

Panmixia. The principle of panmixia is that indifferent organs, since they are of no consequence to the individual, will be mingled at random during reproduction and so will have an effect on the evolution of the species. Certainly these organs can have no effect in determining the survivors of a generation or the propagation of the species, and so they cannot be the basis for natural selection. It is equally certain that they will be inherited, regardless of their state of development; the greatly reduced will be mingled with the highly developed and the result will be an averaging of development of the organ in the entire species. This is, in effect, a reduction in the case of formerly useful organs, but it can hardly bring about an extreme degree of modification.

In 1904 Weismann wrote of this theory: ". . . I was obliged to seek for some other factor in modification, which should be sufficient to effect the degeneration of a disused part, and for a time I thought I had found this in panmixia, that is, in the mingling of all together, well and less well equipped alike. This factor does certainly operate, but the more I thought over it the clearer it became to me that there must be some other factor at work as well, for while panmixia might explain the deterioration of an organ, it could not explain its decrease in size, its gradual wearing away, and ultimate total disappearance. Yet this is the path followed, slowly indeed, but quite surely, by all organs which have become useless." We now recognize this process, which is nothing more than the amphimixis of modern biology, as a means of diversification, or distribution of the various degrees of variation of specific characters, and of maintaining a reasonably definite association of such characters.

Germinal Selection. Weismann's realization of the weakness of the theory of panmixia led him to formulate under the name of germinal selection a theory which now seems of no more value than the former, yet it was based on some conceptions of living substance which are worthy of consideration. An examination of these conceptions leads one to the feeling that Weismann very narrowly missed some important conclusions. In judging his work we must remember that biology was scarcely beyond its infancy during the early years of his scientific activity, and that the great modern discoveries in genetics were little more than begun when he died in 1914, at the age of eighty years.

Weismann assumed as a starting point for his theory that the germ plasm is composed of different living units or "determinants" which control the development of different parts of the organism.

Although his interpretation of these determinants is not exactly the same as our modern understanding of genes, it is a tribute to his insight that he came so near to the fundamentals of the modern theory. He cautions his readers that nothing can be learned *directly* of the intimate structure of the germ plasm; this difficulty has, of course, been partly removed by modern cytology, although we are still unable to see the genes in which we now believe so firmly.

The next step in his theory assumes that, since these particles in the germ plasm are living matter, they are nourished, grow, and multiply. We know no more about this matter than Weismann did, but we must still admit that his opinion is soundly logical. However minute the gene may be, or whatever may be the nature of its organization, we cannot fail to regard it as living matter, subject to the same fundamental laws that have been observed in larger units.

Weismann then assumed that irregularities occur in the nourishment of the determinants and that they differ in their capacity to make use of available food. As a result he supposed that some would increase in vigor at the expense of others, and attain greater expression of the structures which they produced in the body. Such a modification of determinants would necessarily be cumulative, for weakening would still further reduce the possibility of a part's securing adequate nourishment. The result would be a gradual modification of the characters of the organism, some increasing and some decreasing in development.

Criticism of Germinal Selection. At this point the unsoundness of the theory becomes apparent. We must consider that most of the organisms with which we are concerned are normal and lead a normal existence. There is every reason to believe that in such organisms every part of the body is supplied with adequate nourishment and that competition between the various parts does not occur. Even in starvation the whole body starves together; distribution of the available nourishment is made for the greatest common benefit. When we consider the capacity of individual determinants for securing available nourishment we are dealing with hereditary qualities; the very existence of a living unit would demand sufficient ability to meet existing conditions.

It is impossible to see how this theory can account for evolutionary changes in organisms except on the basis of varying hereditary qualities in the germinal determiners. With our modern knowledge of heredity and the theory of the gene we realize that germ cells do contain the genetic foundations of the new individual and that the genes are probably subject to variation like the characters which they bring to expression. This does not, however, explain the process of change.

Three points in Weismann's theory should be borne in mind: 1. The idea of determinants foreshadows the well-established theory of the gene, although it was necessarily vague and imperfect in detail. 2. Weismann's emphasis upon nourishment brings out the importance of environment. This is all the more suggestive when we consider that he did not believe in the activity of the environment in evolutionary processes. 3. The idea that determinants, although hypothetical structures, obey the same laws as the larger units of living matter which can be observed and subjected to experiment is an important and necessary interpretation which we can apply profitably to the gene.

Roux's Intraselection. Prior to the formulation of Weismann's theories, Roux proposed a theory of intraselection which was fundamentally similar in that it assumed struggle among parts of an organism for opportunity to develop as a basis for the evolution of its parts. The theory assumes that conditions within an organism include mechanical relationships such as contact and pressure which exert an influence upon the development of the parts involved. For example, pressure on a bone is supposed to result in a greater deposition of bony tissue to meet the stress. Objection has been made to this theory on the ground that it involves a distinctly Lamarckian point of view, for changes so induced would be responses to the internal environment. Many biologists are still unwilling to accept this point of view. In any case the theory is of minor importance, and must be looked upon as a subsidiary theory. It has not been wholly condemned, although there are excellent reasons for only partial acceptance.

Coincident Selection. This theory, variously expressed, has been widely used in an attempt to utilize Lamarckian characters without recourse to Lamarckian principles. Proponents of the theory recognize the value of individual adaptations, but explain their effect as merely the preservation of the organism until the appearance of germinal variations capable of producing similar adaptations in the entire species.

"It has been objected to this theory, that since the individually acquired modifications possess the main selective value in these instances, there is no reason why the corresponding germinal variations should be fostered at all. The individuals with the right, but slight, congenital variations would have no special advantage over their fellows who show no such coincident variations. Nor is there any ground to assume that the individuals with the greatest amount of plastic modification in a given direction will tend to exhibit similar innate variations to a greater degree than those individuals not possessing this plasticity." (Herbert.)

The Theory of Isolation. In the discussion of natural selection Wagner's theory of isolation was briefly mentioned. It is without doubt one of the valuable theories of species-formation, although, as previously indicated, its fundamental principle of isolation may also be an initial stimulus for natural selection.

Isolation may be due to any factor that prevents the mingling of individuals for purposes of reproduction. Such factors may be purely biological, and they may be purely physical. The writer once noted the flight of a small butterfly, *Plebeius melissa*, over a period of several years, and found that the males were abundant six weeks before the appearance of the first females. The fortuitous destruction of such delicate insects must be great, and the chance of the earliest males to mate correspondingly slight. Such seasonal distribution amounts to isolation through biological factors alone, and must have some influence on the development of the species.

The best example of physical isolation is found in the separation of land areas by barriers. In oceanic islands isolation of this kind is often extreme, and in such a group as the Galápagos Islands the animals may show definite correlation of characters and distribution which is apparently the result of isolation. Beebe's account of the mocking birds, in *Galápagos, World's End*, is an excellent exposition of this correlation. It is as follows:

". . . Without doubt we have here birds which were once living on a single land mass—connected long ago with the mainland, probably along the Costa Rican-Panamanian latitude, and later insulated on a single large Galápagos island. This in turn has, through subsidence, been reduced to a few volcanic peaks, on all of which *Nesomimus* still survive, and which have begun to differ slightly among themselves. This is due probably to causes far other than resulted in the general, generic wing reduction and leg increase, and causes of relatively less importance, whether initiated as internal variations or in response to external conditions we know not. But however these were initiated and preserved, we have a splendid example of evolution in the three planes of space.

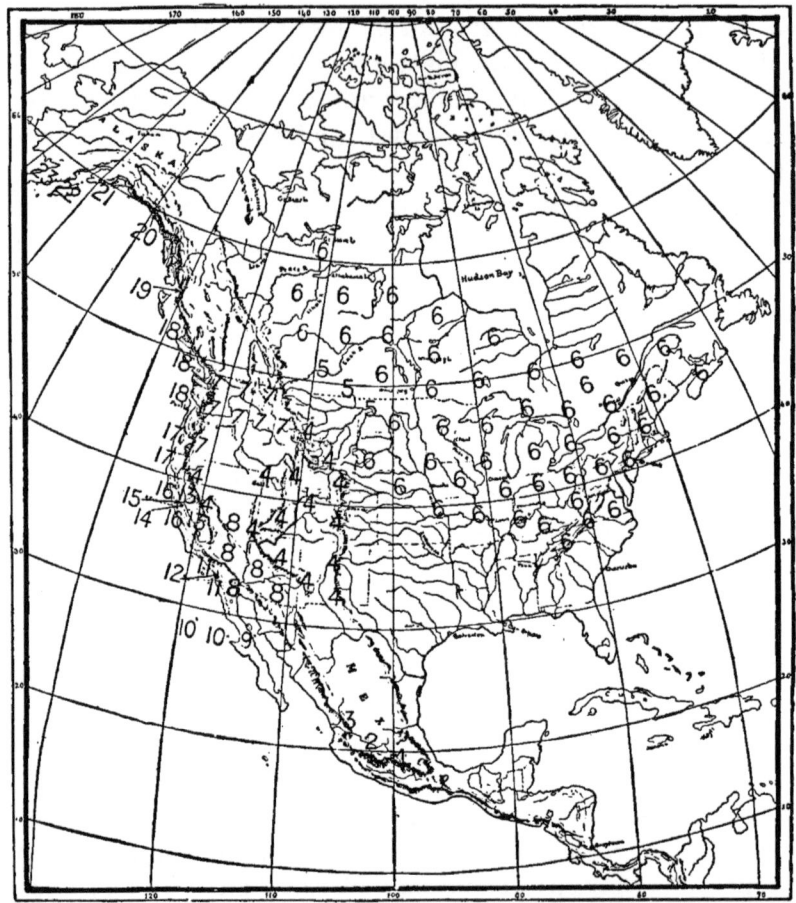

Fig. 209.—The geographical distribution of song sparrows. Each number indicates the habitat of a distinct subspecies. (From Woodruff after Chapman.)

Hence the facts so annoying to the taxonomist of a mocker in Tower Island, far in the northeast of the archipelago, which has a black moustache like the birds in Chatham, seventy miles to the southeast, a black back like those in Bindloe, thirty miles due west, a large beak corresponding to the beaks of the Abingdon birds, forty-five miles northwest, and yet which in all characters is almost identical with the *Nesomimus* of little Culpepper, a speck nearly a hundred and fifty miles to the northwest.

" . .. if we conceive that the ancestors of the birds on Wenman, Tower and Indefatigable were identical, and gradual subsidence separated the three, causal variation would go on with a slightly greater emphasis on one character or another, and with no possibility of indication of specifically separate relationship. . ."

Such is the action of isolation upon the course of evolution. Merely the separation of individuals occupying different positions in the range of variation of the species is enough to account for differences in succeeding generations in the separate localities. It makes no difference whether the characters concerned are important or indifferent; the mere fact that they are a part of the heritage of the species is enough to make them a source of future difference in the isolated races. The distribution of song sparrows in North America is a similar case which has been expressed graphically in Figure 209.

Although Darwin placed little emphasis upon this theory, recently scientists have recognized that it is certain to be a potent factor in the shaping of species. It almost seems axiomatic, in view of our knowledge of geological changes, the possibility of accidental dispersal beyond normal barriers, and the ordinary range of variation, that this should be the case. Quoting Newman, "If natural selection may be said to be the prime factor in producing adaptations, isolation may be said to be the prime factor in species differentiation, guided only within moderate limits by natural selection."

Common Objections to Germinal Theories. In conclusion we should note several characteristics of these theories of evolution which place them on an equal basis with the Lamarckian theories to be treated in the next chapter.

1. No theory of germinal evolution accounts for the origin of variations.

2. Only the mutation theory offers an adequate explanation of change beyond the isolation of pure lines.

3. Specific change has been accomplished very gradually in so many cases that the mutation theory, in spite of its obvious value, seems to be limited in application.

Summary. The limitations of natural selection as an explanation of the origin of species were clearly indicated by Johannsen's study of pure lines. Other theories which have been proposed to supplement or replace that of Darwin are the mutation theory of De Vries, Weismann's theories of panmixia and germinal selection, Roux's intraselection, coincident selection, and the theory of isolation. The occurrence of mutations is well established and seems adequate to explain the formation of species in some cases alone and in some through a process of selection of minor mutations. Weismann's theories are now given little attention but they deserve to be considered for several valuable points even though they must be discarded as theories of evolution. Isolation is apparently an important factor in the development of variations of all kinds into specific characters. The remaining theories are interesting chiefly in the fact that they emphasize the association of changes with external influences. Since their authors were avowedly against Lamarckian principles this tendency is doubly suggestive of the inevitability of such association. Even the writings of Darwin are crowded with Lamarckian explanations, in spite of his early antipathy to Lamarck's views.

REFERENCES

MORGAN, T. H., *Evolution and Adaptation*, 1903.
WEISMANN, A., *The Evolution Theory*, 1904.
DE VRIES, H., *Species and Varieties, Their Origin by Mutation*, 1905.
HERBERT, S., *The First Principles of Evolution*, 1913.
LULL, R. S., *Organic Evolution*, 1917.
NEWMAN, H. H., *Readings in Evolution, Genetics, and Eugenics*, 1921.
WALTER, H. E., *Genetics*, revised edition, 1923.
ELDRIDGE, S., *The Organization of Life*, 1925.

CHAPTER XXIV

THE LAMARCKIAN THEORY

The Lamarckian conception of evolutionary processes has had a more varied career than any of the other existing theories. It has been treated by most evolutionists as the antithesis of Darwin's theory, and most works on this subject devote themselves in some degree to the support of one view and the denial of the other. Under the common designation, the inheritance of acquired characters, the ideas credited primarily to Lamarck have usually suffered in these encounters, and have been dismissed with the conclusion that they are still unproved. Common emphasis upon other theories of descent has resulted in their being much more widely accepted. It is true, nevertheless, that an occasional defender has come to the support of Lamarckian views, and that others have lent their support more or less unwittingly to them in dealing with difficult questions of evolution. Weismann, for example, as pointed out in the last chapter, made use of an outstanding environmental factor in expressing a theory of germinal evolution. At present there is a marked tendency among scientists to feel that Lamarck's views have more to commend them than has been admitted, but they have so far not received so convincing support as Darwin's theory of natural selection and the mutation theory.

Lamarck's Theory of Evolution. Lamarck was not the first scientist to express belief in the shaping of organisms by their environment. He was preceded in this by Buffon and Erasmus Darwin, but like Charles Darwin, he was the first to give the matter adequate presentation to impress it upon the scientific world. That his theory was so largely ignored by his contemporaries was due to a number of factors, among them the opposition of his influential associate, Cuvier. It is abundant testimony to the importance of his theory that his name and Darwin's are preëminent in the literature of evolutionary theories.

Lamarck's theory, when first published in 1809, consisted of two laws which have been translated as follows:

"*First law:* In every animal which has not exceeded the term of its development, the more frequent and sustained use of any organ gradually strengthens this organ, develops and enlarges it, and gives it a strength proportioned to the length of time of such use, while the constant lack of use of such an organ imperceptibly weakens it, causing it to become reduced, progressively diminishes its faculties, and ends in its disappearance.

"*Second law:* Everything which nature has caused individuals to acquire or lose by the influence of the circumstances to which their race may be for a long time exposed, and consequently by the influence of the predominant use of such an organ, or by that of the constant lack of use of such part, it preserves by heredity and passes on to the new individuals which descend from it, provided that the changes thus acquired are common to both sexes, **or to those which have given origin to these new individuals.**"

To these laws he added later the idea that necessity in the organism gives rise to new organs. Other corollaries expressed his belief in various modifying factors, but essentially his theory involves the belief that change springs from the action of the environment upon organisms. He supposed that this action was direct in the case of plants, but that in animals it involved response of the organisms to the environmental condition. His belief in the inheritance of such changes as are obviously produced in correlation with environment is the point which has always been the chief subject for opposition.

Criticism of Lamarck's Laws. *Use and Disuse.* Lamarck's first law concerning the results of use and disuse in organisms is apparently sound. Every individual during his life furnishes an example of the effects of functional activity. If we exercise a muscle repeatedly it gains in development, usually to a visible degree, and whatever may be the effect on its bulk, the power exerted by its contraction is certain to increase. Conversely failure to use muscles results in diminution of muscular strength. A man who leads a sedentary life is not expected to have the powerful muscles of a laborer, and a muscle which for any reason cannot be used at all, may atrophy to an extreme degree.

In considering the effects of use and disuse we need not confine ourselves to the obvious cases of physical activity. Tolerance for poisons is an expression of an obscure functional power of the body, but it is no less subject to increase or decrease. Increase of individual tolerance for the habit-forming drugs is a case commonly known. Ordinary poisons such as strychnine and arsenic have a similar influence. They are used in minute quantities in medicine, and if these doses, tolerated by the normal human body, are gradually increased, a dose can ultimately be administered without causing harm which would be sufficient to kill a number of men if administered without previous preparation. If such an extreme dosage is suddenly discontinued, the effects are as serious as its sudden administration, but it can be reduced gradually until the body is again able to get along without it.

Functions of such organs as the eye are usually thought of in terms of normal exercise and excess. It is true that excessive use of the eyes frequently results in their impairment, but not less true that reasonable use of the eyes results in increased power of vision. Biologists who constantly use the microscope are likely to use one eye to the exclusion of the other, either because of natural preference or convenience in drawing or writing while making observations, and it has often been noted that the eye habitually used becomes much more effective for microscopic observation. The converse cannot be demonstrated satisfactorily in this case, for eyes when present, cannot often be wholly unused. Such cases as the rudimentary eyes of moles and cave animals are suggestive, but they require interpretation.

The evidences of biology therefore support the first of Lamarck's laws. We may definitely conclude that the use of functional power brings about its increase, and disuse it decrease. The matter of size is not necessarily a corollary of functional power, and so is of limited significance.

The Inheritance of Acquired Characters. In formulating his second law Lamarck made an unjustifiably positive statement. No matter how confident we may feel that changes which appear in individuals in response to environmental conditions may exert an influence over future generations, we must admit that evidence of such action is not available. The facts of palaeontology often show such a gradual transition of phylogenetic changes in correlation with a changing environment that they almost convince us of the potency of the environment in bringing about the evolution of organisms, but careful consideration of such individual acquisitions as are frequently produced will show that every generation develops them anew; at birth there is no trace of the characters which the parents had acquired. Such structures as heritable callosities are obviously similar to the calluses which can be developed in an individual, but relationship between the two has not been proved. The association of characters of these two types has long been one of the most vexatious problems in biology, and has been the foundation for many experimental studies.

Necessity in the Production of Organs. The idea that need in an organism may bring about the development of a new organ is scarcely worthy of scientific attention. On a purely logical basis it may be refuted, for organisms, in order to exist, must have all of the organs necessary for successful existence. In human experience need may be interpreted in a different way; in the biological sense only those things are needed which enable the organism to live successfully and to perpetuate the species. In an examination of the facts of biology, and in particular of adaptation, it seems rather that organisms have made use of such things as they have possessed than that they have developed organs to meet preëxisting needs.

The Unity of the Organic World. Finally Lamarck's opinion that the environment acts directly on plants and indirectly through the nervous system on animal structures expresses a naïve disregard for the fundamental unity of the organic world which could hardly fail to weaken the theory of which it is a part. A liberal interpretation shows that it is probably based on the much more intimate association of plants with the physical environment. We now recognize that whatever is true of the fundamental evolutionary processes of one kingdom is in all probability true of the other.

Lamarck's theory therefore narrows down to one critical point of controversy, viz., the inheritance of acquired characters. All other premises are either well established or universally denied, but this one point has baffled scientists. Of the many attempts which have been made to prove or disprove it experimentally a few are significant. The results obtained have generally failed to establish even the possibility of such inheritance, but positive disproof has also been lacking. The question still arises and still finds ardent defenders as well as opponents.

Acceptance of the Inheritance of Acquired Characters. The proponents of Lamarck's view include all kinds of thinkers, among them men of philosophical mind who were content with a purely logical analysis of the question and able biologists whose opinions should have some weight as scientific deductions. We can learn very little therefore from the controversial material now available. One side presents as imposing authority as the other; the scientists are by no means united against the inheritance of acquired characters, nor are its supporters lacking in scientific training and ability.

Since ancient times many men have accepted the idea in connection with human evolution without any critical attempt to establish its validity. The close association of human progress with the activities of past generations is sufficient to justify this feeling, but far from sufficient as yet to prove that any change has actually been produced in the heredity of the human race as a result of human activities. If we could *know* that education gradually modifies the human mind, it would be of the greatest value to us. This is, however, merely *supposed* to occur by those who favor the Lamarckian point of view. The only thing definitely known to us in this connection is that human experience is perpetuated through records and teaching, so that future generations may profit by it as a part of their environment, even though they may have received no inheritance from it.

Herbert Spencer stands out among those who have believed in the inheritance of acquired characters, and like others, his attitude was an outcome of such keen appreciation of the intimate relationship existing between an organism and its environment that he could not imagine an evolutionary process in which environment was wholly passive.

Circumstantial Evidence. Much evidence has been brought to the support of this point of view from the study of existing animals. The similarity of such things as heritable callosities and the callosities which appear in individuals as acquired characters in response to friction is one such case. It has been regarded as wholly logical to conclude that inherent callosities located where they resist extreme friction may well be the result of the long-continued action of such a stimulus. However convincing one may find this type of evidence, he must admit that it is not proof that the character in question was so developed. In the one case we deal with a definitely germinal character in the usual sense; in the other we deal with a character which appears anew with every generation only when the proper environmental stimulus favors its development.

The student should note that questions of this kind may be very intricate and may permit apparently sound conclusions which are, in reality, of no value whatsoever. No better example of the difficulty is available than a recent treatment of this same question. The skin on the soles of the feet is thickened in individuals in proportion to the friction between the feet and the substratum. Walter points out that even in the mud-puppies (*Necturus*, a genus of tailed amphibia) the skin of the soles is thickened, although these animals always live in the water and therefore cannot exert much pressure on the soles of the feet. He adds: "Nor is it *reasonable* to suppose that it ever had any ancestor who did so for the hands and feet of the Amphibia are the most primitive and ancient hands and feet to be found in the animal kingdom without any known ancestral types. The thickening of the skin on the sole of the mud-puppies' feet *must* be due, *therefore*, to germinal determiners and is in no way an acquisition through use." Detlefsen quotes this passage, inserting the italics which are here reproduced, and asks: "But is it really proven that the ancestors never habitually applied pressure or friction to the soles of their feet? Or do we know that the small amount of friction which *Necturus* develops as it crawls along the bottom or behind rocks and stones, would not develop some skin thickening?" These two opinions on one point are an excellent illustration of individual fallibility, for as we examine the evidences of evolution we find that the pentadactyl appendage first appears as an adaptation to terrestrial life. If this generally accepted point is true, then no animal with pentadactyl appendages or their derivatives can fail to have had terrestrial or amphibious ancestors. At some time in the past, therefore, *Necturus* must have had ancestors in which the feet supported the body on solid earth during at least a part of their lives, and here is the desired source of friction! It is therefore just as inaccurate to say that the thickened soles of *Necturus* cannot be due to the individual acquisitions of long ago as to say definitely that they were so produced. Such evidence is simply inconclusive.

Evidences of this kind may be explained just as readily by the Darwinian theory. It is quite as accurate to say that a certain useful character appears fortuitously in some individuals and is perpetuated by natural selection as to say that it developed in the individual as an acquired character and because of its value

THE LAMARCKIAN THEORY 419

finally became a part of the heritage of the species. Either view has its hypothetical aspects; neither is an adequate explanation of the gradual development which is so clearly shown in evolutionary series.

Experimental Evidence. The conditions established for experimentation are that a character must appear in response to a definite environmental condition which does not ordinarily occur in the species under consideration, and second, that this character must persist in some degree in later generations produced after the cessation of the environmental influence.

Experiments dealing with the inheritance of acquired characters have dealt with four kinds of environmental effects, viz., effects upon the soma, effects upon the germ plasm, parallel induction or simultaneous effects upon both soma and germ plasm, and somatic induction or effects upon the soma which are later transmitted to the germ cells.

While these groups are significant in the classification of experimental results, it is obvious that not all of them bear directly upon the problem in hand. The first, effects upon the soma, is obviously of no importance unless some connection with the germplasm can be demonstrated, for we have seen that it is only through the chromosomes of the germ cells that hereditary characters are transmitted. Effects upon the germ plasm have been demonstrated in animals treated with X-rays, alcohol, and other agents. Notable among these are Muller's recent results with *Drosophila* already cited. It is clear that the germ cells can be modified directly, but characters arising from these modifications are in no sense the acquired characters with which evolution is concerned. Parallel induction is no more than a combination of these two. The only remaining category, somatic induction, expresses the condition which must occur if significant acquired characters are to be proved inherited. The character must appear in the body of the individual organism and then affect the germ plasm in such a way that it will be reproduced in succeeding generations.

Weismann's Mice. One of the most famous experiments with somatic modifications was carried out by Weismann. He cut off the tails of young mice generation after generation, breeding the tailless mice and removing the tails which inevitably appeared in their offspring, without affecting in the slightest the development of the tail. At the end of the experiment he produced mice of the same kind as the original generation. This experiment and other evidence furnish abundant proof that mutilations are not inherited.

Parallel Induction in Insects. A number of experiments with insects have shown that conditions of excessively low or high temperature or relative humidity produce melanism. The offspring of insects in which melanism has thus been produced have in some cases also shown melanic color, and these have been interpreted as due to parallel effects upon the germ plasm of the melanic parents. The cases are open to a variety of interpretations, however, none of which is conclusive. In any case, if parallel induction could be proved it would have little significance since the effect on the germ plasm would be direct and the characters therefore not of the type significant in evolution. If parallel induction were well established and frequent in occurrence its standing would be different, for then we might conclude that the soma and germ plasm were normally susceptible to the same stimuli.

Schröder's Insects. Schröder experimented with the caterpillars of a willow moth which usually form a case by rolling the tip of a willow leaf. By cutting off the tips of the leaves on which they were allowed to feed, he forced a number to roll the sides of the leaves in order to conceal themselves. These conditions were repeated with a second generation, and of the third generation of caterpillars four out of nineteen continued the acquired habit even when given entire leaves.

In another experiment he dealt with beetles which live on a smooth-leaved species of willow. Their eggs were removed to pubescent leaves for three generations and at the end of that time he reported that the pubescent leaves were preferred even when free choice of the two varieties was allowed.

Detlefsen objects to Schröder's conclusions on the ground that they are based on too few cases and that the normal range of variation is not taken into consideration. These objections seem sound, but another question which he raises, viz., "whether the modifications would persist or disappear gradually if a reasonable number of additional generations were followed under normal conditions," seems pointless. Either habit may be looked upon as positive. The fact that one prevails under normal conditions may well be taken to indicate that normal conditions favor it, hence a return to the old habits might be only another case of acquisition.

The Experiments of Guyer and Smith. The experiments of Guyer and Smith are widely cited in this connection. These scientists produced an antiserum of fowl blood by injecting a preparation of the lenses of rabbits' eyes. The antiserum was injected into pregnant rabbits without affecting the eyes of the adult, but the young when born showed several defects. Individuals selected from the defective progeny transmitted eye defect through several generations without further treatment. The precaution has been taken to test the transmission of the characters through the male in order that no question of prenatal influence could enter.

These experiments are the most significant yet reported on the transmission of acquired characters. They have been subjected to severe criticisms by several biologists and have been repeated and confirmed by their authors. Other experimenters have failed to secure similar results, although the significance of their failure is reduced by the fact that they did not follow the same conditions. The present status of the experiments is such that they must be grouped with other evidence as inconclusive, although they are highly suggestive.

Castle and Phillips' Ovarian Grafts. One of the most striking experiments relating to the inheritance of acquired characters is that performed by Castle and Phillips by transplanting the ovaries of black guinea pigs to white individuals from which the ovaries had been removed. One female in which the operation was successful was later mated with a white male. Six young produced by this union of white parents were all black.

It is known that black color is dominant over white in the guinea pig, hence this experiment shows conclusively that the germ cells produced by the female were produced by the "black" ovaries and were not influenced by the white body that nurtured them. The experiment was undertaken to determine whether the body, as the immediate environment of the germ cells, exerts any influence on them and has been regarded as absolute proof that this is not the case. However it is not absolute proof. In the first place, not merely the germ cells but ovaries of the black female remained in the white individual, so that the environment of the germ cells was not wholly albino. In the second place, we do not know enough of inheritance to say that the blood stream contains only certain things in connection with one character and other things for its allelomorph. It is at least possible that the same constituents are present in all cases, hence, no matter what the character of a body, cells introduced into it might be expected to avail themselves of the nourishment so provided without losing their own integrity. It has been suggested also that albinism may be due to the absence of a factor, and that a negative condition in the soma could hardly be expected to exert control over the germ cells.

Neo-Lamarckian Theories. The opposition of Neo-Darwinian evolutionists to the Lamarckian theory has been such as to concentrate attention upon the supposed integrity of the germ plasm. It has come to be a very commonly accepted principle that the demonstrable continuity of the germ plasm through successive generations is evidence that it is a thing apart from the soma, influencing it but not influenced in turn. The soma is regarded as a derivative of the germ plasm which develops anew in every generation but fails to make any contribution to the heritage of the species.

The Fallacy of Germinal Continuity. In organisms which always reproduce by sexual processes continuity of the germ plasm is readily demonstrable. The germ cells produced by an individual are obviously derived from the zygote which produced it, and in turn contribute to the zygote from which the germ cells and bodies of the next generation arise. In roundworms of the genus *Ascaris* this continuity has been demonstrated with unusual clarity, for after a few cleavages one cell of the segmenting zygote is set aside as the parent of all germ cells while the rest develop into the body of the worm. Continuity of this evident kind is not often noted, however, and in some of the plants and lower animals we find evidence of continuity of other tissues and even of discontinuity of the germ plasm. Many plants can be developed from fragments of somatic tissue. In such cases we may have continuity of root, stem or leaf tissue from generation to generation. Animals such as *Hydra* display a similar continuity and in the flatworms Child has demonstrated the development of germ cells from somatic tissues, in addition to the somatic continuity made evident by the asexual reproduction of these worms (Fig. 210).

The prevalence of sexual reproduction among the more familiar

THE LAMARCKIAN THEORY

plants and animals gives the idea of germinal continuity a plausibility which it does not deserve. In the light of the facts cited briefly above, it is evident that continuity is merely a corollary of reproductive function, and anything which is capable of bridging the gap between generations will be continuous.

The idea has, nevertheless, influenced the thought of modern supporters of Lamarckian evolution. It has appeared necessary to explain the reciprocal association of the soma and germ plasm whereby the individual might make some contribution to the heritage of the species, and various hypotheses have been advanced with this end in view.

Pangenesis. Darwin, while he was at first vigorously opposed to the views of Lamarck, cited the effects of use and disuse frequently in his

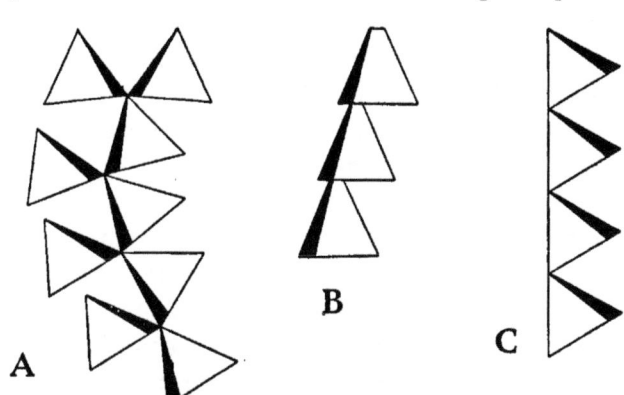

FIG. 210.—Diagrams to illustrate the behaviour of the germ plasm in a succession of generations. Black indicates the germ plasm, the outlined triangle the soma. The succession is from top to bottom. A, continuity of the germ plasm in ordinary sexual reproduction, involving the confluence of two lines at each generation; B, continuity of both germ plasm and soma in fission; C, continuity of soma and discontinuity of germ plasm in plants and flatworms.

writings on evolution. Like his followers he looked upon the body as a thing more or less definitely separated from the germ cells that it produces, and to meet the difficulty of associating somatic characters with the heritage he proposed the theory of pangenesis. This theory assumed that every part of the body constantly gave rise to minute units characteristic of itself. These units were supposed to be taken up by the blood stream and transmitted throughout the body. The units, called gemmules, were supposed to collect in the germ cells, thus influencing them according to the development attained by the somatic structures so that they in turn might carry this influence to the next generation. The acquired characters of Lamarckian theory, according to this hypothesis, would be provided with a vehicle whereby they might definitely influence the heredity of the species. The hypothesis is so unsatisfactory that it is of no more than historical interest. The assumption of purely hypothetical structures for which absolutely no biological support is available is never a convincing basis for scientific thought.

The Mneme Theory. More recently Semon proposed the mneme theory, which is based on the supposition that organic activity and conditions surrounding the organism leave definite traces in the form of engrammes. The engrammes are supposed to accumulate if the condition producing them is continued, and to affect heredity. It is similar to pangenesis in proposing a purely hypothetical vehicle for the recording of the character of the organism, and so is of no more value.

Centro-epigenesis. Rignano's theory of centro-epigenesis is expressed in an ample and excellent treatise on the inheritance of acquired characters. In this work, for the first time, we seem to be very near to a logical consideration of the factors involved. Rignano emphasizes the importance of functional activity in the organism in connection with environmental stimuli as the cause of evolutionary change. His analysis of characters which have erroneously been considered as disproof of Lamarckian views and those which are really entitled to consideration is logically admirable and his work seems worthy of more general consideration than it has received.

Kinetogenesis. A third theory, proposed by Cope, is of limited application. It is fundamentally similar to the last in that it is based upon power of functional response to environmental conditions. This theory, in brief, assumes that the effects of repeated motion or impact, as in the striking of the feet of cursorial animals upon the ground, finds expression in greater development of the parts involved. It would account for the development of elongated feet in the unguligrade animals in this way. However, it has been shown that stimuli of pressure may result in the limitation of bone development, as well as increased deposition of bone, so that the theory is applicable to both of the opposed processes. It may be looked upon as one of the valuable subsidiary theories of evolution, although it is distinctly Lamarckian and to that extent is without actual proof.

The Value of Lamarck's Theory. The foregoing pages are a brief statement of the variegated career of the Lamarckian theory of evolution. It is fifty years older than Darwin's theory, yet it bobs up like a specter even today. In spite of the vigorous opposition of the Darwinian school of thought and the fact that the inheritance of acquired characters is devoid of sound proof, we must recognize that many biologists have such a favorable attitude toward it that they would be glad to see it proved, or at least given as sound a basis for acceptance as the germinal theories of evolution.

It is significant that this condition can exist. Science is not prone to cling to useless theories when once a question is settled, hence this situation is very suggestive. When we consider all of the theories advanced to explain evolutionary processes, we find that none of them is adequate. Natural selection is logical and is commonly accepted as a factor in evolution, but as originally stated it meets a serious obstacle as an explanation of species-formation. The mutation theory obviously explains many changes in organisms, and supplies a valuable corollary for natural selection. In minute mutations we may well have the permanent modifications for natural selection to perpetuate in the gradual differentiation of species, but if we accept this as an adequate explanation of species-formation we must then explain the occurrence of mutations. They are due to chromosomal modification, but what modifies the chromosomes? Are all such changes haphazard? Is the wonderful adjustment of organisms to their environment entirely due to the selection and preservation of random characters? Reason seems to tell us that there is more to the story than this. A satisfactory explanation is not yet available but the very persistence of the Lamarckian idea leads us to consider evolutionary processes with open minds. Partisan views and investigations have failed to attain the desired end, and in their failure we may find a lesson for modern evolution.

Summary. Lamarck's theory of evolution was first stated in 1809 as two laws, known as the law of use and disuse and the law of inheritance of acquired characters. Laws which he stated later are of no moment. Of these two fundamental laws, use and disuse is now commonly known to have an effect upon individual development, but the inheritance of characters produced in this way has never been demonstrated. For many centuries the idea has been accepted by some individuals without proof. Many biologists have also favored it, and have accumulated evidence of a circumstantial nature to support their views. Experimental evidence is also abundant, but experiments have invariably given inconclusive results. The idea of germinal continuity advanced by Weismann has turned attention to the establishment of a reciprocal connection between soma and germ plasm; this tendency colors much of the literature for the Lamarckian view. In spite of the fact that it competes in its unproved state with germinal theories which have more or less convincing support, the idea persists that acquired characters contribute to the heritage. This very persistence is suggestive and valuable in that it forces us to recognize that the question of evolutionary processes is still an open one.

REFERENCES

CASTLE, W. E., and PHILIPS, J. C., "On Germinal Transplantation in Vertebrates," *Carnegie Inst. Wash.*, Publication 144, 1911.

RIGNANO, E., *On the Inheritance of Acquired Characters*, translated by Harvey, 1911.

CHILD, C. M., *Senescence and Rejuvenescence*, 1915.

GUYER, M. F., and SMITH, E., "Studies on Cytolysins. II. Transmission of Induced Eye Defects," *Journ. Exp. Zoöl.* XXXI, 171, 1920.

NEWMAN, H. H., *Readings in Evolution, Genetics and Eugenics*, 1921.

WALTER, H. E., *Genetics*, revised edition, 1923.

CONKLIN, E. G., *Heredity and Environment*, 6th edition, 1924.

DETLEFSEN, J. A., "The Inheritance of Acquired Characters," *Phys. Rev.* V, 244–278, 1925.

CHAPTER XXV

EVOLUTION TODAY

At no time in the history of biology has evolution occupied a more important place in scientific thought than now. It permeates the entire fabric of organized scientific knowledge and furnishes the guiding principle in much, if not all, biological research. The value of its established facts is recognized, but more than that, the value of accurate understanding of the methods by which it works impresses itself upon our minds. No step in biological progress at the present time could be a greater contribution to human knowledge.

We have now completed our survey of the field of evolution. We have considered the growth of the idea from its earliest conception, the evidences of relationship in existing organisms, the evidences of evolution drawn from living things and from the sciences of geology and palaeontology. We have learned of the contributions of genetics to our understanding of the transmission of characters from generation to generation. Finally we have examined the theories which have been proposed to explain the association of all of these facts in the production of the organisms which populate the earth. And we have found that in spite of his wonderful accumulation of facts, man still knows very little of the forces which have placed him where he stands today.

The failure of existing theories of evolution shows very clearly that something has been amiss in the past treatment of the subject. It is difficult to avoid the prejudices of past training. If we bear in mind that we are neither Neo-Darwinians nor Neo-Lamarckians, and that we have no predilection for continuity of the germ plasm or the inheritance of acquired characters, we can hope to learn still more of evolutionary processes. If we remember that we are merely scientists, seeking for clear information and logical deductions, we find that there is a firm and satisfying basis of fact in modern biology for the examination of the problems of evolution, and we can conclude in no more satisfactory way than at this threshold of the evolution of the future.

The Errors of the Past. *Darwin's Theory.* In our discussion of the existing theories we have noted various advantages and shortcomings which are a guidepost to future lines of progress. Darwin's theory has been one of the greatest of all time, yet the investigations of modern biology have cast doubt upon its complete soundness. A careful estimate of the criticisms to which it has been subjected leaves the feeling that the theory is sound, but by no means a complete explanation of evolutionary processes. The occurrence of natural selection can scarcely be doubted, but before it can operate it must have definitely useful or harmful heritable variations with which to work and the theory does not even attempt to explain the occurrence of such variations. This does not rob it of scientific value, but reduces the range of its usefulness; it becomes, as we have noted, chiefly a theory of adaptation.

The Mutation Theory. In connection with natural selection the mutation theory supplies an explanation of the occurrence of such permanent changes as the latter must have to work upon. Some examples of mutations of extreme degree are known which suggest the possibility of even varietal or specific differences appearing suddenly; mutations may therefore be a sufficient cause of the sudden evolution of species in some cases, but it is evident that gradual evolution is more common. Can mutation be a basis for such change? It is obvious in such species as *Drosophila melanogaster* that this is at least a possibility, for here we see that a multitude of slight mutations are constantly appearing, none of them regarded as specific or even varietal differences, but many potentially of importance to the individual. Natural selection, acting upon these minute characters, may readily be supposed to eliminate some and preserve others, to the ultimate modification of the species.

Even when this much is granted the mutation theory, however, we still lack an explanation of the source of evolutionary change. Mutations occur, this nobody denies, but what causes them? We can say that they come spontaneously from the heritage of the species, but what does that mean? An explanation to be of any value must make a thing clear in terms of its more familiar causes or components. Hence the admitted occurrence of mutations of both great and small degree is merely the recognition of a process of change without explanation, just as the recognition of living matter and its properties is not an explanation of life.

Lamarckian Theories. In the work of Lamarck and his supporters we find the only explanation of *how* change occurs in organisms. Lamarck's law of use and disuse and the modern recognition of its principles refer the matter of organic change to definite causes, viz., a definite heritage in living matter, expressing itself according to stimuli received from the surrounding environment. If this were the whole story we might happily become Lamarckians in the strictest sense, but unfortunately it merely carries us into another incomplete conception. We can see change taking place in individual organisms all about us from year to year, sometimes in direct response to environmental conditions, but we have never been able to see the change take place in a species. Here we encounter the serious obstacle which has been called the inheritance of acquired characters and the theory becomes even less useful than the others because it fails in a vital particular and is therefore not even a partial explanation of evolution.

The existing theories of evolution are therefore of little individual value as nuclei for an explanation of the process of evolution. We cannot be selectionists or mutationists or Lamarckians without subjecting ourselves to the limitations which accompany such views, but we can recognize the value which exists in all of them and avoid their limitations readily.

What Must an Adequate Theory Explain? It is improbable that an unbiased point of view can lead at once to a thorough understanding of evolution, but it can at least make for progress along sound lines. Two fundamental points must be met before the processes of evolution can be known to us: (1) The source of variations must be explained, not in terms of axiomatic occurrence as properties of living matter, but in terms of intelligible forces of a more fundamental nature; (2) The perpetuation of some characters in organisms from generation to generation and the elimination of others must be explained by an intelligible process or series of processes. The various fields of biology have contributed so many sound facts to our knowledge of the behaviour of organisms that it seems profitable to begin inquiry with an examination of the fundamental factors in life.

The Factors in the Existence of Living Organisms. This unprejudiced analysis demands first of all an understanding of organic existence. What lies behind the living organism? Why are organisms definitely of this or that species, and why do they live? We need not carry this inquiry as far as the origin of life, for a significant answer is found in the components of existence mentioned from time to time in previous chapters. The organism has definite qualities because they are its heritage. They are not absolutely fixed, for variation is universal, but within the limits of variation they are the same throughout a given species. The organism depends for the expression of these qualities upon things secured from without—from its environment. Its life depends upon its inherent power to secure and use these things. Response to the environment, in brief, is a third factor arising from the other two, and organic existence depends upon the interaction of these three. Without any one of them life must cease. Modification of any one beyond its normal range can only result in abnormal existence or death.

Such an analysis does not explain ultimate causes, nor does it pretend to do so, but in the realization that the three factors are indispensable is disclosed the fallacy of many theories of evolution. We are at present unable to do more than speculate on ultimate causes. Whether life is due to the existence in the organism of some cosmic influence other than those whose existence is proved, or whether it is merely an intricate expression of familiar matter and energy operating in space and time, we do not know. The fact remains, however, that living things are governed by natural laws, whatever may be the activating principle, and it is the immediate task of evolutionists to explain acceptably how organisms are shaped, how species are formed, on the basis of known biological facts.

The heritage has often appealed to biologists as a thing apart, which expresses itself in a definite way in spite of outer influences. It does attain remarkably uniform results in the many individuals composing a species, but we may well ask whether these results are ever, after all, an outcome of the heritage alone. The heritage is sufficiently definite and potent that no one would plant an acorn to raise a peach tree or buy wolves to stock a dairy farm. The egg of a hen can produce nothing but a chick. Keep it in a cool place and it remains an egg—living, but not growing.

Raise it to a high temperature and its substance coagulates—it dies. But keep it at 39° C and development begins. Temperature is one of the common factors of environment, and here we see that the correct temperature is just as essential to the initial development of the chick as the heritage borne by the egg.

In our own bodies we may see the unfolding of hereditary characters without obvious dependence upon the outer world. We grow from day to day throughout childhood. It is true that growth depends to a great extent upon the food that we eat, but some persons fail to grow in spite of sufficient food and others grow to gigantic size. Our mental development increases, but in some individuals this is not the case, although they may have a perfectly sound mental inheritance. The characters of sex appear. We can see in these cases only an indication of the fact already mentioned, that the body itself provides stimuli for the activation and control of its various parts. Some characters develop in response to these stimuli alone.

The environment, therefore, is not merely that which is outside of the individual. Any organ in the body is as truly a part of the environment of other organs as physical factors or living things are a part of the environment of a plant or animal. The environment may properly be subdivided into an external environment, made up of the physical and organic, and an internal environment which is established by the individual's own body.

The essential effects of these environments are the same. Each provides certain stimuli to which organs or organisms may respond according to their inherited capacity. Glands in our bodies, food which we secure from our organic environment and sunlight from the physical environment all contribute to our normal growth. The unfortunate individuals called cretins are mentally and physically retarded because of deficiency of the thyroid gland, but stunted growth and imperfect development may also come from malnutrition and the imperfect bone development called rickets may be due to lack of sunlight or a diet deficient in vitamines.

The Source of Organic Characters. The various parts and functions which characterize a given species are the most definite indication of its heritage, yet it is evident that they are a product not merely of the heritage, but of the heritage responding to environmental stimuli.

In the discussion of acquired characters it has usually been implied that they are the effects of environment impressed upon organisms. This is true of very few characters. Mutilations are, of course, produced in an organism in spite of its inherent qualities, and when the woman of today has her hair curled, no inherent quality of the hair determines the wave, as in the kinky hair of the negro. No one would seriously consider such characters in connection with evolution, although it has been done in the past. The vast majority of acquired characters or modifications are of a very different nature. The formation of calluses, tanning of the skin, muscular development and other so-called modifications are due to individual response to conditions in the external environment, it is true, but can they be produced without the action of a definite heritage?

A simple illustration will show that both heritage and environmental stimulus are as essential in their production as in that of other characters. The ability of the body to deposit pigment is an inherited character. Some people tan readily when exposed to strong sunlight, others burn without tanning and others freckle, but in the case of albinism the heritage is deficient. Albinos can deposit no pigment, therefore no amount of sunlight produces this so-called modification in their bodies. It seems that a character which arises from the response of the heritage to a stimulus from the internal environment may be interpreted as solely a product of the heritage in a broad sense, since the environmental stimulus in this case is another product of the heritage, but in many cases we find that the internal environment merely serves to make available in the proper form certain factors from the external environment. Proper development of the thyroid gland, for example, cannot be attained unless the food contains a sufficient amount of iodine. Even the internal environment is therefore tempered by external environmental conditions. None of the characters which are of interest to us in the study of evolution can be a product entirely of either heritage or environment alone. *The true distinction between the Darwinian and Lamarckian point of view is not the source of the character, but the source of the stimulus which governs its appearance.*

The Rôle of the Gene. A distinction arising from genetics should be clearly understood at this point. The function of the genes as determiners of hereditary characters has been so well established that in speaking of the heritage we must think of the chromosomal complex of a species and the genes which are contained therein. The cytoplasm of somatic cells is the structural material in which the differentiation of tissues is wrought and some biologists have questioned whether the genes might continue to exert an effect on this substance throughout life. Experimental methods for answering such a query have not been devised, but the only logical opinion possible in the light of biological facts is that they do remain active even in somatic cells, for living substance is constantly being torn down and built up through the processes of metabolism, and in enucleate cells constructive metabolism ceases.

It is logical to conclude that the genes which are responsible for the initial production of characters in the developing individual are no less responsible for all of the characters which appear during its life. It is the gene, in the end, which responds to a definite environmental stimulus for the production of a definite character.

The Source of Change. When a certain result can be secured by the proper association of materials and forces we find that an accurate formula is essential to accurate results. The cake turns out well if the proper ingredients are combined in the proper way and baked at the right temperature. Iron and steel take a definite temper if heated to a definite degree and properly quenched. But vary one of these factors and the cake is ruined or the metal does not have the desired quality. If the association of definite factors produces a definite result, it is to be expected that different factors would produce different results.

There is no reason to suppose that living matter is different in relation to these fundamental truths. Every character in a living organism is due to definite qualities of the heritage, reacting to definite stimuli received from the environment. The two are no more separable than batter and heat in the baking of a cake. We must therefore recognize that change in living things cannot fail to result from a change in either of the components.

Scientists have been prone to recognize spontaneous change in the heritage. It would be difficult to prove the truth of such an opinion, for we have seen that living matter is active only in response to environmental stimuli. Moreover, in the analogies cited, change of substance is due to outer forces—either human interference or forces in the physical world. The environment, however, varies according to a vast complex of physical forces which we can explain in part. These forces ultimately resolve themselves into matter and energy operating in space and time, and here we must be contented to stop.

The Tetrakinetic Theory. Osborn expressed views similar to these more than a decade ago. Under the name of the tetrakinetic theory he classified the factors in life as four energy complexes, (1) the inorganic environment, (2) the life environment, (3) the organism, and (4) the heredity-germ. In spite of his masterly analysis the literature of evolution has unfortunately continued along the same old lines.

The interpretation is equivalent to saying that all organic change is based upon changes in the complex environment, referable ultimately to the four entities, matter and energy, space and time. The position is logically tenable, but to satisfy the demands of evolution we must give attention to other details.

Adaptive and Incidental Change. There is every reason to believe that variation of the environment is the immediate stimulus which induces variation in organisms. This does not imply that the environment actually shapes the organism, but *it determines the expression which the organic heritage will attain.* Even identical twins are never really identical. They are genetically the same because they are derived from halves of the same fertilized ovum, but two individuals can never be surrounded by exactly the same conditions. They cannot eat exactly the same food nor the same quantity; they cannot sit in the same chair at the same time; the minute differences in their contacts may even mean life or death, as is true of spatial relations in the traffic of city streets at any time. Such differences as they display must therefore be due to the reaction of their identical heritages to different stimuli.

The suggestion that change in organisms arises in response to change in the environment does not mean that a stimulus always brings about a response which fits the organism to meet the existing environmental condition; in other words adaptations do not necessarily arise as a response to the environmental condition which they enable the organism to meet. Adaptive changes may appear in this way, as when the rays of the sun stimulate the body to deposit pigment for protection, but changes may occur as incidental results. Rickets, although due to lack of sunlight in some cases, unfits an individual for normal life, and the normal skeletal development which is in part due to abundance of sunlight during growth is in no sense an adaptation to life in sunlight. These conditions are, however, no less a result of environmental stimulus than tanning. With respect to adaptation, all changes appearing in response to the internal environment must be incidental.

Incidental changes are a part of the organism just as much as adaptive modifications, hence they are available for other adjustments to the environment. Any incidental modification when it is once developed may affect the organism's relations to other conditions than those which brought about its development. Normal nervous development is partly a response of the heritage to normal thyroid secretion, yet it accomplishes most of our adjustments with the conditions under which we live.

How Has Evolution Progressed? When we examine the phylogenetic series worked out by paleontologists we are able to learn something of the superficial aspects of past evolution, even though we may not be able to fathom the underlying forces. We are able to see evidences of the heritage of long series at different stages of development. Correlation of these characters with the environment may be deduced from a comparison with similar structures of living organisms, but a very definite indication of past environmental conditions, such as climate, is afforded by the fossil remains of other organisms, especially the green plants. We can therefore learn from paleontology not only how the heritage of a given line of descent has changed from age to age but also under what conditions these changes have taken place.

An examination of living organisms is of great value in connection with this study, for even in living things we find conditions similar to those which have preceded evolution in the past. From time to time in the previous chapters of this book organisms have been mentioned which show a distinct capacity for evolutionary change.

The Heritage. Whether in extinct organisms whose descendents are before us or in living things whose possibilities can only be estimated, we find that the heritage must not be limited rigidly to one type of response if evolution is to occur. We cannot take a trout out of water and expect it to live, much less to become a terrestrial animal, but fishes have existed and still exist which might well accomplish this transition in some degree. Fishes with functional lungs can breathe air or secure their oxygen from the water. Among other cases amphibia with gills and lungs breathe in either medium. *Euglena* can carry on either plant or animal metabolism, and the quail can both run rapidly and fly. With the capacity for two or more kinds of activity an organism can carry on either.

Evidences of this kind are so numerous that latitude of hereditary possibility seems a necessary quality in organisms which are to evolve. A species cannot be expected suddenly to acquire something new to meet a given condition; the abundance of extinct species shows that many an organism has failed to meet the conditions surrounding it. If a species has a sufficient range of possibilities to meet the conditions of a changing environment it may survive, as in the past many species have survived. Granting its ability to survive a change of environment, the more valuable characters of the species may persist and be developed while others are being reduced and eliminated. The horse attained a large third digit while losing all others.

The Environment. Like the heritage the environment has always been variable. We witness from season to season and from year to year fluctuations of all of the factors of the physical environment. One year may be wet, another dry. In one region the nights may be cold and the days hot, while in another temperature may be almost constant. Paleontology shows that this is not the limit of fluctuation, but that gradual climatic changes covering thousands of years have occurred. Temperate North America has been both tropical and arctic in the past. Moreover at any given time different parts of the earth's surface present different climatic conditions because of their physiographic features and their varying relations to the sun.

The effects of such fluctuations may be very great. The physical environment is so closely associated with the metabolism of green plants that an exceptionally dry season may mean death to a great many, or if their seeds fall on the wrong kind of soil or in too shady a spot they may perish. The flora of a region is therefore very largely an indication of its physical conditions, and by the flora the fauna may be definitely influenced. Arboreal animals are not found in extensive prairies, for example, nor grazing species in dense forests. Animals may also respond directly to physical conditions; to this relationship is due the fact that amphibia cannot live in deserts nor in the far north.

Environmental change may also be due to the shifting population, as mentioned in Chapter XIII, but whatever its cause, organisms in contact with it must meet the changed conditions or die.

Duration of Environment. When biologists have experimented with environmental effects on organisms in the past they have had to be contented with a study covering relatively few generations and the characters available for study have been relatively unimportant. In long phylogenetic series we find that the gradual development of a species is correlated with a lasting climate, or with a climate which is gradually changing in a way favorable to the existence of the changing organism. The evolution of species has not occurred in a few years nor under frequently changing conditions.

Whether incidental changes or direct adaptive responses in organisms have been the initial step in the evolution of species we cannot say. This is the essence of the conflict between the views of Darwin and Lamarck. If changes have appeared in organisms, giving rise to an assemblage of variations, changing environment may emphasize their relative values in such a way as to cause natural selection. On the other hand it is obvious that some changes come about as direct adaptive response to a new condition. Organisms have unlimited opportunity for contact with new environmental stimuli through migration, climatic change, and fluctuations in the organic environment, hence there is always a chance for the occurrence of new characters of both kinds.

Are Any Characters Permanent? Since all changes in the organism are the product of hereditary genes responding to stimuli from some part of the complex environment, the question arises, how do they differ in permanence? If characters appear in response to conditions of the internal environment, then obviously their reappearance in each generation will be reasonably certain. However even the internal environment is not permanent; it is subject to modification both from within and from without. If the proper condition in the body fails to develop, even apparently hereditary characters fail to appear. A cretin may have a normal mental heritage, but the genes responsible for mental development do not find the proper conditions in its body for their expression. The deficiency may be corrected by the timely administration of thyroid extract and the heritage then responds normally.

Conversely an acquired character or adaptive response will appear in every generation if the proper stimulus is available. As long as the same conditions obtain, the character will apparently be inherited because it will appear in all individuals. It appears during the independent life of the individual, and its appearance is directly associated with an external condition, but these things are no less true of many of our inherent structures. Such a character is therefore fundamentally similar to inherited characters, save only that the heritage in no case provides the stimulus for its development.

Permanence of a character is due to permanence of the contributing factors, heritage and environment. Every character has a basis in the heritage which will be handed down from generation to generation as a part of the normal functions of the living substance, but to the same degree every character depends for its appearance upon the presence of the proper condition in the environment for the normal response of this heritage. How then can a change come about in the heritage?

The Process of Evolution. In answer to this question two interpretations may be based on the facts thus far set down:

1. The power of an organism to develop any character is a part of its heritage. Whatever the nature of the character, it is a product of genes derived from the germ cells of the previous generation and some outer stimulus. Hence even so-called acquired characters may be regarded as an expression of hereditary qualities resident in the germ plasm from which the individual arose. If an existing individual acquires a character, then the germ cells contained genes with the power to produce such a character under the proper stimulus. The parents which produced the germ cells therefore had genes with the same power, and so, step by step, the idea can be carried in either direction through successive generations, even back to ancestral species.

This interpretation, carried to its logical extreme, would include the idea that even the most primitive living matter possessed the inherent ability to produce any of the characters of existing animals in response to the proper stimuli. To a certain extent this must be as true as evolution, but it would require great credulity to believe that an amoeba-like ancestor could develop the characters of man in a generation if given the proper surroundings. The view becomes acceptable only through recognition of the gradual modification of the factors involved, which leads to the second interpretation.

2. It is infinitely more logical to suppose that the heritage in any living organism is subject to change. It has a certain range of power which enables it to respond within limits to different stimuli, and when it has responded in one way it seems probable that its future behaviour and development may be modified as a result. An individual may enter one field of activity after being trained for another, but his earlier training can hardly fail to have an effect on his later work.

This interpretation necessitates the wholly logical belief that each change in an organism through response of its heritage to conditions emanating from any part of the environment may make possible other changes. It recognizes that departure from the one-celled state for colonial or multicellular organization makes possible other steps which the single-celled organism could not directly attain, and that the attainment of triploblastic structure makes possible other specializations impossible to diploblasts. The primitive horses must have had a third digit strong enough to support the weight of the body before the others could be lifted from the ground, but once the latter ceased to bear weight, they would have had a much better chance of reduction than before. In all of these cases it is evident that the characters deal both with the heritage and with environmental conditions.

In spite of the logical aspects of this interpretation we must still meet the old distinction between heritable and acquired characters. The heritage is always expressed through the soma, although it is perpetuated by the germ plasm. This is true of all characters, even those which are admittedly inherited; nevertheless, if so-called acquired characters have any part in evolution, there must be a point of transition where the heritage ceases to find a stimulus from the external environment necessary for its expression and responds to a condition which is normally certain to be supplied by the developing body itself.

Use and Disuse in the Chromosomes. The principle of use and disuse recognizable in organic behaviour has been applied to the chromosomes and their genes as an explanation of such cumulative modification of the heritage through the generations. It is now regarded as probable that the action of a given gene is not limited to the cell in which it lies, but that the genes of a given kind throughout the body may, through the coördination of the individual organism, exert a controlling influence over the part which is capable of expressing them. If such is the case, the spermatogonia and oögonia are no less subject to the coördinating influences. Any increase or decrease in the functional capacity of a gene would therefore be characteristic of the genes in these germ cells as well as of those in the somatic cells expressing the character. The transmission of the change to succeeding generations would be assured. As time went on the effect of continued use might readily bring about a development of the genetic function and an accessory development of associated parts of the internal environment so great that the original stimulus would be a minor matter and its cessation could not be followed by immediate loss of the character whose appearance it had first caused. This theory is in harmony with the known facts, but it remains untried and is therefore only another possible explanation of evolutionary processes.

Evolution and the Internal Environment. Regardless of the association of germinal and somatic genes, we must recognize that the appearance of a new character in an organism, whatever its cause, makes a change in the internal environment. The new character can develop only if it receives the proper support of the remainder of the organism. Consequently if it becomes large or important, its new relationship to the internal environment may become so fundamental that it will be of greater importance in the expression of the character than the stimulus which originally favored its development.

In this secondary adjustment it is evident that structures sometimes take on additional functions which make their persistence necessary to the organism long after the loss of the original use. Man has no further use for pharyngeal pouches as a part of his respiratory system but the first pair have developed into his middle ear and the remainder are associated with the formation of a series of endocrine glands. They persist to the extent necessary for the production of these structures.

Experimental Evolution. The foregoing explanations are obviously not an adequate solution of the process of evolution, nor are they intended to be. They show, however, that it is possible to go much further in a logical evaluation of the factors in evolution than has been done in the formulation of the famous theories of the past. In recognition of the universal importance of the association of heritage and environment in all aspects of life we approach the basis for sound future investigation.

Science cannot be wholly satisfied until its ideas have been demonstrated in the laboratory. If evolution is to reach this happy state of proof, it will obviously not be due to the methods of the past. At present the possibility of finding methods of experimental proof seems remote, for with the facts described in this chapter in mind it is evident that certain conditions must be met.

It would be foolish to expect, as has been done in the past, that the induced change might persist after the removal of the stimulus at the end of a few generations. Even though the heritable power of the organism to respond to a given condition might have changed this could not be expected, for the development of the new response would not necessarily destroy the organism's capacity for the old. Both old and new responses may be looked upon as the positive results of different factors.

If, however, a change can be induced and intensified by gradual modification of the stimulus through many generations until a result of considerable importance in individual life is attained, we shall have an "acquired character" worthy of a test. And if the character in question should develop accessory relationships of value within the organism it will be even better. Under these conditions we may expect the removal of the stimulus, if the organism can still live under the old conditions, to be accomplished without the loss of the character. Even under such conditions we would have to recognize that the change proceeded not from the heritage nor from the environment alone, but from the two acting together. And if such an experiment should fail, we must then look for some unknown property in living matter which makes its heritage in some degree independent of environmental conditions.

It is highly probable in the light of modern knowledge that the explanation of evolution will be based upon the recognized facts of biology. Living things as we see them now are exceedingly varied and complex, but if we could look back over the complete record of their development we should probably find a tale of gradually increasing complexity as each successive stage realized the possibilities of its heritage and thereby made possible still other steps in evolution. Modern thought on this increase of capacity has crystallized under the name of emergent evolution. The name is scarcely necessary for the belief is no more than a logical statement of things long held true. The rôle of sentient powers and inherent directive forces has also gained a prominent place in evolutionary thought, but here we are dangerously near to passing out of the transitional zone into purely philosophical speculation.

Although discouraging it is not surprising that the method of evolution should be so elusive, for it is the most complex and baffling field of research. The genes with which the problem is so intimately concerned are exceedingly small. They have never been seen. Even the chromosomes must be subjected to extensive treatment before they can be examined. Physicists may do as they will with the atom and electron; these units are tractable in comparison with genes, for methods can be devised for studying them without destroying them. Living matter presents more difficulties. When chromosomes are made visible they are dead and the gene remains a hypothetical unit in their densely stained substance.

In spite of these limitations the fact of evolution remains an established principle of biology through all of the investigations and disputes concerning its methods, and, knowing this to be so, we may feel confident that we shall some day solve the problems which are before us today.

Summary. Considering all of the proposed theoretical explanations of evolutionary method we find an almost universal lack of explanation of the origin of variations. Darwin's theory of natural selection and the mutation theory together serve to explain adaptation but must begin with variation. Lamarck's theory offers the idea of interaction of heritage and environment but fails to account for the extension of its results beyond the individual. An adequate theory of evolution must do both things, i.e., it must account for changes and explain how they become the characters of species. A logical analysis of the available facts shows that the organism and all significant organic characters are a product of the heritage reacting to the complex environment, and variations in the latter seem sufficient to account for variations in the resulting living organisms. This view is essentially the tetrakinetic theory of Osborn. We are able to see variation as a process only in the individual, however, and it remains for us to explain how these changes become a part of the species. That they are of the heritage is easily shown. Use and disuse of functional capacity in the hereditary bodies, the genes, has been proposed as an explanation of changes in the functional capacity of these minute bodies. This may account for the gradual development of characters through a long succession of generations under continued or progressive environmental conditions, and when characters have attained great development in response to conditions of the external environment they may well have attained such importance in the internal environment that their persistence will be necessary. Future work in evolution must take into account these things. The difficulties of experimental work are great, but the fact of evolution is so well established that we may expect confidently to solve the riddles of evolutionary method in the future.

REFERENCES

MORGAN, T. H., *Evolution and Adaptation*, 1903.
THOMSON, J. A., *Heredity*, 1909.
OSBORN, H. F., *The Origin and Evolution of Life*, 1918.
MORGAN, C. L., *Emergent Evolution*, 1922.
ELDRIDGE, S., *The Organization of Life*, 1925.
KEPNER, W. A., *Animals Looking into the Future*, 1925.
NOBLE, E., *Purposive Evolution*, 1926.
LINDSEY, A. W., "Factors in Phylogenetic Development," *American Naturalist* XLI, 251–265, 1927.
JENNINGS, H. S., "Diverse Doctrines of Evolution," etc., *Science* LXV, 19–25, 1927.
WASHBURN, M. F., "Purposive Action," *Science* LXVII, 24–28, 1928.

INDEX

Italics indicate pages bearing figures or diagrams.

Abyssal realm, 215; adaptations to, 216; conditions in, 215; fishes of, *216*
Accidental destruction, 392
Acipenser, bony scutes, *69;* pelvic girdle, *78*
Acquired characters, 7, 10, 386; inheritance of, 414, 415, 416
Adaptation, basis of, 247; and heredity, 264; process and result, 210; and external environment, 212; and internal environment, 213; transition from individual to species, 262
Adaptations, abyssal, 215; ambulatory, 217; of anteaters, 232; aquatic, 214; to aridity, 224, *226;* of benthos, 214; of birds, 225, *235;* of carnivorous animals, 233; cursorial, 218; flight, 225; for securing food, 232; fossorial, 221; for gliding, 228; for grazing, 232; to light, 223; of loon, 213; of nekton, 215; to organic environment, 230; to physical environment, 213; of plankton, 215; protective, 234; saltatory, 219; scansorial, 222
Adaptive branching, 252
Adaptive radiation, 27, 252
Adhesive organs, 223
Africa, fossil man in, 199
Agave, *226*
Aggressive colors, 242
Albinos, 432
Algonkian, 124
Alhambra plum, 350
Allantois, *45;* human, *46*
Allelomorphs, 279; multiple, 314
Allosomes, 299
Alluring colors, 242
Alps, 120
Alternation of generations, 23
Altitude, effect on climate, 248
Amber, 127; fossils, *127*

Ambulatory adaptations, 217
Ambulatory animals, 253
Amitosis, 91
Ammonites, 133
Amnion, *45;* human, *46*
Amphibamus, 151
Amphibia, 37, 130; in vertebrate evolution, 150; of Carboniferous, *151*
Amphioxus, 145; cytoplasmic differentiation of egg, *302*
Amphioxus theory of vertebrate descent, 144, 146
Anabolism, 30
Analogies, significance of, 111
Analogy, 27
Anaximander, 5, 6
Ancestral inheritance, law of, 271
Anchitherium, 180, 183
Ancon sheep, 404
Andalusian fowls, 281
Anemones, sea, *214*
Angraecum sesquipedale, 393
Animal associations, see Associations
Animal hybrids, 345
Annelid theory of vertebrate descent, 144, 146
Annelida, 28, 137; artificial parthenogenesis in, 330
Anopheles, head and mouth parts, *234*
Anteaters, *233;* adaptations of, 232
Antennae of mosquito, *234*
Anthropoidea, 188
Ants, and aphids, 231
Aorta, 62
Aortic arches, diagram, *64;* of human embryo, *49*
Apes, half-, 188; man-like, 189; New World, 188; Old World, 188
Aphids, 231; and pure lines, 356; life cycle of, 329, *330*
Appalachian mountains, 120, 258
Appendages, 35

445

446 INDEX

Appendix, human, 88, 398
Apterygota, 136
Aquinas, Thomas, 8
Arachnida, 135
Arboreal origin of man, 191
Arboreal quadrupeds, 191
Archaeopteryx lithographica, *155*, 156
Archenteron, 42
Archeozoic, 124; fossils of, 129
Arches, branchial, see Branchial arches
"Arcturus Adventure," 260
Argonauta, 132
Aridity, adaptations to, in animals, 224; in plants, 224
Arisaema triphyllum, sex reversal in, 336; Siamese twins, *337*
Aristotle, 5, 7
Armadillo, *236*
Armor, of alligator, 236; of armadillo, *236;* of *Stegosaurus*, 235; of *Triceratops*, 235, *236*
Arthropod theory of vertebrate origin, 144
Arthropoda, 28; adaptations of legs, 218; ancestors, 134; aquatic, 135; burrowing, 221; terrestrial, 136
Artiodactyla, 164
Ascaris, germinal continuity, 306, 422
Asexual propagation, 350
Asphalt, 124
Asses, 164, 183, 345
Associations, animal, gregariousness, 237; communal, 239; commensal, 240; symbiosis, 240; parasitism, 240
Asterias forbesii, 265
Auchenia, skull, feet, and teeth, *185*
Augustine, 8
Australian region, 255, *256*
Australopithecus africanus, 201
Autosomes, 299
Autotomy, 237
Aves, see Birds
Axolotl, 95, 211

Babcock and Clausen, 275, 317, 342, 345, 347, 350
Back-cross, *285*
Bacon, 8
Baptanodon, *153*
Barnacles, 214
Barriers, 257

Basques, 207
Bateson, 264, 269, 319
Bats, *227*, 253
Beagle, voyage of, 14
Beaks, of carnivorous birds, 233, *235;* of seed-eating birds, 234, *235;* of insectivorous birds, 234, *235*
Bear animalcules, 137
Bears, polar, 260
Bechuanaland, 199
Beebe, Wm., 2, 14, 216, 244, 254, 260, 394, 409
Bee moth, 347
Bees, evolution in, 397; and fruit, 231
Beetle, Japanese, 251
Beetles, jumping adaptations, 220; running leg of tiger beetle, *218*; swimming leg of gyrinid, *218*; Schroder's experiments with, 420
Belemnites, 133
Belostomatidae, 254
Benedict, a Clydesdale stallion, *352*
Benthos, 214; aërial, 217
Beresovka mammoth, *126*
Bering isthmus, 258, *259*
Biology, 11
Biometry, 15, 275
Bipedality, 219
Birds, 37; adaptations for securing food, 233, *235;* digestion, 227; feathers, 225; flight adaptations, 225; metabolism, 228; origin of, 154; respiration, 228; secondary sexual characters, 334; sex reversal in, 327; skeleton, 225; tail and wing functions, 225; temperature, 228; and warning colors, 242; mocking, 409
Birth rate, 377; decrease in, 377; of the unfit, 378; of mentally normal, 378; and immigration, 380
Bison, 347
Blastocoele, *40*
Blastula, *40, 41*
Blending inheritance, 318
Blood, 113
Blood tests, 95
Blue Andalusian fowl, 281
Body cavity, 35
Boveri, 301
Brachiation, 193
Brachiopoda, 28, 132
Brachiosaurus, *153*

INDEX

Brain, embryonic, 48; of primates, 188; of vertebrates, *86*
Branchial arches, 54
Branchiostoma, 145
Bridges, 314
British Guiana, tinamou, 244
Bryophyta, 28
Bryozoa, 28, 132
Buffon, 9, 413
Bugs, 253
Bull, Ephraim Wales, 344
Bumble bees, and red clover, 230
Burbank, 350
Bursa, seed capsule shape, 317

Cacops, 151
Caenogenesis, 211
Callosities, 417
Cambrian, 123; coelenterates, 130; mollusca, 133; sponges, 130; trilobites, 134
Camels, 184; phylogeny, *185*
Cannabis sativa, 336
Carboniferous, 123; amphibia, *151*
Carnegie, 380
Castes, neuter, evolution of, 397
Castle, W. E., 318, 421
Casts, *125*, 126
Cat, skull, *391*
Catarrhini, 188
Caterpillar, 211; mimicry of twig, *244*
Catocala ilia, *243*
Cattle, 347; selection in, 351; polled Hereford, 404
Cave drawings, 1, *206*, 207
Cebidae, 188
Cells, *30*, *31*; importance in genetics, 272.
Cellulose, digestion in termites, 240
Cenozoic, 122
Cephalopoda, 37, 132; concealing discharges, 217, 237
Cercopithecidae, 188
Cereus giganteus, *226*
Chamaeleon, 223
Change, of habits, 392; source of, 433; adaptive and incidental, 434
Characters, acquired, inheritance of, 386, 414, 415; evidence of, 417
Characters, organic, source of, 431; permanence of, 437; useful and harmful, 389; importance to individual, 387

Chat, yellow-breasted, 394
Child, 422
Chilopoda, 136
Chimpanzee, *189*
Chlorophyll, 91
Choloepus didactylus, *223*
Chondrocranium, 54, 68; of elasmobranch, *58*
Chordata, 28, 37
Chorion, *45*; human, *46*
Chromomeres, 291
Chromosomes, characteristics of, 292; organization of, 291; in mitosis, *100*; in gametogenesis, 292, *293*; abnormal behaviour, 299; non-disjunction, *300*; multiplication, 299; in crossing over, *309*; and infertility, 349; and Mendelian inheritance, 303; in gynandromorphs, 333; haploid and diploid number, 292; and sex, 297, *298*; x, y and z, 298; of aphids, 329, *330*; of *Ascaris*, *295*; of *Drosophila*, *291*; of honeybee, 329; of man, 310, 361
Chromosome theory, 289; evidence for, 300
Cicindelidae, cursorial adaptations of, 218
Circulation of human embryo, *48*
Circulatory system, of fishes, 59; of dogfish, *61*; of terrestrial vertebrates, 61
Class, 26
Classification, 19, 25; of animals and plants, 28
Clavicle, of *Sauripterus*, *76*; of terrestrial vertebrates, *77*; of primitive reptile, *79*; of mammals, *80*; of primates, 188
Claws, *85*; of anteaters, 232; of flesh-eating birds, 233, *235*; as weapons, 237
Cleavage, 39; of frog's egg, *41*; of pigeon's egg, *41*
Climate, changes of, 248, *249*; and food supply, 250
Climbing, adaptations for, 222
Clones, 355
Cocoanuts, dispersal by water, 260
Codling moth, *21*
Coelenterata, 28, 130, 215
Coelom, 35
Colonial animals, 23

Coloration, 242
Color-blindness, a sex-linked character, 364; inheritance of, *365*
Columbian elephant, 398, *399*
Comanchean, 122
Combinations, 268
Commensalism, 240
Communal associations, in insects, 239; in man, 240
Comparative anatomy, 3, 68
Competition, 6; and natural selection, 388; and useful characters, 389
Complementary factors, *319*
Concealing discharges, 237
Concealment, 230
Concord grape, 344
Conductivity, 90
Consciousness, 1, 2
Continuity of germ plasm, 422
Continuous variations, 267
Contractility, 90
Convergence, 27
Corals, 214
Coral snake, 242
Corixidae, 254
Corn, origin of, 271; and Mendelian inheritance, 286, 287; hybridization and selection, 342; crosses of, 344; heterosis in, *343;* sex reversal in, 336
Correns, 276
Corythosaurus, 153
Craniata, 144
Crayfish, *134*
Cretaceous, 122; Bryozoa of, 132
Cretaceous Eocene, change of climate during, 248
Cretinism, 95
Crew, 335
Cricotus, 151
Crinoidea, *131*
Crocodilia, 237
Crô-Magnon race, 206; sculpture, *206*, 207; implements, *207;* mental development, 208
Cross fertilization, 231
Crossing over, mechanism of, *309;* multiple, 309, *310;* percentage of, 311, *312;* in Drosophila, *307*, 308
Crossopterygii, fins of, *76*, 149
Crows, 239
Cryptobranchus allegheniensis, skull, *58*

Cryptopsaras coursii, 216
Crypturus variegatus, 244
Ctenophora, 28
Cumulative factors, *318*
Cursorial adaptations, 218
Cuttlefishes, 132
Cuvier, 11, 413
Cyclostomata, 37
Cydninae, 254
Cynodontia, 157
Cytoplasm, 32; rôle in inheritance, 301; differentiation in eggs, 302

Dactylethra, pelvic girdle, *78*
Daphnids, 356
Dark Ages, 8
Dart, Raymond, 199
Darwin, Charles, 9, 12, *13*, 230, 250, 269, 369, 385, 413
Darwin, Erasmus, 9, 10, 370, 413
Darwinian theory, 385, 418, 428
Darwinism, see Darwinian theory
Davenport, C. B., 362
Deaf-mutism, inheritance of, *367*
Debris, floating, as aid to dispersal, 258
Deep-sea fishes, *216*
Deer, protective adaptation, 236; signal marks, 243
Defectives, inbreeding of, 367; segregation of, 382; sterilization of, 382
Defects, inheritance of, in man, 366
Democritus, 5
Dentalium, 302, cytoplasmic differentiation of egg, *302*
Dentition, heterodont, 159; of elephant, 168; of horse, 175, *177*
Dermal bones, 55
Desert, Great Salt Lake, 255
Detlefsen, 418, 420
Devonian, 123, 150; amphibia of, 130; trilobite of, 134
De Vries, Hugo, 12, 276, 402
Dibelodon, 170
Differential birth rate, 377; causes of, 379
Dihybrid, 279; ratios, 281, *282*
Dinosaurs, *153*, 219
Dinotherium, 169
Diploblasts, 34
Diplocaulus, 151
Diplopoda, 136
Dipnoi, 149

INDEX 449

Discontinuous variation, 267
Dispersal, of mammals, 160; aids to, 258; barriers to, 257
Distribution, geographical, see Geographical distribution.
Disuse, see Use and disuse
Divergence, 27
Division of labor, 23
Dogs, ancestry of, 271
Dominance, 280
Dordogne, 207
Draco volans, 230
Drosophila, 419, 428; chromosomes of, *291;* mutation and normal eye, *269;* and Mendelian inheritance, 287; multiple allelomorphs, 314; linkage and crossing over, *307*, 308; multiple crossing over, 309; sex-linkage, 312, *313*, *315;* non-disjunction, 300; reverse mutations, 402; chromosome maps, *311;* and pure lines, 356
Drosophila melanogaster, see *Drosophila*.
Dryopithecus fontani, 198
Dubois, Eugen, 198
Duck mole, shoulder girdle of, *80*
Dugdale, 368

Earthworm, *36*
Echidna, claw of, *85*
Echinodermata, 28, 132
Ectoderm, *40, 41*
Edwards, Jonathan, 370
Ehrlich, 95
Elasmosaurus, *153*
Electrical organs, 237
Elephants, Indian, *165;* African, *165;* foot, *166;* skull, *167;* tooth, *168;* size of, 166; limbs, 166; trunk, 166; tusks, *167;* teeth, 168; Columbian, 398, *399;* evolution of head and teeth, *174;* phylogeny of, 169
Elephas, 172; *antiquus*, 173; *columbi*, 173, *399; imperator*, 173; *meridionalis*, 173; *primigenius*, *126*, 173
Embryology, 39
Embryos, *52*, *53*
Empedocles, 5, 6
Endocrine glands, 94
Endoderm, *40, 41*
English sparrow, 251
Enteron, *40*

Entheus peleus, 22
Environment, 2, 211; modification of, 248; availability of, 252; response to, 251; continuity of, in evolution, 436; duration of, 437; of fossils, 127; effects on soma and germ plasm, 419; and existence, 430, 431; and sex determination, 338; and eugenics, 383; and human life, 372
Eoanthropus dawsoni, 202; phylogeny, 204; skull, *203*
Eocene, 122; elephants, 168; horses, 176, *178*, 179; camels, 185
Eohippus, 176, *178*, 183; feet and teeth, *179*
Ephemera varia, *138*
Epihippus, 179, 183
Equation division, 294
Equus, 182, 183
Eryops, *151*
Estabrook, 368
Ethiopian, 255, *256*
Eubleptus danielsi, *140*
Eugenics, 374; goal of, 375; limitations, 377; and environment, 383; practical aspects, 383
Eugenics Record Office, 359
Euglena, 93
Eutheria, 103, 159
Evidence, for evolution, 105, 119; for inheritance of acquired characters, 417, 419
Evolution, 3, 4; history of, 5; emergent, 442; purposive, 17; experimental, 440; nature of progress in, 435; latitude of heritage in, 435; and environment, 436, 440; of Mollusca, 132; of Arthropoda, 134, *137;* of insects, *142;* of vertebrates, 144; of elephants, 164, *174;* of horses, 175, *183;* of camels, *185;* of man 187
Evolutionists, 9
Excretory system, 65; in chick embryos, *65*
Existence, factors in, 430
Exoskeleton, 82
Eye, development of, 49, *50;* evolution of, 395; of primitive insect, *396;* of squid, *112;* of vertebrate, *113;* loss in burrowing animals, 224
Eyelid, third, 88

Factors, and unit characters, 306; complementary, *319;* cumulative, 316, *318;* duplicate, 316, *317;* lethal, 320, *322;* multiple, 316; supplementary, 319; in existence, 430
Family, 26
Faserstrang, 147
Feathers, 83
Felis, skull, 391
Ferns, *24*
Fertilization, 296
Filial regression, law of, 271, *272*
Fin, pectoral, of dogfish, *75;* of *Sauripterus*, *76;* transitional, 76
Fishes, 37; as vertebrate ancestors, 148; in nekton, 215
Fission, 100
Flea, amber fossil, *127;* jumping adaptations, 220
Flies, adhesive organs of, 223
Flight, theories of origin, 154; adaptations of birds, 154, 225; in insects, 227, *229;* in vertebrates, 227-229
Flippers, 111
Fly, house, 231
Flying dragon, 229, *230*
Flying frog, 229
Flying lemur, 229
Flying squirrel, 229
Foetal membranes, *45*, 102; human, *46*
Food securing, adaptations, 232; of grazing species, 232; of anteaters, 232; of carnivorous animals, 233
Food supply and climate, 250
Forelegs, skeletons of, *81*
Fortuitous variations, 268
Fossils, formation of, 124, *125;* interpretation of, 127; succession of, 129
Fossorial adaptations, 221
Fossorial bugs, 254
Four o'clocks, *281*
Fowls, Andalusian, 281; Sebright bantams, 334; effects of removal of gonads, 334; sex reversal in, 335, *336*
Fox, arctic, 242
Freemartin, 335
Frog, development of egg, *41;* flying, 229; sex reversal in, 327; spermatogonium, *290;* spinal ganglion cell, *290*
Functions, variations in, 265

Galápagos islands, 258
"Galápagos, World s End," 14, 409
Galen, 8
Galeopithecus, 229
Galton 16, 271, 369
Galton s laws, 271
Gametes, 101; individuality of, 328; of *Sphaerella*, *326*, 327; see Germ cells
Gametogenesis, 292, *293*
Gametophyte, *24*
Gastrostomus bairdii, *216*
Gastrula, *40*, 41
Gastrulation, 42; in frog, *41;* in pigeon, *43*
Gelastocoridae, 254
Gemmules, 423
Gene, theory of, see Chromosome theory
Genes, 305; localization of, 310; in ontogeny, 322; rôle of, 432
Genetics, 4, 17, 262; foundations of, 264; practical value, 341
"Genetics in Relation to Agriculture," 275
Genotype, 279
Genotypic selection, 353
Genus, 26
Geographical distribution, 254; realms, 255, *256;* barriers, 257; aids to dispersal, 258
Geological table, 122
Geological time, 120; divisions of, 122
Geology, 119
Germ cells, development, 292, *293;* human, *296;* see Gametes
Germ plasm, 305; continuity of, 422
Germinal continuity, 422
Germinal selection, 406
Germ layers, 32, *40*, *41*
Gerridae, 254
Gibbon, 189
Gila monster, 242
Gill slits, 37
Gipsy moth, 251
Girdles, diagram, *77;* pelvic, *78;* pectoral, *79*, *80*
Glands, endocrine, 94; and internal environment, 431
Gliding, adaptations for, 228
Gnathostomata, 147
Goddard, 376

INDEX   451

Goldschmidt, 323
Gonads, effects of removal, 334
Goniatites, 133
Gorilla, 189, *190;* skeleton, *192*
Grasshopper, 220, 243; jumping leg of, *218*
Great Salt Lake desert, 255
Greek philosophers, 5
Gregariousness, 238
Gregory, W. K., 188
Gregory of Nyssa, 8
Guinea-pigs, inheritance in, 282, *283;* ovarian grafts in, 421
Guyer, 382, 421
Gynandromorphs, 332, *333*

Hair, and scales, 83, *84;* persistent, 87; form of, 267
Halobates, 254
Handlirsch, 136, 140
Hapalidae, 188
Hare, prairie, 242
Harmful characters, 389
Harvey, 8
Heart, of vertebrates, 61; diagram, *62;* human, development of, *63*
Hefner, 364
Heidelberg man, 202; jaw of, *202*
Height, inheritance of, 271
Hellgramites, 211
Helm, W. B., 266
Hemimetabola, 136
Hemiptera, adaptive branching in, 253
Hemp, sex reversal in, 336
Herbert, 409
Hereditary bridge, 274
"Hereditary Genius," 369
Heredity, 15, 264; chromosome theory of, 280; in man, 359
Heritability, importance in genetics, 270
Heritage, 2; latitude of, in evolution, 435; and existence, 436; subject to change, 439
Hermaphrodites, 332
Hermit crab, 240, *241*
Hesperopithecus haroldcooki, 201
Heteroceras, 133
Heterosis, in plants, 342; in animals, 348
Heterozygous, definition of, 279
Himalayan mountains, 257
History of evolution, 5

Hipparion, 181, 183
Hippidion, 182, 183
Hoatzin, 157; young, climbing, *156*
Holmes, 368, 369
Holometabola, 136
Hominidae, 188
Homo, see man; *heidelbergensis,* 202; jaw, *202;* phylogeny 204; *neanderthalensis,* 202, 204; skull and skeleton, *205;* phylogeny, 204; restoration, 200; *sapiens,* 26; *sapiens,* Crô-Magnon race, 206; relation to Neanderthal man, 206; restoration, 200; phylogeny, 204
Homologies, significance of, 111
Homology, 27
Homozygous, definition of, 279
Homozygous crosses, 356
Honey-bee, *21;* sting, *238;* organization of colony, 239; hybridization in, 347; selection in, 351
Hoofs, *85*
Hooke, 8
Hormones, 94; and secondary sexual characters, 314, 334; effects of insufficiency, 334
Hornbills, 396
Horns, as weapons, 237
Horse, adaptive structure, 175; teeth, *177;* skull, *177;* evolution of, 176; protective adaptations, 236; orthogenetic variation in, 268; selection in, 351, *352;* South American horses, 182; North American horses, 182; extinction of, 184; and mule, 345
Humming bird, ruby-throat, 394
Huxley, 15
Hybridization, 274; practical value of, 341; of domestic animals, 345; limitations of, 348
Hybrids, heterosis in, 342; asexual propagation of, 350; infertility of, 349
Hydra, 34, 51, *55;* asexual reproduction in, 273, *274;* clones in, 356; and continuity of germ plasm, 422; variation in, 265
Hylobates, 189
Hyoid arch, 54, 75
Hypohippus, 180, 183
Hypopharynx, of *Anopheles, 234*
Hyracotherium, 176, 183

Ice floes, 258
Ichthyosaur, *153*
Idiacanthus ferox, 216
Idioplasm, 291
Immigration, 380
Immune sera, 95
Implements, 1
Inbreeding, in man, 367
Induction, parallel, 419; in insects, 420; somatic, 419
Infertility, interspecific, 348; of hybrids, 349
Inheritance, of acquired characters, 414, 415; acceptance of, 416; circumstantial evidence for, 417; alternative, 280; blending, 280; mosaic, 280; in four o'clocks, *281;* in Andalusian fowls, 281; in Guinea-pigs, 282, *283;* in peas, 278; of more than three characters, 284; of eye color, 361; of hair color, 361; of skin color, 362
Insects, 136; adaptations for securing food, 233; adaptive branching in, 253; diversity of, 138; metamorphosis of, 136; mouth parts, 138, *139;* mimicry, 245; Paleozoic, *140;* Schröder's, 420; wings, 139
Instincts, and natural selection, 396
Insulin, 95
Intraselection, 408
Irises, 345
Irish Elk, 389
Irritability, 90
Isolation, biological, 409; physical, 409; theory of, 409; spatial, 387

Jack-in-the-pulpit, sex reversal in, 336; Siamese twins, *337*
Japanese beetle, 251
Java, 198
Java ape-man, 199
Jaws, 73
Jelly-fishes, *213*, 214
Johannsen, 354
Jones, F. W., 193, 195, 197
Jukes family, 368
Jumping, 219; value of, 220; in insects, 220; in vertebrates, 219
Jurassic, 122; squids, 133
Just (and Lillie), 301

Kallikak Family, 370
Kangaroo, 219

Kant, 8
Katabolism, 30
Kea parrot, 267
Kellicott, 367
Kiang, 183
Kidneys, 66
Kingdom, 26
Kingsley, 75
Krapina, 203

Labium, of *Anopheles*, 234
Labrum epipharynx of *Anopheles, 234*
Lamarck, 9, 10, *11*, 115, 385; and variation, 269
Lamarck's laws, 11, 414
Lamarckian theory, 413; errors of, 428; value of, 424
Lance, 122
Leeuwenhoek, 8
Legs, of insects, *218*, 233
Lemur, flying, 229
Lemuroidea, 188
Lethal factors, 320, *322*
Lichens, symbiosis, 240
Life, 2, 3
Light, adaptations to, in plants, 223; in animals, 224; pigmentation, 224; loss of eyes, 224
Lillie, 323, 335
Lillie and Just, 301
Limb buds, 47
Limbs, of elephants, 166
Limestone, 120
Line selection, 353
Linkage, 285, 306; effect on dihybrid ratios, 286; and sex, 286; in *Drosophila*, *307;* in man, 310; in other organisms, 310
Linnaeus, 9, 388
Linophryne lucifer, 216
Llama, 184; skull, feet, teeth, *185*
Lobster, *36*
Locomotion, in terrestrial organisms, 217
Locy, 10
Loon, adaptations of, 213
Lou Dillon, *352*
Loxodonta, 173
Lull, 122, 127, 133, 150, 160, 164, 166, 168, 176, 217, 257, 388
Luminous organs, 216
Lung fishes, 130
Lush, 347

INDEX 453

MacFarland, E. M., 226
Maggot, 136
Malpighi, 8
Malthus, 10, 14, 250, 385
Mammae, of primates, 188; supernumerary, in man, 87
Mammalia, 37
Mammals, ancestors of, 157; adaptations of, 159; adaptive radiation in, 253; classification of, 159; dispersal of, 160; evolution of, 157; secondary sexual characters of, 334
Man, 26; systematic position, 187; structural plan, 187; arboreal origin, 191; structural adaptations, 196; skeleton, *192;* evolution of, 187; fossil remains, 198, 202; significance of geological record, 201; climatic factors in evolution, 201; phylogeny, *204;* ancestral species, restored, *200;* course of evolution, 208; loss of hair, 197; heredity in, 359; heritable characters, 360; unit characters, 361; chromosomes, 361; symphalangism, *363;* polydactyly, *266,* 363; sex-linkage, 364; inheritance of color blindness, *365;* inheritance of defects, *366;* pedigrees, 366; inheritance of deaf-mutism, *367;* inbreeding, 367; and environment, 372; desirable traits, 374; importance of mental development, 375; Heidelberg, *202;* Neanderthal, *200,* 204, *205;* Piltdown, *203;* Crô-Magnon, *200,* 206
Mandibles, of *Anopheles, 234*
Mangro, 363
Mantis, raptorial foreleg of, *218*
Maple seeds, 260
Marsh, 164
Marsupials, 159
Mass selection, 351
Mastodon, 170, *171;* skull and tooth, *174*
Matthew, 179, 197
Maturation divisions, 294
Maxillae, of *Anopheles, 234*
Maxillary palpi of *Anopheles,* 234
May-fly, *138*
McClung, 308, 333
Meckel's cartilage, 54, 73, 110; of embryo kitten, *60*
Mendel, 16, *17,* 276

Mendelian hybrids, of plants, 342
Mendelian inheritance, 276; and chromosomal behaviour, 303; fundamental principles, 279; monohybrid ratio, 277; dihybrid ratio, 281; practical importance, 287
Mental defects, and eugenics, 376; inheritance of, 370
Mental traits, 370
Merychippus, 181, 183; teeth, *181*
Mesembryanthemum, 226
Mesohippus, 179, 183; teeth, *180;* feet, *180*
Mesonephric tubule, *65*
Mesonephros, 66
Mesozoic, 122; reptiles of, *153*
Metabolism, 30, 90, 114; of animals, 92; of plants, 91; other types, 93
Metameres, 35
Metamorphosis, 24; of insects, 136
Metanephros, 66
Metatheria, 102, 159
Method, of evolution, 16
Methods, of selection, 351
Mice, field, 230
Mice, Weismann's, 419
Migration, 251, 254, 392
Milk, 102
Milkweed seeds, 260
Mimicry, 241, 245; and classification, 26
Mind, brachiation and, 193; environment and, 194
Miocene, 122; apes, 198; camels, 185; climate, 180; elephants, 169; horses, 180
Miohippus, 179, 183
Mirabilis jalapa, 281
Mississippian, 123
Mitochondria, functions in heredity, 303
Mitosis, *100*
Mocking birds, 409
Modifications, 268
Moeritherium, 168; skull and tooth, *174*
Molds, *125,* 126
Mole, 253
Mole cricket, 221; burrowing foreleg of, *218*
Mole, European, *221;* adaptations, 221
Mollusca, 28, 37, 132; in nekton, 215; in plankton, 215

Mongolian race, hair of, 267
Monkeys, and warning colors, 242; prehensile tails, 223
Monohybrid ratio, 277, *279;* modification of, 280, *281*
Monohybrids, of peas, 277
Monotremata, 157
Montanian-Coloradian, 122
Moron, 376
Morphological variation, 265
Moth, gipsy, 251
Moth, geometrid larva, 244
Moth, willow, 420
Mountains, as barriers to dispersal, 257; effect on climate, 248
Mouth parts, of insects, 138, *139*
Mouths of insects, 233; piercing and sucking, *234;* mandibulate, *139*
Mulattos, 362
Mule, 345; fertility of, 349
Muller, H. J., 314, 320, 402, 419
Multiple allelomorphs, 314
Mutants, in cultivated plants, 342
Mutations, cause of, 405; in Drosophila, *269*, 404; and natural selection, 404; and pure lines, 357; in Oenothera, 402; theory, 428; reverse, 402
Mutilations, 386; source of, 432

Nails, *85;* of primates, 188
Narcissus, 345, *346*
Natural history, 5
Natural Selection, 385; answers to objections, 395; examples of, 393; in *Angraecum sesquipedale*, *393;* objections to theory, 395; origin of theory, 385; statement of theory, 385; summary of, 392; underlying principles, 386
Nautilus, 132
Neanderthal man, 204; cranial capacity, 204; mental development, 205; phylogeny, 204; speech, 205; skull and skeleton, *205;* stature and posture, 205
Nearctic, *256;* realm, 255
Nebraska, fossil man in, 201
Necturus, 418; and inheritance of acquired characters, 418; pectoral girdle, 80; pelvic girdle, 77, *78;* sacrum, 73
Negro, hair of, 267; white crosses, 362

Nekton, 215; aërial, 217
Nemathelminthes, 28
Neo-Lamarckian theories, 422
Neolaurentian, 124
Neoteny, 211; in axolotl, 211
Neotropical, 255; *256*
Nepidae, 254
Nesomimus, 409
Neural groove, 44; in chick embryo, *43*
Neural tube, *44*
Neurenteric canal, 42, *44*, 47
Newman, 152, 411
Nilsson-Ehle, 316
Nilsson, Hjalmar, 353
Non-adaptive changes, 247
Non-adaptive characters, 212, 434
North America, climate of, 248
Notholaena, 225
Notochord, 37, 51, *56*, *57*
Notonectidae, 254
Nucleus, 32; rôle in inheritance, 301
Nuttall, 95

Obelia, *20*, 23
Oceans as barriers to dispersal, 258
Ocellus, *110*
Octopus, 132
Octoroon, 363
Oenothera lamarckina, scintillans, oblonga, lata, 402, *403; gigas, nanella, 403*
Oligocene, 122; amber fossils of, *127*; elephants, 169; horses, 179; camels, 185
Ommatidium, *111*
Onohippidion, 182, 183
Ontogeny, 109
Onychophora, 136
Oöcyte, 292
Oögonia, 292
Opthalmochlus duryi, 141
Orang, 189
Orchesella rufescens var. *pallida*, eye, *396*
Order, 26
Ordovician, 123; fishes, 130, 144; annelids, 132; Brachiopoda, 132; vertebrates, 144
Organic environment, 230
Organic world, unity of, 416
Organisms, as agents in dispersal, 261; relationships of, 232

INDEX

Organogeny, 47
Organs, complex, and natural selection, 395; electrical, 237; necessity in production of, 416; vestigial, and natural selection, 397
Oriental region, 255, *256*
"Origin of Species," 12, 15, 212, 385
Orohippus, 179, 183
Orthogenetic variations, 268
Orthoptera, jumping adaptations in, 220
Osborn, H. F., 6, 198, 199, 202, 206, 207, 252
Os coccyx, 87
Ostriches, 219
Ovarian grafts, 421
Overproduction, 250
Overspecialization, 398
Ovum, *31*, 40, 102, 292; of birds, 295; human, *296*
Oyster, rate of reproduction, 234, 388

Paedogenesis, 211
Palaearctic realm, 255, *256*
Palaeomastodon, *169*; skull and tooth, *174*
Palaeopithecus sivalensis, 198
Palaeostraca, 135
Paleolaurentian, 124
Paleontology, 19, 119
Paleozoic, 123; fossils of, 130; echinoderms, 132; insects, *140*
Pan pygmaeus, 189
Pangenesis, 423
Panmixia, 405
Parahippus, 180, 183
Paramecium, 91; binary fission in, 273; and pure lines, 357
Parasitism, 240
Parental care, 234
Parrot, kea, 267; use of beak in climbing, 223
Parthenogenesis, 328; and pure line equivalents, 355
Pass-for-white, 363
Paurometabola, 136
Peas, garden, 277; sweet, 319
Pectoral girdle, 54, 79; of dogfish, *75;* of *Sauripterus*, *76;* of primitive reptile, *79;* of human embryo, *80;* of duck-mole, *80*
Pedigrees, human, 366
Pelvic girdle, 54, 77, *78*

Pennsylvanian, 123
Pentadactyl appendage, *77*, 79; and locomotion, 217
Peonies, 345
Perissodactyla, 164
Permian, 123; reptiles of, 152; change of climate during, 248
Petrogale xanthopus, *220*
Pharyngeal clefts, 37
Phaseolus vulgaris nana, 354
Phenotype, 279
Phenotypic selection, 350
Phillips, 421
"Philosophie Zoologique," 11
Photostomias guernei, *216*
Photosynthesis, 91, 230
Phylum, 26
Physiological variation, 265
Physiology, 90
Piltdown man, 203; skull, *203;* phylogeny, 204
Pisces, 37
Pithecanthropus erectus, 198; skull, *199;* restoration, *200;* relation to man, 199
Pituitary gland, 94
Placoid scales, *82*
Plankton, 215; aërial, 217
Plant-animals, 92
Plants, adaptation to aridity, 224, *226;* dispersal, 260; heterosis in, 342; variation in, 265
Platyhelminthes, 28
Platyrrhini, 188
Plebeius melissa, 409
Pleistocene, change of climate during, 248; glaciation, 202; elephants, 173; horses, 182
Plesiosaur, *153*
Plesippus, 183
Plica semilunaris, 88
Pliny, 8
Pliocene, 122; climate of, 201; Bryozoa, 132; elephants, 172; horses, 182; camels, 185
Pliohippus, 182
Pliopithecus antiquus, 198
Poëbrotherium, skull, feet and teeth, *185*
Poetaz narcissus, *346*
Poeticus narcissus, *346*
Polar bodies, 294
Polyanthus narcissus, *346*

Polydactyly, 265, *266*
Polypterus, ribs, *71;* pelvic girdle, *78*
Pongo, 189
Porifera, 28, 130
Portuguese man-of-war, 23
Prawn, deep-sea, 217, 237
Precipitin tests, 96
Prehensile appendages, for climbing, 222; of primates, 191; for handling objects, 194; effect on evolution of primates, 196
Prenatal influence, 7
Primates, characteristics of, 187; classification of, 188; effects of terrestrial life on, 195
Primitive man, 3
Primordial germ cells, 292
Prjevalsky horse, *352*
Proboscidea, 164
Procamelus, skull, feet and teeth, *185*
Proctodaeum, 48
Pronephric tubule, *65*
Pronephros, 66
Pronuclei, 297
Proserpinaca palustris, leaves of, *390*
Protective adaptations, 234
Protective coloration, 10, 242
Proterozoic, 124; fossils of, 129; change of climate during, 248
Proteus anguinus, *224*
Protohippus, 181, 183
Protoplasm, 29
Prototheria, 102, 157, 159
Protozoa, 28, 32, *33*, 130, *131;* clones in, 356; symbiosis with termites, 240
Protylopus, skull, feet and teeth, *185*
Psychological variations, 267
Psychozoic, 122
Pteridophyta, 28
Pterodactyl, 153, 229, *232*
Pteropus, *227*
Pterosaurs, 230
Pterygoquadrate, 54, 73; of elasmobranch, *58*
Pterygota, 136
Pure lines, *354;* equivalents of, 355; of wheat, *356;* selection in, 356; as limits of selection, 357
Pytonius, *151*

Quadroon, 362
Quaternary, 122
Quinto Porto, *347*

Rabbits, 219; inheritance of color in, *321;* inheritance of ear length in, 318; Guyer and Smith's experiments with, 421
Recapitulation theory, 110
Recessive, 279, in back-cross, 285
Reciprocal cross, 308
Reduction division, 294
Regression, filial, Galton's law of, 355, 401
Relationship, significance of, 105; of individuals, 106; of species, 107; of organism and environment, 261
Reproduction, 90; of cells, 99; of individuals, 101; accessory functions of, 101; in unicellular organisms, 100, 272, *273;* in multicellular organisms, 273; as passive protection, 234; and change of environment, 250
Reptiles, cursorial, 219; in vertebrate evolution, 152; of Mesozoic, *153*
Reptilia, 37
Rhacophorus, 229
Rhamphorhynchus phyllurus, *232*
Ribs, 71, 72
Rickets, 431, 434
Rignano, 424
Rocks, formation of, 119
Rocky Mountains, 120, 257
Roses, 345
Rotatoria, 28, 137; elimination of male in, 329
Roux, 291, 408

Sacrum, 72; human, *73*
Saint-Hilaire, 9, 12
Salamander, blind, *224;* cell from peritoneum, *290*
Saldidae, 254
Saltatory adaptations, 219
Sambo, 363
Sandstone, 120
San Jose Scale, 251
Sauripterus taylori, pectoral girdle and fin, *76*
Scale, San Jose, 251
Scales, 83
Scansorial adaptations, 222; and adaptive radiation, 253
Scaphyrhynchus, pelvic girdle of, *78*
Schaffner, 336
Schistosoma haematobium, *23*

Schoetensack, 202
Schuchert, 120, 248
Scientific methods, development of, 8
Sciuropterus volucella, *231*
Scott, 172, 182
Sea anemone, *214;* and hermit crab, *241*
Sea urchins, artificial parthenogenesis in, 330; cytoplasmic differentiation in ovum of, *302*
Secondary sexual characters, 22, 314, 334
Secretions, repellent, 237
Sedum, *226*
Seeds, dispersal of, 260
Segregation, of unit characters, 279; and chromosomes, 303; and natural selection, 387; of defectives, 382
Selection, 271, 274; practical value and methods of, 351; in pure lines, 356, *357;* limits of, 357; phenotypic, 350; natural, 385; sexual, 394; germinal, 401, 406; coincident, 408; in bees, 397; results of, in horses, *352*
Semon, 424
Serosa, *45*
Sex, chromosomes, 298; determination of, *298*, 325; reversal, 327, 335, 338; reversal in fowl, *336;* control in *Arisaema*, *337;* and the heritage, 327; purpose of, 331
Sexes, differentiation of, 331; contributions to heritage, 300
Sex-linkage, in Drosophila, 312, *313*, *315;* in man, 314, 364, *365*
Sexual colors, 244
Sexual forms, 22
Sexual selection, 10, 394
Shale, 120
Sheep, Ancon, 404; selection in, 351
Shepherd's purse, 316; inheritance in, *317*
Shore bugs, 254
Shrews, 253
Shull, 316, 342, 380
Sierra Nevada mountains, 257
Silurian, 123, 150; lung-fishes of, 130; change of climate during, 248
Simia, 189
Simiidae, 188
Siwalik Hills, 198

Skeleton, appendicular, **77**; of lower fishes, 51; of higher fishes, 55; above fishes, 55; visceral, 73, *74;* of man and gorilla, *192;* of Neanderthal man, 205
Skull, ganoid stage, 68, *69;* dermal bones of, 69; of *Cryptobranchus*, *58;* in Amphibia, 70; above Amphibia, 71; human, 71; of elephants, *167;* of *Moeritherium*, *168;* of Proboscidea, *174;* diagram of mammalian, *59;* of horse, 177; of *Smilodon*, *391;* of *Felis*, *391;* of Java ape-man, *199;* of Piltdown man, *203;* of Neanderthal man, *205*
Sloth, two-toed, *223*
Sloths, 222
Smilodon, skull of, *391*
Smith, E., 421
Social animals, 22
Sociology, 376
Soma, 306
Somatoplasm, 305
Song sparrows, distribution of, *410*
Sparrow, English, 251; song, distribution of, *410*
Specialization, 93; limitations of, 252
Species, 19, 25
Spencer, Herbert, 15, 417
Spermatids, 294
Spermatocyte, 292; of P*roteus*, *290*
Spermatogenesis, 292; in *Ascaris*, *295*
Spermatophyta, 28
Spermatozoa, 292; human, *296*
Sphaerella, 91; life history, *326*
Spinal column, 72
Spirula, 133
Spondylomorum, *34*
Sponges, 214
Spontaneous generation, 6
Spores, 325; of *Sphaerella*, *326*
Sporophyte, *24*
Spy, 203
Squids, 132
Squirrels, as arboreal animals, 229; flying, 229; in adaptive radiation, 253
Starfishes, artificial parthenogenesis in, 330
Stegocephalia, *151*, 152
Stegodon, 172; tooth of, *174*
Steinmetz, 380
Stenodictya lobata, *140*

Stenoma schlaegeri, 245
Sterility, 397
Sterilization of defectives, 382
Stings, of insects, 237; of honey-bee, *238*
Stomodaeum, 48
Struthiomimus, *153*
Sturtevant, 314
Styela partita, eggs of, 302
Stylopid, *141*
Subspecific forms, 21
Supplementary factors, 319
Survival of the fittest, 7, 386
Swammerdam, 8
Sweet peas, 319
Swimming legs, in bugs, 254
Symbiosis, 240
Symmetry, 34, 214
Symphalangism, *363*
Synapsis, 292, *293*

Tail, in man, 87
Tails, as weapons, 237
Talpa europaea, *221*
Tarpan, 183
Tatu novemcintus, *236*
Teeth, 82; abnormal growth in woodchuck, *212*, 213; adaptations, 83; adaptive radiation in, 253; development of, *82;* of beaver, 210; of carnivores, 159, 233; of cynodonts, 157; of elephants, *168;* heterodont dentition, 159; of herbivores, 159; of horse, 210; of mammals, *158;* and scales, 82; of squirrels, 210
Termite, 271
Termites, symbiosis with protozoa, 240
Terrestrial adaptations, 217
Terrestrial life, demands upon vertebrates, 148; forms transitional to, 148
Tertiary, 122; climate, 160; mammals, 160; region of Bering Straits during, *259*
Tetrad, 294; in *Ascaris*, *295*
Tetrakinetic theory, 434
Tetralophodon, 170, *172*
Thallophyta, 28
Theophrastus, 5, 7
Theories, adequate, conditions for, 429; Amphioxus, 144–146; Annelid, 144, 146; Arthropod, 144;

Centroepigenesis, 424; coincident selection, 408; chromosome, 289; Darwinian, 385, 428; Germinal continuity, 422; germinal selection, 406; Goldschmidt's, 323; intraselection, 408; isolation, 409, 428; kinetogenesis, 424; Lamarckian, 413; Mneme, 424; mosaic vision, 112; mutation, 402; natural selection, 385; origin of flight, 154; pangenesis, 423; panmixia, 405; sexual selection, 394; tetrakinetic, 434; use and disuse in chromosomes, 439
Thymus, 94
Thyroid, 94
Tiger, sabre-tooth, 390; skull, *391*
Tiger-beetle, running leg, *218*
Tinamou, 244, 294
Trapa natans, leaves, *390*
Tree frogs, adhesive organs of, 233; climbing, 223; flying, 229
Triassic, 122; changes of climate, 248; mollusca, 133
Triceratops, *236*
Trihybrid, 279, 283, *284;* in Guineapig, 283; ratios, 284
Trilobites, 134, *135*
Trilophodon, 169, *170;* skull and tooth, *174*
Trinacromerion, *153*
Trinil race, 199
Triploblasts, 34
Trout, 243
Trunk, 166
Tschermak, 276
Tulips, 345
Tusks, *167*, 168, *174*

Ungulata, 164; cursorial, 219
Unguligrade appendage, 253
Unit characters, 279; and chromosomes, 303; and factors, 306; of man, 361
Use and disuse, 414

Variability, of organisms, 115; of environment, 116; cause and effect in, 117
Variation, 247; and natural selection, 386
Variations, kinds of, 265; source of, 269

INDEX 459

Veliidae, 254
Venom, 237
Vertebrae, 56; development of, diagram, *59*
Vertebrates, 37; evolution of, 144, *161*; emergence of terrestrial, 147
Vesalius, 8
Vespertilio noctula, *227*
Vestigial structures, 85
Vilmorin, 353
Vitamines, 431
Vitis labrusca, 344; *V. vinifera*, 345
Volant animals, in adaptive radiation, 253; dispersal, 260
Volvox, 51, *54*

Wagner, Moritz, 388, 409
Wallaby, 219, *220*
Wallace, 14, 250, 385, 393
Walter, 274, 356, 418
Warning colors, 242
Water, as environment, 148
Water lilies, 260
Weapons, origin of, 194
Web of life, 230

Weismann, 15, 269, 291, 405
Wheat, 316; pure lines of, *356*
Whetham, 367
Whip-poor-will, 242
Wilder, H. H., 69, 72, 198
Wings of insects, 139, *229*
Winship, 370
Wisdom teeth, 88
Wolves, 239
Woodchuck, abnormal growth of incisor, *212*
Wrens, nesting habits of, 396

Xerophytes, 224, *226*
X-rays, 116, 320, 419

Yolk, 40
Yolk sac, *45*; human, *46*
Yucca, *226*

Zebra, 164, 183
Zebu, 347
Zoögeographical realms, 255, *256*
Zygote, 326